THE LANGUAGE OF GAME THEORY

Putting Epistemics into the
Mathematics of Games

World Scientific Series in Economic Theory
(ISSN: 2251-2071)

Series Editor: Eric Maskin *(Harvard University, USA)*

Published

Forthcoming

World Scientific Series in Economic Theory – Vol. 5

THE LANGUAGE OF GAME THEORY

Putting Epistemics into the Mathematics of Games

Adam Brandenburger

New York University, USA

World Scientific

NEW JERSEY · LONDON · SINGAPORE · BEIJING · SHANGHAI · HONG KONG · TAIPEI · CHENNAI

Published by

World Scientific Publishing Co. Pte. Ltd.

5 Toh Tuck Link, Singapore 596224

USA office: 27 Warren Street, Suite 401-402, Hackensack, NJ 07601

UK office: 57 Shelton Street, Covent Garden, London WC2H 9HE

Library of Congress Cataloging-in-Publication Data
The language of game theory : putting epistemics into the mathematics of games / by Adam Brandenburger (New York University, USA).
pages cm. -- (World scientific series in economic theory, ISSN 2251-2071 ; vol. 5)
Includes bibliographical references and index.
ISBN 978-9814513432 -- ISBN 9814513431
1. Game theory. I. Brandenburger, Adam.
HB144.L36 2014
519.3--dc23

2013038870

British Library Cataloguing-in-Publication Data
A catalogue record for this book is available from the British Library.

In-house Editor: Alisha Nguyen

Typeset by Stallion Press
Email: enquiries@stallionpress.com

Printed in Singapore

"Over the past decade epistemic game theory has emerged as one of the principled alternatives to more traditional approaches to economic interactions and Adam Brandenburger has played a central role in that emergence. For anyone interested in epistemic game theory, or game theory in general, this book is a *must have*. But even more important is the opportunity this volume, and epistemic game theory in general, presents to empirical scientists. As Brandenburger notes in his Introduction, until now epistemic game theory has been a theoretical discipline. This volume should make it clear, however, that it could be — and likely soon will be — an empirical undertaking. Anyone interested in behavioral, psychological, or neurobiological studies of how we make decisions during strategic play will find in this volume a profoundly fascinating set of empirically testable hypotheses just waiting to be examined."

Paul Glimcher
New York University, USA

"Adam Brandenburger has played a leading role in developing the program of epistemic game theory, the goal of which is to provide a deeper and clearer foundation for game theory as a whole. This volume collects a remarkable body of work by Brandenburger and his collaborators, in which penetrating conceptual analysis and the development of a rich mathematical theory go hand in hand. The work offers much of great interest to computer scientists, who will see many connections with their study of recursive and corecursive structures, of processes and their logics, and of multi-agent systems; and to mathematicians and logicians interested in making precise models of the reflexive structures inherent in systems containing rational agents who can reason about the system of which they form a part. I hope that this timely collection will help to stimulate cross-disciplinary work on these fundamental topics."

Samson Abramsky
Oxford University, UK

"Games are playgrounds where players meet and interact, guided by streams of information and opinion. Adam Brandenburger's work has been instrumental in creating a new rich epistemic framework doing justice to both games and their players. This timely book will help a broader audience learn and appreciate the resulting theory."

Johan van Benthem
University of Amsterdam, The Netherlands
Stanford University, USA

"Economics, so grounded in the notion of equilibrium, has required substantial foundational work on reasoning about reasoning — epistemics — in interacting situations (games). Yet, if plain reasoning is difficult enough, just imagine epistemics. Adam Brandenburger, as is evident from the elegant and clear chapters of this book, is a master of the trade. His highly regarded research, always subtle and deep, is of the kind that establishes milestones while at the same time opening up vistas to new, and unexpected, frontiers. This book is specialized, certainly, but it is a must."

Andreu Mas-Colell
Universitat Pompeu Fabra, Spain

To my daughter
Lucy Brandenburger
12 years old and who happily explained epistemic game theory to me while
we were playing cards one day.
Adam: "Lucy, hold up your cards. I can see your hand."
Lucy: "Dad, it doesn't matter if you see my cards. You don't know my
strategy."

To my late mother
Ennis Brandenburger
who passed on to me her love of language in all its forms.

And to my wife
Barbara Rifkind
my beloved partner in the game of life.

Foreword

Game theory attempts to predict how agents (people, firms, governments, etc.) will behave in strategic situations. And far-and-away the most important predictive concept in game theory's evolution has been Nash equilibrium, a configuration of strategies from which no agent gets a higher payoff by deviating unilaterally.

On the face of it, Nash equilibrium seems a plausible prediction: why, after all, would a rational agent *want* to deviate if his payoff doesn't rise as a result? But this question assumes that the agent knows that other agents will be playing *their* Nash equilibrium strategies, an assumption that, on reflection, may seem too strong.

The job of epistemic game theory is to provide justifiable and secure foundations for behavioral predictions. In that role, it places limits on what an agent believes about others' behavior only on the basis of rationality and beliefs about rationality. It has been a remarkably fruitful and interesting enterprise.

Adam Brandenburger is one of epistemic game theory's pioneers. In particular, he and Eddie Dekel were the first to show that, under suitable conditions, rationalizability, iterated elimination of dominated strategies, and subjective correlated equilibrium — three central concepts in game theory — all amount to the same thing, a result that is the starting point for almost all subsequent work.

I am extremely pleased that Adam has agreed to put some of his fundamental papers on epistemic game theory — the article with Eddie and seven others — together in one volume, along with a lucid, simple, and beautifully written introduction to the subject.

Eric Maskin
Editor-in-Chief
World Scientific Series in Economic Theory

Contents

Contents

About the Author

Photo by Jerllin Cheng

Adam Brandenburger is the J.P. Valles Professor at the NYU Stern School of Business, New York University. He was a professor at Harvard Business School from 1987 to 2002. Adam Brandenburger works in the areas of game theory and, more recently, quantum information, and applies his work in game theory to the area of business strategy. Within game theory, he was one of the pioneers in developing what has come to be called "epistemic game theory" — an approach that studies the effect of how players in a game reason about one another (including their reasoning about other players' reasoning) on how the game is played. He also developed (with Harborne Stuart) the concept of a "biform game" — a model of strategy as moves that affect the amount of value created, and its division, among the players in a game. His book *Co-opetition* (with Barry Nalebuff, Doubleday, 1996) applies this idea to business strategy. Recently, he has worked on the implications for game theory of giving players access to quantum rather than classical information resources. Adam Brandenburger is an Associated Faculty Member at the Center for Data Science, New York University, and is currently Vice Dean for Innovation at the NYU Stern School of Business. He received his B.A., M.Phil., and Ph.D. degrees from the University of Cambridge.

Acknowledgments

I am deeply grateful to my co-authors Bob Aumann, Larry Blume, Eddie Dekel, Amanda Friedenberg, and Jerry Keisler for the enormous privilege of working with them, and also for permission to include joint papers in this volume. I am also deeply grateful to Gus Stuart for the many hours spent discussing topics in this volume and for the privilege of writing (other) papers with him. Samson Abramsky, Jonathan Haidt, Rena Henderson, Jessy Hsieh, Alex Peysakhovich, Barbara Rifkind, Paul Romer, Zvi Ruder, Natalya Vinokurova, Shellwyn Weston, and students in my NYU class The Project provided important input. I learned at the feet of many teachers. I note, with especial affection, Kenneth Arrow, Margaret Bray, Frank Hahn, David Kreps, Louis Makowski, and Andreu Mas-Colell. Eric Maskin very generously invited me to contribute this volume to his series. Zvi Ruder, Alisha Nguyen, and the editorial team at World Scientific provided unstinting support throughout.

I would like to acknowledge the following journals for granting their permissions to reproduce the chapters in this volume:

Econometrica

Chapter 3: "Rationalizability and Correlated Equilibria," with Eddie Dekel, *Econometrica*, 55, 1987, 1391–1402.

Chapter 5: "Epistemic Conditions for Nash Equilibrium," with Robert Aumann, *Econometrica*, 63, 1995, 1161–1180.

Chapter 6: "Lexicographic Probabilities and Choice under Uncertainty," with Larry Blume and Eddie Dekel, *Econometrica*, 59, 1991, 61–79.

Chapter 7: "Admissibility in Games," with Amanda Friedenberg and H. Jerome Keisler, *Econometrica*, 76, 2008, 307–352.

Journal of Economic Theory

Chapter 2: "Hierarchies of Beliefs and Common Knowledge," with Eddie Dekel, *Journal of Economic Theory*, 59, 1993, 189–198.

Chapter 4: "Intrinsic Correlation in Games," with Amanda Friedenberg, *Journal of Economic Theory*, 141, 2008, 28–67.

Chapter 8: "Self-Admissible Sets," with Amanda Friedenberg, *Journal of Economic Theory*, 145, 2010, 785–811.

Studia Logica

Chapter 1: "An Impossibility Theorem on Beliefs in Games," with H. Jerome Keisler, *Studia Logica*, 84, 2006, 211–240.

Introduction

Adam Brandenburger

It is never easy, if even possible, to put a precise date on the beginning of a new field. Still, if one were to choose a year for the origin of game theory, it would surely be 1928, with von Neumann's fundamental paper of that year (von Neumann [1928]). von Neumann wanted to find the best way to play a game of strategy:

> n players $S_1, S_2, \ldots, S_n$ are playing a given game of strategy, $\mathcal{G}$. How must one of the participants, S_m, play in order to achieve a most advantageous result? (von Neumann [1928, p. 13])

To formulate the question in its essential form, von Neumann introduced the concept of a strategy — i.e., a complete plan of play — for each player. The task for player S_m is now to pick one of his strategies "without knowing the choices of the others" (von Neumann [1928, p. 17]). von Neumann said that this lack of knowledge was built in:

> [I]t is inherent in the concept of "strategy" that all information about the actions of the participants and the outcomes of "draws" [i.e., moves by Nature] a player is able to obtain or infer is already incorporated in the "strategy." Consequently, each player must choose his strategy in complete ignorance of the choices of the rest of the players and of the results of the "draws." (von Neumann [1928, p. 19])

Players might observe moves by other players and by Nature during the course of a game, but, by definition, they cannot observe other players' strategies.

One might think that this feature of strategic analysis creates an insurmountable obstacle. In all but the simplest games, the best strategy for one player to choose varies with the strategies the other players choose.

If the latter choices are a matter of "complete ignorance," then there seems to be no clear way for the first player to make a good choice.

von Neumann proposed a solution to this problem: This is his famous maximin criterion, which goes as follows. Ann chooses a strategy that yields her the highest guaranteed payoff. Suppose there are two players: Ann and Bob. For each strategy that she can choose, Ann supposes that Bob will, on his side, choose a strategy that is the worst for her. She then chooses a strategy that yields the highest such worst payoff — in other words, her highest guaranteed payoff. Bob chooses a strategy according to the same criterion.

By the time we come to von Neumann and Morgenstern (1944), von Neumann's maximin criterion had been extended to a complete theory for n-player general-sum games. Each subset ("coalition") of players jointly chooses strategies, under the worst-case assumption about how the complementary set of players chooses. This is the cooperative game model of von Neumann and Morgenstern.

Nash (1951) took us back to the game model (the noncooperative model) in which players choose strategies individually, not jointly. Within this setting, he took game theory down a very different road. He supposed that Ann's strategy is one that she might optimally choose, if she actually knew Bob's chosen strategy. Likewise, Bob's strategy is one that he might optimally choose, if he actually knew Ann's chosen strategy. (This, of course, is the concept of Nash equilibrium.) Nash banished the uncertainty about strategies, which was, in von Neumann's view, a fundamental characteristic of games.

Epistemic game theory (EGT) is concerned with noncooperative game theory, in which players choose strategies individually. But, whereas mainstream noncooperative game theory has followed Nash, EGT puts uncertainty about strategies center-stage. In this way, it really goes back to von Neumann's view of games and reconstructs noncooperative game theory accordingly. We will come back to cooperative theory later.

Epistemic Game Theory

In EGT, in line with von Neumann's approach, a game is pictured as a collection of decision problems under uncertainty, one for each player. Unlike von Neumann's approach, however, a player's decision criterion is typically not maximin, but the less conservative one of maximization of expected payoff. Under this criterion, each player formulates a probability distribution over which strategy each of the other players chooses, and chooses a strategy to maximize his expected payoff under his distribution. We will say more about other decision criteria later.

This approach raises a key question: What constraints (if any) should be placed on the probability distributions the players employ in their expected-payoff calculations? A good first definition of EGT is that it studies various answers that one might give to this question, and works out the consequences for how games are played.

For example, we might put no restrictions on the probability distributions. This will be the case if we assume that each player is rational, but nothing more. Each player has his own entirely unrestricted probability distribution over the strategies chosen by the other players, and maximizes with respect to this distribution. Such a distribution represents the player's own subjective (what Savage [1954, p. 3] called "personalistic") view of the likelihood of various choices that the other players can make.

But, EGT begins to take on its characteristic form when we go further and assume not only that each player is rational, but also that each player thinks the other players are rational. That is, we now ask that Ann not only forms a probability distribution over Bob's strategy choices (and chooses her strategy accordingly), but also that, for each of Bob's strategies on which she puts positive probability, she produces a probability distribution over her own strategy choices under which the strategy in question for Bob is optimal for him. Notice that, at this point, we have arrived at the need for a mathematical structure for talking about probability distributions (what Ann thinks about Bob's choice of strategy) and probability distributions over probability distributions (what Ann thinks Bob thinks about her choice of strategy).

We can go further in this direction: Ann is rational, thinks Bob is rational, and thinks Bob thinks she is rational. And so on. The limiting case — the infinite conjunction of all such finite-length "Ann thinks Bob thinks...is rational" conditions — is called the assumption of "rationality and common belief in rationality" in EGT. To date, much of the focus of the field has been on answering the question of which strategy choices by players are consistent with this assumption. The question turns out to be much trickier then one might first think.

Theory or Language?

Let's pause and ask a question about the nature of EGT. Is it a full-fledged theory? a language? something else? At this point in its development,

EGT seems best described as a mathematical language. It provides a way of capturing certain intuitive concepts (rationality, beliefs, beliefs about beliefs, ...) so that they can be put to work in formulating and analyzing strategic situations. But, as yet, EGT is devoid of empirical content. For this reason, it cannot yet be considered a full theory. For progress on this front, we must look beyond the mathematics.

One very promising potential source of empirical content for EGT is the area of social neuroscience called "theory of mind" (Premack and Woodruff [1978]). This area studies the ability of humans and other species to represent the mind of a fellow member of their species in terms of the latter's desires, intentions, and beliefs. (See Singer [2009] for more on the definition of this concept and for a survey of the current state of knowledge concerning the underlying brain mechanisms. Adolphs [2010] has an interesting discussion of links between social neuroscience and other disciplines, including economics and computer science.) Human infants acquire the ability to form beliefs about what other people believe (Wimmer and Perner [1983]). Call and Tomasello (2008) survey the research that has followed the pioneering animal study by Premack and Woodruff (1978).

Data of this kind can be expressed in the EGT formalism, which will then begin to take on the shape of a full theory. This has not been the flavor of the field in its early phase, which has focused on the limiting case of common belief rather than on finite-order beliefs. The language can express these latter cases, too. Interestingly, there are indications that studying these cases may be harder and can lead to quite different answers. Brandenburger and Friedenberg (2013) show that, for a given game, the implication of imposing rationality and mth-order belief in rationality, however large m might be, can differ from the implication of imposing rationality and common belief in rationality. There is a 'discontinuity at infinity.'

Finding solid connections between EGT and empirics — in a number of areas, no doubt, and not just in social neuroscience — is both a timely and an exciting avenue. We will say a little more about empirics in the final section of this Introduction.

Limits in Principle

It turns out that even if we lift all practical limits on levels of reasoning in games, there are limits in principle. Chapter 1 in this book shows that any model of beliefs, beliefs about beliefs, ..., of a certain kind must have a 'hole' in it; not all possible beliefs can be present. A so-called complete belief model does not exist.

Such results — on what is impossible even in principle — arise in various fields. In naive set theory, there was Russell's Paradox ("the set of all sets that are not members of themselves"). Indeed, the nonexistence of a complete belief model can be viewed as a game-theoretic analog of Russell's Paradox. In computer science, there is the distinction between computability and computational complexity. The first topic considers the limits to what can be computed in principle, even with unbounded resources. The second studies what can be computed tractably — say, in polynomial time. (Papadimitriou [1993] is a standard introduction.)

In both these other fields, the impossibility results came first. In EGT, the order was reversed. First, various models of all possible beliefs, beliefs about beliefs,..., were constructed. The earliest such models were built by Armbruster and Böge (1979) and Böge and Eisele (1979), followed by Mertens and Zamir (1985). Chapter 2 presents another early model of all possible beliefs, beliefs about beliefs,....

The work in Chapter 1 came later. It makes clear that in these models of all possible beliefs, beliefs about beliefs,..., the word "all" must be understood in a restricted fashion, since impossibility is avoided.

Belief models borrow the concept of "type" that Harsanyi brilliantly introduced in his pioneering study (Harsanyi [1967–68]) of uncertainty in games. In EGT, each player has a set of possible types, where each type is associated with a probability distribution over the possible strategies and types for the other players. Beliefs, beliefs about beliefs,..., can be deduced from types, which makes the latter a very convenient element of the language of game theory.

Without Harsanyi's insight, it would have been much harder for EGT to get off the ground. But, even though he came very close, Harsanyi did not quite initiate EGT. His interest was in uncertainty about the structure of the game — principally, about the payoffs of the players. Like Nash, he was not interested in uncertainty about strategies. In fact, Harsanyi's concepts were more powerful than the use to which he himself put them.

Fundamental Theorem of Epistemic Game Theory

Once a foundation for belief models (albeit, restricted belief models) was established, it became possible to tackle the question: What does the epistemic condition of rationality and common belief in rationality (RCBR) imply for how a game will be played? We already mentioned the importance of other — less 'extreme' — epistemic conditions. But, the field began with the extreme case, and this is how a lot of the language has been developed.

At first blush, the question has a clear answer. There are layers to the question, though, as we shall see. Here is the first answer. If each player is rational, then none will play a so-called dominated strategy (since no probability distribution can make a dominated strategy optimal). But, then, if each player also believes that the other players are rational, no player will play a strategy that is dominated only after all dominated strategies have been deleted from the original game. And so on. Under RCBR, then, each player will choose an iteratively undominated (IU) strategy.

In fact, this is not quite a complete answer, even as a first answer, because we want a converse, too. A simple converse would go something like this: Given a game and an IU strategy for each player, we can build a belief model so that these strategies are consistent with RCBR in the model. With this, we can say that the epistemic condition of RCBR not only implies that IU strategies will be played, but actually identifies this set (and no smaller set) of strategies. We get a characterization of the epistemic condition of RCBR.

This characterization result — which we can call the Fundamental Theorem of EGT — has been proved a number of times in a variety of forms. Results in Chapter 3 yield an early proof (albeit with a pre-epistemic interpretation of the mathematics). Friedenberg and Keisler (2011) give state-of-the-art versions of the characterization result.

Here is another layer to the question. To explain, let's consider three players: Ann, Bob, and Charlie. Charlie thinks that Ann and Bob choose their strategies independently. (Remember: We are doing noncooperative game theory.) Does this mean that Charlie's probability distribution over Ann's strategy choice and Bob's strategy choice must be independent — i.e., must be a product distribution? The answer is no, because this neglects the possibility of common-cause correlation. Even though Ann and Bob choose independently, there could be variables affecting their choices that are correlated. Indeed, in EGT, such variables are right in front of us: they are Ann's and Bob's beliefs about the game. Charlie can think that Ann's strategy choice and Bob's strategy choice are correlated, if he thinks that Ann's and Bob's beliefs are correlated.

This leads to the question: What does the epistemic condition of RCBR imply for how a game will be played, when we allow for common-cause correlation? Chapter 4 shows that the answer, this time, is not the IU strategies, but a strictly smaller set of strategies. In other words, there are IU strategies that rely on too much correlation — on correlation that cannot arise through the mechanism of correlated beliefs. What exactly is this smaller set? At the moment, the question is open. We do not know what the Fundamental Theorem of EGT looks like under this kind of correlation.

This is a good indication of how young a field EGT is. This is just one of a good number of very basic questions in EGT to which we do not yet have answers.

Epistemic vs. Ontic Views

Return for a moment to Nash's development of game theory. If EGT is the broad language for describing and analyzing strategic situations we said it is, then, in particular, we should be able to express the idea of Nash equilibrium in the language. Chapter 5 is an attempt to do exactly this.

Here is the specific question: What epistemic conditions on a game imply that the players will play a Nash equilibrium? There is an easy answer for pure strategies and a deeper answer for mixed strategies. For pure strategies, the answer is essentially the one give earlier — that each player knows the strategies chosen by the other players. In epistemic belief models, the condition becomes: Each player has a correct belief about the other players' strategy choices. Formally, each player assigns probability 1 to the actual strategies chosen. Of course, there is also the condition that each player is rational.

These conditions apply to pure strategies. They could also apply to mixed-strategy equilibria, if we think of mixed strategies classically — that is, as choices from the set of randomizations over pure strategies. But, there is a natural epistemic question: Can a mixed strategy be turned on its head, so to speak, and viewed as representing other players' uncertainty about that player's choice of pure strategy?

Let's borrow some very useful terminology from physics — specifically, from the area of quantum mechanics. The "ontic" view of a quantum state is that it describes an objective reality of a physical system, while the "epistemic" view is that the state describes the state of knowledge of an observer of the system (see Spekkens [2007]).

What is the epistemic vs. ontic view of mixed strategies? As in the Fundamental Theorem of EGT, there is, again, a big distinction between games with two players and games with three or more players. One reason is the same as before: Charlie can think that Ann's and Bob's choices of strategy are independent or correlated. Another reason, different from before, is that two players can agree or disagree about the probabilities they assign to a third player's choice of strategy: Ann and Bob can agree or disagree about Charlie. To obtain Nash equilibrium, we impose epistemic conditions that yield independence and agreement. This allows us to assign to each player, in a well-defined way, a commonly held probability distribution by all other players over that player's choice of strategy, where these distributions are combined in an independent manner. Add an appropriate epistemic condition on rationality, and these distributions, viewed as mixed strategies, constitute a Nash equilibrium.

A clear lesson of the epistemic analysis of equilibrium is that the conditions involved are very restrictive. Correlation and disagreement are the general case in EGT. In order to obtain Nash equilibrium, we have to rule them out, by finding epistemic conditions that do so.

Conventional and epistemic game theory proceed very differently. The starting point of the conventional theory is a so-called solution concept — almost always Nash equilibrium or one of its many variants. Usually said to be the embodiment of 'rational behavior' in some way or other, these solution concepts are used to analyze game situations. In EGT, the primitives are more basic. The very idea of rational behavior has to be defined. (Moreover, we have seen that, once it is carefully defined, rationality alone does not yield Nash equilibrium. Several additional conditions are involved.) We also have to specify what each player believes about other players' strategy choices, beliefs, rationality, etc. EGT is much more explicit than is conventional game theory about what is being assumed about the players in a game.

It is an intriguing fact that the ontic vs. epistemic issue arises both in game theory and in foundations of quantum mechanics (QM). Even more, von Neumann (1932) was the first to analyze the issue in QM. Also, somewhat like the use of games in mathematics itself — think, for example, of the topic of determinacy in set theory — games have become a useful device in quantum foundations. (The "Guess Your Neighbor's Input" game is a notable example; see Almeida *et al.* [2010].) Since EGT looks specifically at the ontic vs. epistemic issue, perhaps, game theory done this way will turn out to be particularly applicable to QM.

Invariance and Admissibility

Return to the Fundamental Theorem of EGT. The question was: What does the epistemic condition of rationality and common belief in rationality

(RCBR) imply for how a game will be played? What we said earlier on this question still only scratches the surface.

Battigalli and Siniscalchi (2002) made a major advance by extending investigation of the question beyond game matrices to game trees. In EGT, game trees pose a whole new level of complexity because players not only hold beliefs, beliefs about beliefs, . . . , but also revise these beliefs as the play of the game proceeds. They do so by conditioning on the observed events in the tree. But there is more. In probability theory, conditional probabilities are defined up to their value on probability-0 events. In EGT, however, the probability-0 events cannot be ignored because what one player may view as chance might be under the control of another player.

For example, Bob might assign probability 1 to Ann's making a move that does not reach him (i.e., does not give him the next move). In this case, his conditional distribution over what Ann might subsequently do, if she does, in fact, give Bob the next move, is undefined. This is not a problem from Bob's perspective, as his expected payoff does not depend on his move. Now, though, suppose that Ann thinks that Bob assigns probability 1 to her not giving him the next move. She still has to decide whether or not to give him the next move — the event is under her control. Her decision will depend on how she thinks Bob would react, and this depends on how she thinks Bob will update his probability distribution, conditional on the probability-0 event that she did give him the move. In sum, Ann needs a model of how Bob forms conditional probabilities even on events to which he assigns probability 0.

We have spelled this point out in detail, because it indicates the need for an extended non-Kolmogorovian probability theory in game theory. It turns out that there is a tailor-made such theory, due to Rényi (1955). (Rényi's motivations were from statistics and physics. The game theory of his time had not been developed to the point where the motivation from this direction could have been apparent.) This is the theory that Battigalli and Siniscalchi beautifully developed and applied to games.

All of this is about game trees, where players make observations during the play of the game and revise their beliefs. But, remember that, for von Neumann, the concept of strategy was basic. By listing each player's possible strategies, we reduce a game tree to a game matrix, and then conduct the analysis (in von Neumann's case, maximin analysis) on the matrix.

Actually, it is not immediately obvious that von Neumann's approach can work. In general, more than one game tree reduces to a given game matrix. Perhaps, crucial information is lost in considering only the matrix. Implicitly, at least, von Neumann was assuming that no 'strategically relevant' information is lost.

This view was made explicit in the highly influential paper by Kohlberg and Mertens (1986), who proposed an invariance principle. They argued that if two trees reduce to the same matrix, then the analysis of the two trees must be the same. Not only is no 'strategically relevant' information lost in working with the matrix, but it also is essential to work with the matrix, they said. Otherwise, we risk having the analysis depend on which of several strategically equivalent trees we happen to have in front of us.

So, how does one perform an invariant analysis? The key turns out to be the admissibility (weak dominance) concept. A strategy is admissible in the matrix if and only if it is rational in any tree that reduces to that matrix. (For a precise statement and proof of this claim, see Brandenburger [2007, p. 488].) Therefore, to build EGT in a manner that satisfies Kohlberg–Mertens invariance, we assume that each player adheres to an admissibility requirement.

Let us examine admissibility more closely. It is a standard result in decision theory that a strategy for Ann is admissible if and only if it maximizes her expected payoff under some probability distribution that puts strictly positive probability on each of Bob's possible strategy choices. This creates a puzzle, first observed by Samuelson (1992). Suppose that some of Bob's strategies are inadmissible (for him). Then, it seems that if Ann thinks that Bob adheres to the admissibility requirement, she will assign probability 0 to these strategies. The assumption that Ann adheres to the admissibility requirement (so that she assigns positive probability to all of Bob's strategies) appears to contradict the assumption that Ann thinks that Bob adheres to the admissibility requirement (so that she assigns probability 0 to some of Bob's strategies). Can we, in fact, do game theory with admissibility?

Chapter 6 offers a resolution of this puzzle. Ann is equipped not with one probability measure over Bob's strategy choices, but with a sequence of probability measures, which we called a lexicographic probability system (LPS). An LPS is used lexicographically in determining an optimal strategy: Pick those strategies that maximize expected payoff under the first probability measure. From this set, pick those strategies that maximize expected payoff under the second probability measure. And so on. The strategies for Bob that receive positive primary probability are considered infinitely more likely than the strategies that receive positive secondary probability, and so on. (This can be made into a formal statement, using infinitesimals.) This is the resolution of the puzzle. Ann need not assign probability 0 to any of Bob's strategies, but, she can consider some of them infinitely less likely than others.

This approach is given a foundation by starting with the standard Anscombe–Aumann (1963) axiomatization of the expected payoff criterion, and modifying the axioms in order to obtain an LPS representation. Note that, again, we extend Kolmogorov probability theory in order to do game theory — this time, in order to do game theory in an invariant manner.

Chapter 7 defines an epistemic condition that is a lexicographic analog to the basic RCBR condition. Rationality relative to an LPS is defined in the way just explained. We also need a lexicographic analog to the concept of "assigns probability 1 to," which is how belief is formalized in ordinary probability theory. Say that Ann "assumes" (our term for the lexicographic case) an event if she considers all states in the event infinitely more likely than all states not in the event. (Of course, the definition for infinite spaces needs more care.) With these ingredients, we can formulate the epistemic condition of "rationality and common assumption of rationality" (RCAR) and characterize it in terms of which strategies can then be played.

The characterization identifies a new solution concept on the matrix — call it a "self-admissible set" (SAS). Chapter 8 studies this solution concept in its own right, showing how it behaves in three very famous games — Centipede (Rosenthal [1981]), the Finitely Repeated Prisoner's Dilemma, and Chain Store (Selten [1978]) — and establishing various general properties.

Back to the main story: An important asymmetry has arisen at this point. The basic epistemic condition of RCBR is characterized by the set of iteratively undominated (IU) strategies. This is the Fundamental Theorem of EGT. The (more advanced) epistemic condition of RCAR is characterized by the SAS concept. But, by analogy with the first result, would we not expect RCAR to be characterized by the set of iteratively admissible (IA) strategies? The intuition is analogous (given that we have solved the Samuelson puzzle). If each player adheres to the admissibility requirement, then none will play an inadmissible strategy. But then, if each player also assumes that the other players adhere to the requirement, no player will play a strategy that is inadmissible after all inadmissible strategies have been deleted from the original game. And so on. Under RCAR, each player will choose an IA strategy.

This turns out to be wrong. The IA set constitutes one of the SAS's in a game, but, usually, a game has other (distinct) SAS's, too. One way to explain why this is so is that, unlike dominance, the admissibility concept involves a fundamental nonmonotonicity.

To get the IA strategies from within EGT, we need to formulate the RCAR condition not in an arbitrary belief model, but in one of the 'large' belief models we talked about in the earlier section "Limits in Principle." There is more. A result in Chapter 7 shows that in a large belief model (where the idea of "large" is suitably captured), for any finite integer m, the condition of "rationality and mth-order assumption of rationality" is characterized by $(m + 1)$ rounds of elimination of inadmissible strategies. (One round is for the rationality assumption, and m rounds are for mth-order assumption of rationality.) However, it is impossible to have RCAR (infinite levels) in such a structure. In some sense, the players are being asked to think about too much. It appears that EGT imposes another limit in principle to reasoning in games. Having said this, some recent papers (see, in particular, Keisler and Lee [2011]) test the robustness of this impossibility result. This is another very open area in EGT.

Questions and Directions

As EGT has developed, it has led to some provocative questions about very basic ingredients of game theory. One such question is: What — really — is a game model?

Conventional game theory says that a game model is a game matrix or tree. EGT says no — this is only a partial model. A full game model consists of a matrix or tree together with a belief model — i.e., a space of possible beliefs, beliefs about beliefs, ..., for each player. In other words, in EGT, a game model consists of a game in the classical sense, but also a model containing the beliefs, beliefs about beliefs, ..., that each player might hold. EGT respects the 'trilogy' of decision theory: strategies, payoffs, and probabilities. A game model should include the players' probabilities over what they are uncertain about, in addition to their strategy sets and payoff functions.

What can be thought of as an output in conventional game theory becomes an input in EGT. Conventionally, we start with a Nash equilibrium — say, one in which Ann chooses the strategy *Up* and Bob chooses the strategy *Left*. We can associate beliefs with this pair of strategies: Ann assigns probability 1 to Bob's choosing *Left* and Bob assigns probability 1 to Ann's choosing *Up*. (The definition of Nash equilibrium does not require us to make this association, but it is a natural one to make.) The solution concept is interpreted as telling us what beliefs the players hold. In short, beliefs emerge as an output of the analysis.

In EGT, we cannot even begin analysis until we have specified the players' beliefs. Only once we have done so can we narrow down the strategy choices. This time, beliefs are an essential input into the analysis.

Where do we look for this input? What is underneath a belief model? For this, it seems we should look to outside factors, such as the environment in which strategic interaction takes place, the prior experiences of the players, their personalities, etc. Moving in this direction makes game theory much less self-contained as a discipline than it traditionally has been.

This causes mixed reactions. To some people, it is a big defect of EGT: It is not a complete theory. To others, it is a welcome end to a dubious enterprise of trying to build a complete theory of strategic interaction on the basis of 'pure reason' alone. Such aspects as context and history, among others, are now seen to matter. This issue has divided game theorists from early on. Von Neumann and Morgenstern (1944) were firmly in the second camp:

> [W]e shall in most cases observe a multiplicity of solutions. Considering what we have said about interpreting solutions as stable 'standards of behavior' this has a simple and not unreasonable meaning, namely that given the same physical background different 'established orders of society' or 'accepted standards of behavior' can be built.... (von Neumann and Morgenstern [1944, p. 42])

(Admittedly, this was written in the context of cooperative theory, but, remember that cooperative theory was their general theory of n-player games.) If we understand the term "physical background" to refer to the game matrix (or tree), then von Neumann and Morgenstern believed that game theory should not attempt to give a unique answer based on this model alone. Nash (1950) took the exact opposite view, and asked precisely for a unique answer from the matrix:

> We proceed by investigating the question: what would be a 'rational' prediction of the behavior to be expected of rational [ly] playing the game in question? By using the principles that a rational prediction should be unique, that the players should be able to deduce and make use of it, and that such knowledge on the part of each player of what to expect the others to do should not

lead him to act out of conformity with the prediction, one is led
to the concept of a solution [viz., Nash equilibrium] defined before.
(Nash [1950])

We can sum up the issue this way: Game theory built on the matrix alone is
indeterministic (von Neumann–Morgenstern) or deterministic (Nash). EGT
is clearly a theory of the first kind.

Here is another question about the basic ingredients of game theory:
What is a player? Or, more precisely put, what is the appropriate unit of
behavior? Once again, von Neumann–Morgenstern and Nash took different
paths. In the von Neumann–Morgenstern cooperative theory, the notion of
joint action by several players is basic. In Nash's noncooperative theory,
individual action is basic. The unit of behavior is, in the first case, a group,
and, in the second case, an individual. EGT suggests that the boundary
between the noncooperative and cooperative theories is more blurry than
one might think. In the section "Fundamental Theorem of Epistemic Game
Theory," we explained that individual actions can become correlated via
correlation of beliefs. We also said that the iteratively undominated (IU)
strategies — a very basic object in noncooperative game theory — contain
more correlation than can be explained this way. What is this extra
correlation? It is precisely joint action by players. (Chapter 4 gives more
detail on this point.) Noncooperative game theory turns out to include in it
what is usually thought of as a distinguishing feature of cooperative theory.
The boundary between the two branches of game theory appears to be less
clear-cut than previously thought.

Let us now come back to the matter of the decision criterion that
is attributed to the players in EGT. So far, the field has almost always
assumed expected-payoff maximizing behavior. The main reason for this is a
"one new thing at a time" philosophy. EGT is radically different from classical game theory in its rejection of the use of equilibrium as the starting point
of analysis. It would be hard to get a clear view of the effect of this shift,
if EGT made other changes at the same time. Since expected-payoff maximization is the assumption almost always made in the classical theory, EGT
holds this part fixed. We can think of the approach as rather like conducting
a controlled experiment, in which one variable is changed while the others
remain fixed. Even maintaining the classical decision criterion, EGT turns
out to be very different from classical game theory — as we have already
glimpsed and as will become even clearer in the remainder of this book.

Still, it would be highly desirable to explore what happens if a
nonclassical decision criterion is used instead. Fortunately, EGT looks to
be rather modular with respect to the decision criterion used. The hope is

that much of the architecture of EGT — the concept of type, the inductive definitions of rationality, belief in rationality, etc. — should continue to work even if we 'plug in' a different decision criterion. Such an extension of EGT will be another way in which empirical content can be added to the field — by using decision theories that have better empirical underpinnings.

At various points in this Introduction, we have pointed to possible directions in which EGT could further develop. We have noted some very basic open questions in the field, which is still very young. We have also seen some unexpected consequences of the approach: EGT has prompted us to ask — or re-ask — some fundamental questions about the whole architecture of game theory. We believe this to be another virtue of the epistemic approach.

References

Adolphs, R (2010). Conceptual challenges and directions for social neuroscience. *Neuron*, 65, 752–767.

Almeida, M, J-D Bancal, N Brunner, A Acín, N Gisin, and S Pironio (2010). Guess your neighbors input: A multipartite nonlocal game with no quantum advantage. *Physical Review Letters*, 104, 230404.

Anscombe, F and R Aumann (1963). A definition of subjective probability. *Annals of Mathematical Statistics*, 34, 199–205.

Armbruster, W and W Böge (1979). Bayesian game theory. In Moeschlin, O and D Pallaschke (Eds.), *Game Theory and Related Topics*. Amsterdam: North-Holland.

Battigalli, P and M Siniscalchi (2002). Strong belief and forward-induction reasoning. *Journal of Economic Theory*, 106, 356–391.

Böge, W and Th Eisele (1979). On solutions of Bayesian games. *International Journal of Game Theory*, 8, 193–215.

Brandenburger, A (2007). The power of paradox: Some recent developments in interactive epistemology. *International Journal of Game Theory*, 35, 465–492.

Brandenburger, A and A Friedenberg (2013). Finite-order reasoning. Working Paper.

Call, J and M Tomasello (2008). Does the chimpanzee have a theory of mind? 30 years later. *Trends in Cognitive Sciences*, 12, 187–192.

Friedenberg, A and HJ Keisler (2011). Iterated dominance revisited. Available at www.public.asu. edu/~afrieden.

Harsanyi, J (1967–68). Games with incomplete information played by "Bayesian" players, I–III. *Management Science*, 14, 159–182, 320–334, 486–502.

Keisler, HJ and BS Lee (2011). Common assumption of rationality. Available at www.math.wisc.edu/~keisler.

Kohlberg, E and J-F Mertens (1986). On the strategic stability of equilibria. *Econometrica*, 54, 1003–1038.

Mertens, J-F and S Zamir (1985). Formulation of Bayesian analysis for games with incomplete information. *International Journal of Game Theory*, 14, 1–29.

Nash, J (1950). Non-cooperative games. Doctoral dissertation, Princeton University.

Nash, J (1951). Non-cooperative games. *Annals of Mathematics*, 54, 286–295

Papadimitriou, C (1993). *Computational Complexity*. Reading, MA: Addison-Wesley.

Premack, D and G Woodruff (1978). Does the chimpanzee have a theory of mind? *Behavioral and Brain Sciences*, 1, 515–526.

Rényi, A (1955). On a new axiomatic theory of probability. *Acta Mathematica Academiae Scientiarum Hungaricae*, 6, 285–335.

Rosenthal, R (1981). Games of perfect-information, predatory pricing, and the chain store paradox. *Journal of Economic Theory*, 25, 92–100.

Samuelson, L (1992). Dominated strategies and common knowledge. *Games and Economic Behavior*, 4, 284–313.

Savage, L (1954). *The Foundations of Statistics*. New York, NY: Wiley.

Selten, R (1978). The chain store paradox. *Theory and Decision*, 9, 127–159.

Singer, T (2009). Understanding others: Brain mechanisms of theory of mind and empathy. In Glimcher, P, C Camerer, E Fehr and R Poldrack (Eds.), *Neuroeconomics: Decision Making and the Brain*, pp. 251–268, Amsterdam: Elsevier.

Spekkens, R (2007). Evidence for the epistemic view of quantum states: A toy theory. *Physical Review A*, 75, 032110.

von Neumann, J (1928). Zur Theorie der Gesellschaftsspiele. *Mathematische Annalen*, 100, 295–320. (Bargman, S (1955). English translation: On the theory of games of strategy. In Tucker, A and RD Luce (Eds.), *Contributions to the Theory of Games, Volume IV*, pp. 13–42. Princeton, NJ: Princeton University Press.)

von Neumann, J (1932). *Mathematische Grundlagen der Quantenmechanik*. Berlin: Springer. (Translated (1955) as *Mathematical Foundations of Quantum Mechanics*. Princeton, NJ: Princeton University Press.)

von Neumann, J and O Morgenstern (1944). *Theory of Games and Economic Behavior*. Princeton, NJ: Princeton University Press.

Wimmer, H and J Perner (1983). Beliefs about beliefs: representation and constraining function of wrong beliefs in young children's understanding of deception. *Cognition*, 13, 103–128.

Chapter 1

An Impossibility Theorem on Beliefs in Games

Adam Brandenburger and H. Jerome Keisler

A paradox of self-reference in beliefs in games is identified, which yields a game-theoretic impossibility theorem akin to Russell's Paradox. An informal version of the paradox is that the following configuration of beliefs is impossible:

Ann believes that Bob assumes that
Ann believes that Bob's assumption is wrong

This is formalized to show that any belief model of a certain kind must have a "hole." An interpretation of the result is that if the analyst's tools are available to the players in a game, then there are statements that the players can think about but cannot assume. Connections are made to some questions in the foundations of game theory.

Originally published in *Studia Logica*, 84, 211–240.

Keywords: belief model; complete belief model; game; first order logic; modal logic; paradox.

Financial support: Harvard Business School, Stern School of Business, National Science Foundation, and Vilas Trust Fund.

Acknowledgments: We are indebted to Amanda Friedenberg and Gus Stuart for many valuable discussions bearing on this chapter. Samson Abramsky, Ken Arrow, Susan Athey, Bob Aumann, Joe Halpern, Christopher Harris, Aviad Heifetz, Jon Levin, Martin Meier, Mike Moldoveanu, Eric Pacuit, Rohit Parikh, Martin Rechenauer, Hannu Salonen, Dov Samet, Johan van Benthem, Daniel Yamins, and Noson Yanofsky provided important input. Our thanks, too, to participants in the XIII Convegno di Teoria dei Giochi ed Applicazioni (University of Bologna, June 1999), the Tenth International Conference on Game Theory (State University of New York at Stony Brook, July 1999), the 2004 Association for Symbolic Logic Annual Meeting (Carnegie Mellon University, May 2004), and seminars at Harvard University, New York University, and Northwestern University, and to referees.

1. Introduction

In game theory, the notion of a player's beliefs about the game — even a player's beliefs about other players' beliefs, and so on — arises naturally. Take the basic game-theoretic question: Are Ann and Bob rational, does each believe the other to be rational, and so on? To address this, we need to write down what Ann believes about Bob's choice of strategy — to decide whether she chooses her strategy optimally given her beliefs (i.e., whether she is rational). We also have to write down what Ann believes Bob believes about her strategy choice — to decide whether Ann believes Bob chooses optimally given his beliefs (i.e., whether Ann believes Bob is rational). Etc. Beliefs about beliefs about... in games are basic.

In this chapter we ask: Doesn't such talk of what Ann believes Bob believes about her, and so on, suggest that some kind of self-reference arises in games, similar to the well-known examples of self-reference in mathematical logic. If so, then is there some kind of impossibility result on beliefs in games, that exploits this self-reference?

There is such a result, a game-theoretic version of Russell's paradox.[1] We give it first in words. By an *assumption* (or strongest belief) we mean a belief that implies all other beliefs. Consider the following configuration:

Ann believes that Bob assumes that
Ann believes that Bob's assumption is wrong.

To get the impossibility, ask: Does Ann believe that Bob's assumption is wrong? If so, then in Ann's view, Bob's assumption, namely "Ann believes that Bob's assumption is wrong," is right. But then Ann does not believe that Bob's assumption is wrong, which contradicts our starting supposition. This leaves the other possibility, that Ann does not believe that Bob's assumption is wrong. If this is so, then in Ann's view, Bob's assumption, namely "Ann believes that Bob's assumption is wrong," is wrong. But then Ann does believe that Bob's assumption is wrong, so we again get a contradiction.

The conclusion is that the configuration of beliefs in bold is impossible. But, presumably, a model of Ann's and Bob's beliefs that contained all beliefs would contain this configuration of beliefs (among many other

---

[1]The informal statement here in the Introduction is a multi-player analog of the Liar Paradox. The formal statement we give later (Section 6) is a multi-player analog of Russell's paradox.

configurations). Apparently, such a model — which we will call a complete belief model — does not exist. Alternatively put, every model of Ann's and Bob's beliefs will have a 'hole' in it; not all possible beliefs can be present.[2]

Formal notions of belief are of central importance in both game theory and modal logic. We think that our impossibility result is best understood if it is formulated in both settings in parallel. Since our results originated from a problem in game theory, we will first formulate our results in a setting which is consistent with the literature in game theory (beginning in Section 3), and then reformulate our results in a modal logic setting (in Section 7).

As we will see in Section 9, the notion of assumption, or strongest belief, is essential for the impossibility result. In the verbal argument, as well as in the formalization to be given later, the statement in italics must be about one particular belief for Bob but all beliefs for Ann. This interpretation seems natural when one speaks of Ann's beliefs and Bob's assumption. In the modal logic setting, belief and assumption are different modal operators. The assumption operator was introduced and analyzed (and called the "all and only" operator) in Humberstone (1987).[3] The assumption operator is also closely related to the modal operator I, which means "has been informed that," in the paper of Bonanno (2005).

We now add a little more precision to the verbal impossibility argument (in the game-theoretic setting). In general, a *belief model* has a set of states for each player, and a relation for each player that specifies when a state of one player considers a state of the other player to be possible. The concepts of assumption and belief have natural definitions in such a model. Given a belief model, we next consider a *language* used by the players to formulate their beliefs about the other players. We then say that a given belief model is *complete* for a language if every statement in a player's language which is possible (i.e., true for some states) can be assumed by the player.[4] Thus, completeness is relative to a language. Completeness for a given language

---

[2]For an earlier — and weaker — impossibility result, see Brandenburger (2003). We recap this result later. Other well-known paradoxes of belief include G. E. Moore's paradox ("It is raining but I don't believe it"), and the Believer's Paradox (Thomason [1980]). Huynh and Szentes (1999) give a one-player impossibility result.

[3]We thank Eric Pacuit for calling this to our attention.

[4]The word "complete" has been used in this way in papers in game theory. A more descriptive term would be "assumption-complete," but we will keep to the shorter version. Completeness in the present sense is not related to the notions of a complete formal system or a complete theory in logic.

is determined by the beliefs for each player about the other player (but not by the beliefs for the players about themselves). Depending on our choice of language, we may get a nonexistence or an existence result for complete models.

The main impossibility theorem will show: *No belief model can be complete for a language that contains first order logic.* That is, every belief model has holes expressible in first order logic, where a hole is a statement about one player that is possible but is never assumed by the other player. In fact, we will show that in any model of Ann's and Bob's beliefs, a hole must occur at one of the following rather simple statements:

0. The tautologically true statement,
1. Bob rules out nothing (i.e., considers everything to be possible),
2. Ann believes that Bob rules out nothing,
3. Bob believes that Ann believes that Bob rules out nothing,
4. Ann believes that Bob's assumption is wrong, and
5. Bob assumes that Ann believes that Bob's assumption is wrong.

The informal argument given in this introduction shows that Ann cannot assume, or even believe, statement 5.

Section 2 gives some more motivation for the chapter, by explaining connections between the completeness concept and some questions that have arisen in game theory. We return to these connections at the end of the chapter, in Section 11.

The formal argument in the game-theoretic setting is developed in Sections 3–5. Section 3 gives the definition of a belief model, and the mathematical notions of a set being assumed and being believed. Section 4 contains the notion of a belief model being complete for a language. Section 5 contains our main impossibility theorem, Theorem 5.4. The proof of Lemma 5.6 in that section matches the informal argument.

In Section 6, we prepare the way for a modal logic version of our impossibility results by presenting a basic "single player" modal logic with an assumption operator. This logic is a simple special case of the logic of Bonanno (2005) for belief revision. Bonanno gave an axiom set and completeness theorem for his logic. To shed some light on the behavior of the assumption operator, we restate his theorem in our simpler setting. An earlier and different axiom set and completeness theorem for modal logic with an assumption operator was given by Humberstone (1987).

In Section 7, we present a modal logic with assumption operators for two players, and reformulate our impossibility result in that logic. For a more detailed modal logic analysis of our impossibility result (which refers to an earlier version of this paper) see Pacuit (2006).

In Section 8, we introduce belief models with additional structure, strategies, which arise naturally in game theory. These *strategic models* are not needed in our main impossibility theorem, but are of interest in the application of our results to game theory. The next two sections contain cases where complete belief models do exist. In these positive results we get strategic models.

In Section 9, we obtain two existence theorems which suggest that the notion of a statement being assumed is an essential ingredient in our main impossibility result, and that the problem is intrinsically multi-player, i.e., game-theoretic, in nature. Section 10 provides some positive results on completeness relative to restricted (but still interesting) languages. We show there are strategic models which are complete for the fragment of first order logic that is closed under finite conjunction and disjunction, the universal and existential quantifiers, and the belief and assumption operators, but not under negation. To get these models, we construct strategic models that are topologically complete, i.e., that have a topological structure and in which every nonempty compact set of states can be assumed.

2. The Existence Problem for Complete Belief Models

Belief models and languages are artifacts created by the analyst to describe a strategic situation. A long-standing question in game theory is whether these artifacts can or should be thought of as, in some sense, available to the players themselves. (For a discussion and references to other discussions of the issue, see, e.g., Brandenburger and Dekel [1993, Section 3].)

Arguably, since we, the analysts, can build belief models and use a language, such as first order logic, these same tools should indeed be available to the players. Unless we want to accord the analyst a 'privileged' position that is somehow denied to the players, it is only natural to ask what happens if a player can think about the game the same way. But then our impossibility theorem says: *If the analyst's tools are available to the players, there are statements that the players can think about but cannot assume. The model must be incomplete.* This appears to be a kind of basic limitation in the analysis of games.[5]

---

[5]See Yanofsky (2003) for a formal presentation of the idea that mathematical paradoxes indicate limitations of various systems.

This limitation notwithstanding, the question of the existence of a complete belief model turns out to be very relevant to what is known as the "epistemic program" in game theory. One aim of this program is to find conditions on the players — specifically, on their rationality, belief in one another's rationality, etc. — that lead to various well-known solution concepts (iterated dominance, Nash equilibrium, backward induction, and others). Completeness of the belief model has been found to be needed in at least two of these analyses. Battigalli and Siniscalchi (2002) use completeness in their epistemic conditions for extensive-form rationalizability (a solution concept due to Pearce [1984]). Brandenburger, Friedenberg, and Keisler (2006) use completeness in their epistemic conditions for iterated admissibility (iterated weak dominance). These solution concepts are of independent interest, but they both also give the backward-induction outcome in perfect-information games.[6] So, we see that completeness is also very relevant (at least under these two analyses) to giving a firm foundation for the fundamental concept of backward induction in games.

Of course, given our impossibility result, both Battigalli–Siniscalchi and Brandenburger–Friedenberg–Keisler must restrict the language that the players can use to formulate their beliefs. They (effectively) do this by making various topological assumptions on the belief models they use. We will show in Section 10 that topological assumptions can yield complete belief models for "positive languages."

Separate from the topological approach, though, this chapter does pose the following open question in game theory. Can we find a logic $\mathcal{L}$ such that: (i) complete belief models for $\mathcal{L}$ exist for each game; (ii) notions such as rationality, belief in rationality, etc. are expressible in $\mathcal{L}$; (iii) the ingredients in (i) and (ii) can be combined to yield various well-known game-theoretic solution concepts, as above?

Section 11 comes back to some related papers in game theory, after our formal treatment.

3. Belief Models

In this section, belief models for two players are introduced, which are designed to allow us to state our impossibility results in the simplest form. In a belief model, each player has a set of states, and each state for one player has beliefs about the states of the other player. (Extending the definition to finitely many players would be straightforward.)

---

[6]Under assumptions ruling out certain ties among payoffs. See the cited papers for details.

Definition 3.1. A *belief model* is a two-sorted structure

$$\mathcal{M} = (U^a, U^b, P^a, P^b, \ldots)$$

where U^a and U^b are the nonempty universe sets (for the two sorts), P^a is a proper subset of $U^a \times U^b$, P^b is a proper subset of $U^b \times U^a$, and P^a, P^b are serial, that is, the sentences

$$\forall x \exists y P^a(x, y), \quad \forall y \exists x P^b(y, x)$$

hold. A belief model may also contain zero or more additional relations represented by the three dots. The set of relations $\{P^a, P^b, \ldots\}$ is called the *vocabulary of* $\mathcal{M}$. We place no restriction on the size of the vocabulary.

To simplify notation, we will always use the convention that x is a variable of sort U^a and y is a variable of sort U^b. We say x *believes* a set $Y \subseteq U^b$ if $\{y : P^a(x, y)\} \subseteq Y$, and x *assumes* Y if $\{y : P^a(x, y)\} = Y$. We also use the analogous terms with a, b and x, y reversed.

Thus, assumes implies believes. The members of U^a and U^b are called *states for* Ann and Bob, respectively, and the members of $U^a \times U^b$ are called *states*. P^a and P^b are called the *possibility relations*. Intuitively, $P^a(x, y)$ means that state x for Ann considers state y for Bob to be possible. So x assumes the set of all states that x considers possible, and x believes the sets which contain all the states that x considers possible.

Note that every state for Ann assumes a unique subset of U^b, and every state for Bob assumes a unique subset of U^a. By the definition of a belief model, every state for Ann assumes a nonempty subset of U^b, and some state for Ann assumes a proper subset of U^b. Likewise for Bob and U^a.

Remark 3.2. This shows that the notions of belief and assumption do not collapse to the same notion under further conditions. There must be a state for Ann that assumes a proper subset of U^b, and this state believes U^b but does not assume U^b.

There is no equality relation between elements of different sorts, so we can always take the universe sets U^a and U^b of a belief model to be disjoint. That is, every belief model is isomorphic as a two-sorted structure to a belief model with U^a, U^b disjoint.

The belief models which arise naturally in game theory have additional structure, including strategies. As we explained in the Introduction, strategies will not be needed in our main impossibility theorems, so for clarity we will postpone their treatment until Section 8.

A belief model as defined here does not specify beliefs for Ann about Ann or beliefs for Bob about Bob. That is, it does not include a relation saying when a state for Ann considers another state for Ann to be possible. However, since additional relations are allowed in the vocabulary of a belief model, one can form belief models with additional possibility relations on $U^a \times U^a$ and $U^b \times U^b$ which specify beliefs for Ann about Ann and for Bob about Bob. One can also add relations on $U^a \times U^a \times U^b$ and $U^b \times U^a \times U^b$ to specify beliefs about states. Our framework allows these extra relations, but they do not play a role in the impossibility result.

4. Complete Belief Models

Given a belief model, the next step is to specify a language used by the players to think about beliefs. We will then be able to talk about the completeness of a model, which is relative to the language. That is, a model will be complete if every statement in a player's language which is possible (i.e., true for some states) can be assumed by the player. (Otherwise, the model is incomplete.)

Conceptually, the language for a player should be a set of statements that the player can think about. We will be concerned with the family of subsets of U^b that Ann can think about, and the family of subsets of U^a that Bob can think about. The exact definition of a language will not matter much, but it will be convenient to take the statements to be first order formulas. This will give us a lot of flexibility because we are allowing a belief structure to have extra predicates in its vocabulary.

Let us first consider an arbitrary structure $\mathcal{N} = (U^a, U^b, \ldots)$, which may or may not be a belief structure. By the *first order language for* $\mathcal{N}$ we mean the two-sorted first order logic with sorts for U^a and U^b and symbols for the relations in the vocabulary of $\mathcal{N}$. Given a first order formula $\varphi(u)$ whose only free variable is u, the set *defined by* φ in $\mathcal{N}$ is the set $\{u : \varphi(u)$ is true in $\mathcal{N}\}$.

In general, by a *language for* $\mathcal{N}$ we will mean a subset of the set of all formulas of the first order language for $\mathcal{N}$. Given a language $\mathcal{L}$ for $\mathcal{N}$, we let $\mathcal{L}^a, \mathcal{L}^b$ be the families of all subsets of U^a, U^b, respectively which are defined by formulas in $\mathcal{L}$.

Remark 4.1. (i) If the vocabulary of $\mathcal{N}$ is finite or countable, then any language $\mathcal{L}$ for $\mathcal{N}$ has countably many formulas, so the sets $\mathcal{L}^a$ and $\mathcal{L}^b$ are at most countable.

(ii) If $\mathcal{M}$ is obtained from $\mathcal{N}$ by adding additional relations to the vocabulary, then any language for $\mathcal{N}$ is also a language for $\mathcal{M}$.

We now define the notion of a belief model which is complete for a language.

Definition 4.2. Let $\mathcal{M}$ be a belief model, and let $\mathcal{L}$ be a language for $\mathcal{M}$. $\mathcal{M}$ is *complete for* $\mathcal{L}$ if each nonempty set $Y \in \mathcal{L}^b$ is assumed by some $x \in U^a$, and each nonempty $X \in \mathcal{L}^a$ is assumed by some $y \in U^b$.

In words, a belief model is complete for a language if every nonempty set of Bob's states which is definable in the language is assumed by one of Ann's states, and vice versa.

Proposition 4.3. *Suppose two belief models $\mathcal{M}$ and $\mathcal{K}$ are elementarily equivalent, that is, they satisfy the same first order sentences. Then any language $\mathcal{L}$ for $\mathcal{M}$ is also a language for $\mathcal{K}$, and $\mathcal{M}$ is complete for $\mathcal{L}$ if and only if $\mathcal{K}$ is complete for $\mathcal{L}$.*

Proof. Being complete for $\mathcal{L}$ is expressed by the set of first order sentences

$$\exists y \varphi(y) \to \exists x \forall y [P^a(x, y) \leftrightarrow \varphi(y)]$$

for each formula $\varphi(y) \in \mathcal{L}$, and similarly with a, b and x, y reversed. $\qquad\square$

In general, a language for a belief model $\mathcal{M}$ will contain formulas involving the possibility relations P^a, P^b as well as the symbols of the reduced structure $\mathcal{N} = (U^a, U^b, \ldots)$. In the very special case that the relations P^a, P^b do not occur in the formulas of $\mathcal{L}$, and in addition that $\mathcal{L}$ is not too large, we get an easy example of a complete belief model.

Example 4.4. Let $\mathcal{N} = (U^a, U^b, \ldots)$ be a structure where the universe sets U^a and U^b are infinite and the vocabulary (indicated by $\ldots$) is at most countable. Let $\mathcal{L}$ be the first order language of $\mathcal{N}$. Then there are relations P^a, P^b such that $\mathcal{M} = (U^a, U^b, P^a, P^b \ldots)$ is a complete belief model for the language $\mathcal{L}$.

To see this, we note that since $\mathcal{L}^a$ and $\mathcal{L}^b$ are at most countable, one can choose surjective mappings $f \colon U^a \to \mathcal{L}^b \backslash \{\emptyset\}$ and $g \colon U^b \to \mathcal{L}^a \backslash \{\emptyset\}$, and let $P^a(x, y)$ iff $y \in f(x)$ and $P^b(y, x)$ iff $x \in g(y)$.

More generally, the above example works if the universe sets U^a, U^b are infinite and of cardinality at least the number of symbols in the vocabulary of $\mathcal{N}$.

Our main result (Theorem 5.4) will show that no belief model $\mathcal{M}$ can be complete for its own first order language, regardless of the size of the vocabulary. For this reason, one is led to consider belief models which are complete for various subsets of the first order language, as in Section 10.

5. Impossibility Results

As a warm-up, we review an earlier impossibility result from Brandenburger (2003), which shows that a belief model cannot be complete for a language when the family of definable sets is too large. Given a set X, the power set of X is denoted by $\mathcal{P}(X)$, and the cardinality of X is denoted by $|X|$.

Proposition 5.1. *No belief model $\mathcal{M}$ is complete for a language $\mathcal{L}$ such that $\mathcal{L}^a = \mathcal{P}(U^a)$ and $\mathcal{L}^b = \mathcal{P}(U^b)$.*

Proof. Given a belief model $\mathcal{M}$, we have $|U^a| \leq |U^b|$ or $|U^b| \leq U^a$, say the former. Since $U^a \times U^b$ has nonempty proper subsets, we must have $|U^b| > 1$. Then by Cantor's theorem, $|U^b| < |C|$ where C is the set of all nonempty subsets of U^b. It follows that $|U^a| < |C|$. Let $f \colon U^a \to C$ be the function where $f(x)$ is the set that x assumes. There must be a set $Y \in C \backslash range(f)$. Then $\emptyset \neq Y \in \mathcal{L}^b$ but no x assumes Y, so $\mathcal{M}$ is not complete for $\mathcal{L}$. $\qquad\square$

More generally, the above argument shows that if $|U^a| < |\mathcal{L}^b| - 1$ (that is, $\mathcal{L}^b$ is too large in cardinality), then $\mathcal{M}$ cannot be complete for $\mathcal{L}$.

In this chapter, only the two variables x and y will be needed in formulas. We now introduce some notation that will make many formulas easier to read.

Definition 5.2. If $\varphi(y)$ is a statement about y, we will use the formal abbreviations

$$x \text{ believes } \varphi(y) \quad \text{for } \forall y[P^a(x,y) \to \varphi(y)],$$
$$x \text{ assumes } \varphi(y) \quad \text{for } \forall y[P^a(x,y) \leftrightarrow \varphi(y)].$$

Similarly with a, b and x, y reversed.

Note that "x believes $\varphi(y)$" and "x assumes $\varphi(y)$" are statements about x only.

Definition 5.3. The *diagonal formula* $D(x)$ is the first order formula

$$\forall y[P^a(x,y) \to \neg P^b(y,x)].$$

This is our formal counterpart to the intuitive statement "Ann believes Bob's assumption is wrong." Note that the intuitive statement contains the word "believes", but the diagonal formula is not of the form x *believes* $\varphi(y)$ in the notation of Definition 5.2.

Here is our main impossibility result, which works for countable as well as large languages.

Theorem 5.4. *Let* $\mathcal{M}$ *be a belief model and let* $\mathcal{L}$ *be the first order language for* $\mathcal{M}$. *Then* $\mathcal{M}$ *cannot be complete for* $\mathcal{L}$.

The theorem is an easy consequence of the next two lemmas.

Lemma 5.5. *In a belief model* $\mathcal{M}$, *suppose* $\forall y P^a(x_1, y)$ *and*

$$x_2 \text{ believes } [y \text{ believes } [x \text{ believes } \forall x \, P^b(y, x)]].$$

Then $D(x_2)$.

Proof. We must show that

$$\forall y [P^a(x_2, y) \rightarrow \neg P^b(y, x_2)].$$

Suppose not. Then there is an element y_2 such that

$$P^a(x_2, y_2) \wedge P^b(y_2, x_2).$$

It then follows in turn that

$$y_2 \text{ believes } [x \text{ believes } \forall x \, P^b(y, x)],$$
$$x_2 \text{ believes } \forall x \, P^b(y, x),$$
$$\forall x \, P^b(y_2, x),$$
$$\forall x \, [x \text{ believes } \forall x \, P^b(y, x)],$$
$$x_1 \text{ believes } \forall x \, P^b(y, x),$$
$$\forall y \forall x \, P^b(y, x).$$

This contradicts the hypothesis that P^b is a proper subset of $U^b \times U^a$. □

Lemma 5.6. *Suppose* $\mathcal{M}$ *is a belief model. Then no element* x_0 *of* U^a *satisfies the formula*

$$x \text{ believes } (y \text{ assumes } D(x)) \tag{$*$}$$

in $\mathcal{M}$.

The proof of this lemma will closely match the argument given in the Introduction; the statement in bold type there is the informal version of the formula $(*)$. The informal version in the Introduction is a two-player analog of the Liar Paradox. It is a semantic statement, since it involves the notion of an assumption being "wrong." Lemma 5.6, on the other hand, is a formal result, which is a two-player analog of Russell's Paradox. In full unabbreviated form, formula (*) is

$$\forall y[P^a(x,y) \to \forall x(P^b(y,x) \leftrightarrow \forall y[P^a(x,y) \to \neg P^b(y,x)])].$$

Proof. We suppose that an element x_0 satisfies formula (*) in $\mathcal{M}$ and arrive at a contradiction. We ask whether $D(x_0)$.

Case 1. $D(x_0)$. Since $\forall x \exists y P^a(x,y)$, we may choose y_0 such that $P^a(x_0, y_0)$. Since $D(x_0), \neg P^b(y_0, x_0)$. But since x_0 satisfies (*), y_0 assumes $D(x)$. Then $P^b(y_0, x_0) \leftrightarrow D(x_0)$, so $P^b(y_0, x_0)$ and we have a contradiction.

Case 2. $\neg D(x_0)$. Choose y_0 such that $P^a(x_0, y_0) \wedge P^b(y_0, x_0)$. Since $P^a(x_0, y_0)$ and x_0 satisfies (*), y_0 assumes $D(x)$. Since $P^b(y_0, x_0)$ and y_0 assumes $D(x)$, we have $D(x_0)$, contradiction. $\square$

Here is an English translation of the above proof, replacing x by "Ann", y by "Bob", the relations P^a and P^b by "sees", and $D(x)$ by "Ann believes that Bob cannot see Ann". This proof, unlike the rough argument in the Introduction, involves only Ann's beliefs about Bob and Bob's beliefs about Ann.

Suppose Ann believes that Bob assumes that Ann believes that Bob cannot see Ann.

Case 1. Ann believes Bob cannot see Ann.

Ann sees (a state for) Bob who cannot see Ann. Bob sees Ann iff Ann believes Bob cannot see Ann. Since Bob cannot see Ann, Ann does not believe Bob cannot see Ann. Contradiction.

Case 2. Ann does not believe Bob cannot see Ann.

Ann sees (a state for) Bob who sees Ann. Bob sees Ann iff Ann believes Bob cannot see Ann. Since Bob sees Ann, Ann believes Bob cannot see Ann. Contradiction.

We are now ready to prove Theorem 5.4. The proof will actually give a sharper result which pinpoints the location of the holes in a belief model. We say that belief model $\mathcal{M}$ has a *hole* at a set Y if Y is nonempty but is not assumed by any element. Thus, a belief model is complete for a language $\mathcal{L}$ if and only if it has no holes in $\mathcal{L}^a$ and no holes in $\mathcal{L}^b$.

Let us also say that $\mathcal{M}$ has a *big hole* at Y if Y is nonempty but is not believed by any element. Thus, $\mathcal{M}$ has a big hole at Y if and only if it has a hole at every nonempty subset of Y.

Theorem 5.7. *Every belief model $\mathcal{M}$ has either a hole at U^a, a hole at U^b, a big hole at one of the formulas*

$$\forall x \; P^b(y, x), \tag{i}$$

$$x \; believes \; \forall x \; P^b(y, x), \tag{ii}$$

$$y \; believes \; [x \; believes \; \forall x \; P^b(y, x)], \tag{iii}$$

a hole at the formula

$$D(x), \tag{iv}$$

or a big hole at the formula

$$y \; assumes \; D(x). \tag{v}$$

Thus, there is no belief model which is complete for a language $\mathcal{L}$ which contains the tautologically true formula and formulas (i)–(v).

This immediately implies Theorem 5.4, since each of the formulas (i)–(v) is first order. Looking back at the list of statements in the Introduction, formulas (i)–(v) are the formal counterparts of the statements 1–5 respectively, and the sets U^a and U^b are counterparts of the tautologically true statement 0.

Proof. Suppose the theorem does not hold for a belief model $\mathcal{M}$. Since $\mathcal{M}$ does not have holes at U^a and U^b, there is an element y_1 which satisfies formula (i) and an element x_1 such that $\forall y P^a(x_1, y)$. Since $\mathcal{M}$ does not have big holes at formulas (i)–(iii), there is an element x_2 that believes formula (i) and thus satisfies (ii), an element y_3 that believes (ii) and thus satisfies (iii), and an element x_4 that believes formula (iii). Then by Lemma 5.5, x_4 satisfies formula (iv). Since there is no hole at (iv), there is an element y_4 which assumes the formula (iv) and thus satisfies (v). But then by Lemma 5.6, $\mathcal{M}$ must have a big hole at (v), and we have a contradiction. □

In Section 9, we will give an example showing that the list of formulas (i)–(v) in the above theorem cannot be shortened to (i)–(iv).

6. Assumption in Modal Logic

In this section, we prepare for a modal formulation of our results by presenting a basic modal logic with an assumption operator in a single-player setting, which we will call *assumption logic*. This logic is related to the modal logic of Humberstone (1987) and is also a simpler special case of Bonanno's modal logic for belief revision in Bonanno (2005). Bonanno's logic has modal operators B_0 for initial belief, B_1 for final belief, I for being informed that, and A for the universal operator, which allows him to get an axiom set and completeness theorem. In our simpler case, the initial belief operator B_0 is the same as the universal operator A, and the operator I will be interpreted as assumption.

We will use the standard symbol $\square$ for the belief operator, and the symbol $\heartsuit$ for the assumption operator.

We refer to Boolos (1993) for an elementary introduction to modal logic. The models for assumption logic are (Kripke) *frames* $\mathcal{W} = (W, P)$ where P is a binary relation on W. The elements of W are called worlds, and P is called the *accessibility relation*. At a world w, $\square\varphi$ is interpreted as "w believes φ", $\heartsuit\varphi$ as "w assumes φ", and $A\varphi$ as $\forall z \varphi$.

The *formulas* of assumption logic are built from a set L of proposition symbols and the false formula $\perp$ using propositional connectives and the three modal operators, $\square$, $\heartsuit$, and A. Note that $\neg \perp$ is the true formula.

In a frame $\mathcal{W}$, a *valuation* is a function V which associates a subset of $V(\mathbf{D}) \subseteq W$ with each proposition symbol $\mathbf{D} \in L$. For a given valuation V, the notion of a formula φ being *true* at a world w, in symbols $w \models \varphi$, is defined by induction on the complexity of φ. For a proposition symbol $\mathbf{D}$, $w \models \mathbf{D}$ if $w \in V(\mathbf{D})$. The rules for connectives are as usual, and the rules for the modal operators are as follows:

$$w \models \square\varphi \text{ if for all } z \in W, \quad P(w, z) \text{ implies } z \models \varphi.$$
$$w \models \heartsuit\varphi \text{ if for all } z \in W, \quad P(w, z) \text{ if and only if } z \models \varphi.$$
$$w \models A\varphi \text{ if for all } z \in W, \quad z \models \varphi.$$

A formula is *valid for V in $\mathcal{W}$* if it is true at all $w \in W$, and *satisfiable for V in $\mathcal{W}$* if it is true at some $w \in W$.

Note that if the valuation assigns a first order definable set to each proposition symbol, then for each modal formula φ, $w \models \varphi$ is expressible by a formula with one free variable in the first order language of $\mathcal{W}$.

To shed more light on the behavior of the assumption operator in the modal logic setting, we restate the axioms and completeness theorem of Bonanno (2005) with the simplification that comes from eliminating the initial belief operator B_0.

Rules of Inference for Assumption Logic

Modus Ponens: From φ, $\varphi \rightarrow \psi$ infer ψ.
Necessitation: From φ infer $A\varphi$.

Axioms for Assumption Logic

All propositional tautologies.
Distribution Axioms for $\square$ and A:

$$\square(\varphi \rightarrow \psi) \rightarrow (\square\varphi \rightarrow \square\psi), \quad A(\varphi \rightarrow \psi) \rightarrow (A\varphi \rightarrow A\psi).$$

S_5 Axioms for A:

$$A\varphi \rightarrow \varphi, \quad \neg A\varphi \rightarrow A\neg A\varphi$$

Inclusion Axiom for $\square$:

$$A\varphi \rightarrow \square\varphi$$

Axioms for $\heartsuit$:

$$\heartsuit\varphi \wedge \heartsuit\psi \rightarrow A(\varphi \leftrightarrow \psi), \quad A(\varphi \leftrightarrow \psi) \rightarrow (\heartsuit\varphi \leftrightarrow \heartsuit\psi),$$
$$\heartsuit\varphi \wedge \square\psi \rightarrow A(\varphi \rightarrow \psi), \quad \heartsuit\varphi \rightarrow \square\varphi$$

Proposition 6.1. (*Bonanno* [2005] *Soundness and Completeness*)
   (*i*) *Assumption logic is sound, that is, every provable formula is valid in all frames.*
   (*ii*) *Assumption logic is complete, that is, every formula which is valid in all frames is provable.*

7. Impossibility Results in Modal Form

In this section, we reformulate our two-player impossibility result in a modal logic setting. For each pair of players cd among Ann and Bob, there will be an operator $\square^{cd}$ of beliefs for c about d, and an operator $\heartsuit^{cd}$ of assumptions for c about d. We first define the models of our modal logic for two players.

Definition 7.1. An *interactive frame* is a structure $\mathcal{W} = (W, P, U^a, U^b)$ with a binary relation $P \subseteq W \times W$ and disjoint sets U^a, U^b, such that $\mathcal{M} = (U^a, U^b, P^a, P^b)$ is a belief model, where $U^a \cup U^b = W$, $P^a = P \cap U^a \times U^b$, and $P^b = P \cap U^b \times U^a$.

In an interactive frame, the states for both a and b become members of the set W of worlds. P is the *accessibility relation*.

This definition makes no restrictions on the part of P in $U^a \times U^a$ and $U^b \times U^b$. Thus, beliefs for players about themselves are allowed in the interactive frame, but the corresponding belief model does not depend on them. With this setup, it will be apparent that our impossibility phenomenon is not affected by the players' beliefs about themselves.

The requirement that $\mathcal{M}$ is a belief model means that the sets U^a, U^b are nonempty, and the relations $P \cap U^a \times U^b$ and $P \cap U^b \times U^a$ are serial proper subsets.

We are using frames with a set of states for each player but only a single accessibility relation. This is convenient for a study of the beliefs for one player about another. Another approach would be to use frames with an accessibility relation for each player, as in Lomuscio (1999). This approach gives a modal logic with a belief operator for each player about the state of the world.

We now introduce the formulas and semantical interpretation for the modal logic of interactive frames.

Interactive modal logic will have two distinguished proposition symbols $\mathbf{U}^a, \mathbf{U}^b$ and a set L of additional proposition symbols. By a *modal formula* we mean an expression which is built from proposition symbols and the false formula $\perp$ using propositional connectives, the universal modal operator A, and the modal operators $\square^{cd}$, $\heartsuit^{cd}$ where c and d are taken from $\{a, b\}$.

As before, a *valuation* V associates a subset of $V(\mathbf{D}) \subseteq W$ with each proposition symbol $\mathbf{D}$ in the set L.

Given a valuation V on $\mathcal{W}$, the notion of a world w being true at a formula φ, in symbols $w \models \varphi$, is defined by induction on the complexity of φ as follows: $w \models \mathbf{U}^a$ if $w \in U^a$, and similarly for b. That is, U^a is true at each state for Ann, and $\mathbf{U}^b$ is true at each state for Bob. The rules for connectives are as usual, and the rules for the modal operators for each pair of players $c, d \in \{a, b\}$ are:

$$w \models \square^{cd}\varphi \text{ if } (w \models \mathbf{U}^c \wedge \forall z[(P(w,z) \wedge z \models \mathbf{U}^d) \rightarrow z \models \varphi]).$$
$$w \models \heartsuit^{cd}\varphi \text{ if } (w \models \mathbf{U}^c \wedge \forall z[(P(w,z) \wedge z \models \mathbf{U}^d) \leftrightarrow z \models \varphi]).$$
$$w \models A\varphi \text{ if } \forall z \ z \models \varphi.$$

Validity and satisfiability are defined as in the preceding section.

We again note that if the valuation assigns a first order definable set to each proposition symbol, then for each modal formula φ, $w \models \varphi$ is expressible by a formula of the first order language of $\mathcal{W}$.

In the notation of Section 5 where x has sort U^a and y has sort U^b,

$$x \models \Box^{ab}\varphi \text{ says "x believes $\varphi(y)$",}$$
$$x \models \heartsuit^{ab}\varphi \text{ says "x assumes $\varphi(y)$",}$$

and similarly with a, b and x, y reversed.

In classical modal logic, one often adds axioms such as $\Box\varphi \to \varphi$ or $\Box\varphi \to \Box\Box\varphi$, which are reasonable hypotheses for beliefs about one's own beliefs. In the two-player setting, the analogue of $\Box\varphi \to \varphi$ would be

$$\Box^{aa}\varphi \to \varphi, \quad \Box^{bb}\varphi \to \varphi.$$

The analogue of $\Box\varphi \to \Box\Box\varphi$ would be

$$\Box^{aa}\varphi \to \Box^{aa}\Box^{aa}\varphi, \quad \Box^{ab}\varphi \to \Box^{ab}\Box^{ab}\varphi,$$

and similarly with a and b reversed.

However, similar properties which involve only a player's beliefs about the other player cannot be valid in an interactive frame. It is easy to see that each of the formulas

$$\Box^{ab}\mathbf{U}^b \leftrightarrow \mathbf{U}^a, \quad \Box^{ba}\mathbf{U}^a \leftrightarrow \mathbf{U}^b, \quad \Box^{ab}\mathbf{U}^a \leftrightarrow \bot, \quad \Box^{ba}\mathbf{U}^b \leftrightarrow \bot$$

is valid in all interactive frames. Since the sets U^a and U^b are nonempty and disjoint, it follows that the formulas

$$\Box^{ab}\mathbf{U}^b \to \mathbf{U}^b, \quad \Box^{ab}\mathbf{U}^b \to \Box^{ba}\Box^{ab}\mathbf{U}^b, \quad \Box^{ab}\mathbf{U}^b \to \Box^{ab}\Box^{ab}\mathbf{U}^b$$

are *never* valid in an interactive frame.

We now restate the results of Section 5 in the modal setting. The universal operator A will not be needed in these results, but is included in the language because, as in the preceding section, it makes it possible to state the properties of interactive frames as axioms in the modal language. The modal operators such as $\Box^{aa}$ of beliefs and assumptions for players about themselves will also not be needed, and are mentioned only to clarify the overall picture. What we do need are the operators $\Box^{ab}$, $\heartsuit^{ab}$ and their counterparts with ba of beliefs and assumptions for one player about the other.

Definition 7.2. For the remainder of this section we will always suppose that $\mathcal{W}$ is an interactive frame, $\mathbf{D}$ is a proposition symbol (for diagonal), and V is a valuation in $\mathcal{W}$ such that $V(\mathbf{D})$ is the set

$$D = \{w \in W : (\forall z \in W)[P(w,z) \to \neg P(z,w)]\}.$$

"Satisfiable" means satisfiable in $\mathcal{W}$ at V, and similarly for "valid".

Lemma 7.3. ($= Lemma\ 5.5$) *If $\heartsuit^{ab}\mathbf{U}^b$ is satisfiable then*

$$[\Box^{ab}\Box^{ba}\Box^{ab}\heartsuit^{ba}\mathbf{U}^a] \to \mathbf{D}$$

is valid.

Lemma 7.4. ($= Lemma\ 5.6$) *$\neg\Box^{ab}\heartsuit^{ba}(\mathbf{U}^a \wedge \mathbf{D})$ is valid.*

We say that $\mathcal{W}$ with V has a *hole* at a formula φ if either $\mathbf{U}^b \wedge \varphi$ is satisfiable but $\heartsuit^{ab}\varphi$ is not, or $\mathbf{U}^a \wedge \varphi$ is satisfiable but $\heartsuit^{ba}\varphi$ is not. A *big hole* is defined similarly but with $\Box$ instead of $\heartsuit$. An interactive frame $\mathcal{W}$ with valuation V is *complete* for a set $\mathcal{L}$ of modal formulas if it does not have a hole in $\mathcal{L}$.

Theorem 7.5. ($= Theorem\ 5.7$) *There is either a hole at $\mathbf{U}^a$, a hole at $\mathbf{U}^b$, a big hole at one of the formulas*

$$\heartsuit^{ba}\mathbf{U}^a, \quad \Box^{ab}\heartsuit^{ba}\mathbf{U}^a, \quad \Box^{ba}\Box^{ab}\heartsuit^{ba}\mathbf{U}^a,$$

a hole at the formula $\mathbf{U}^a \wedge \mathbf{D}$, or a big hole at the formula $\heartsuit^{ba}(\mathbf{U}^a \wedge \mathbf{D})$. Thus, there is no complete interactive frame for the set of all modal formulas built from $\mathbf{U}^a, \mathbf{U}^b, \mathbf{D}$.

As a corollary, we see that the impossibility result also holds for the set of modal formulas which have only the assumption operators $\heartsuit^{ab}, \heartsuit^{ba}$, without the belief operators.

Corollary 7.6. *There is a hole at one of the formulas*

$$\mathbf{U}^a, \quad \mathbf{U}^b, \quad \heartsuit^{ba}\mathbf{U}^a, \quad \heartsuit^{ab}\heartsuit^{ba}\mathbf{U}^a, \quad \heartsuit^{ba}\heartsuit^{ab}\heartsuit^{ba}\mathbf{U}^a, \quad \mathbf{U}^a \wedge \mathbf{D}, \quad \heartsuit^{ba}(\mathbf{U}^a \wedge \mathbf{D}).$$

Thus, there is no complete interactive frame for the set of modal formulas built from $\mathbf{U}^a, \mathbf{U}^b, \mathbf{D}$ in which the belief operators do not occur.

Problem 7.7. *Is there an interactive frame $\mathcal{W}$ which is complete for the set of modal formulas φ built from $\mathbf{U}^a, \mathbf{U}^b, \mathbf{D}$ such that the assumption operators $\heartsuit^{cd}$ do not occur in φ (that is, for each such formula φ, $\mathcal{W}$ with V does not have a hole at φ)?*

Following the pattern in Section 6, one can build a set of axioms for interactive modal logic, and then prove a completeness theorem using the methods of Bonanno (2005). There would be axiom schemes analogous to those of the preceding section for the two-player belief and assumption operators, plus a finite set of axioms saying that U^a, U^b are nonempty and partition W, that P^a and P^b are proper, and that $\forall x \, \exists y P^a(x, y)$ and similarly for b.

8. Strategic Belief Models

Strategic models are a particular class of belief models, used in applications to games. In a strategic model, each player has a set of strategies and a set of types, each strategy-type pair is a state for a player, and each type for one player has beliefs about the states for the other player. It is intended that the strategy sets (the sets S^a, S^b below) are part of an underlying game with payoff functions (maps π^a, π^b from $S^a \times S^b$ to the reals), but the payoff functions are not needed in the formal treatment here.

Definition 8.1. Given a pair of nonempty sets (S^a, S^b), an (S^a, S^b)-based *strategic model* is a belief model $\mathcal{M} = (U^a, U^b, P^a, P^b, \ldots)$ where:

(a) U^a and U^b are Cartesian products $U^a = S^a \times T^a, U^b = S^b \times T^b$; members of S^a and S^b are called *strategies*, and members of T^a and T^b are called *types*.

(b) $P^a((s^a, t^a), y)$ depends only on (t^a, y), and $P^b((s^b, t^b), x)$ depends only on (t^b, x).

(c) The vocabulary of $\mathcal{M}$ must include the following additional relations, which capture the sets S^a, T^a, S^b, T^b: The binary relation τ^a on U^a which says that two states in U^a have the same type, for each $s^a \in S^a$ the unary relation $s^a(x)$ on U^a which holds when x has strategy s^a, and the analogous relations with b in place of a.

Thus, a strategic model has the form:

$$\mathcal{M} = (U^a, U^b, P^a, P^b, \tau^a, \tau^b, s^a, s^b, \ldots : s^a \in S^a, s^b \in S^b).$$

In view of condition (b), each type t^a for Ann assumes a nonempty set of states for Bob, and *vice versa*. Also, the extra relations in the vocabulary of a strategic model give us the following useful fact. (Recall that an elementary submodel of $\mathcal{M}$ is a submodel $\mathcal{N}$ such that each tuple of elements of $\mathcal{N}$ satisfies the same first order formulas in $\mathcal{N}$ as in $\mathcal{M}$.)

Proposition 8.2. *If $\mathcal{M}$ is an (S^a, S^b)-based strategic model and $\mathcal{N}$ is an elementary submodel of $\mathcal{M}$, then $\mathcal{N}$ is an (S^a, S^b)-based strategic model.*

Proof. Let $\mathcal{N} = (V^a, V^b, \ldots)$. We must find sets of types T_0^a, T_0^b such that $V^a = S^a \times T_0^a$ and $V^b = S^b \times T_0^b$. We first observe that for each $s^a \in S^a$, the sentence $\exists x s^a(x)$ holds in $\mathcal{N}$ because it holds in $\mathcal{M}$. Since $\mathcal{N}$ is a submodel of $\mathcal{M}$, every $x \in V^a$ satisfies $s^a(x)$ for some $s^a \in S^a$. Moreover, for each $r^a, s^a \in S^a$, $\mathcal{N}$ satisfies the sentence

$$\forall x[s^a(x) \longrightarrow \exists u[r^a(u) \wedge \tau^a(x, u)]],$$

It follows that $V^a = S^a \times T_0^a$, where

$$T_0^a = \{t^a \in T^a : (s^a, t^a) \in V^a)\}$$

We can define T_0^b in a similar way and get $V^b = S^a \times T_0^b$. It is clear that the relations P^a, P^b, τ^a, τ^b have the required properties in $\mathcal{N}$. Therefore, $\mathcal{N}$ is an (S^a, S^b)-based strategic model. $\qquad\square$

Here is a simple example of a strategic model.

Example 8.3. Consider Matching Pennies (Figure 8.1) and the associated strategic model in Figure 8.2. Here, Ann's type is either v^a or w^a; Bob's type is either v^b or w^b. Type v^a of Ann assumes the singleton set $\{(L, v^b)\}$ as depicted by the plus sign in the upper left of Figure 8.2. In words, this type of Ann assumes that Bob chooses the strategy L and is of type v^b. The interpretation of the rest of Figure 8.2 is similar.

Fix the state (U, v^a, R, w^b). At this state, Ann chooses strategy U, and assumes that Bob chooses L and is of type v^b. Bob in fact chooses R (contrary to what Ann assumes), and assumes that Ann chooses U and is of type v^a (which is indeed the situation). This is an example of the kind of game scenario that strategic models are designed to describe.

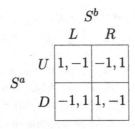

Figure 8.1.

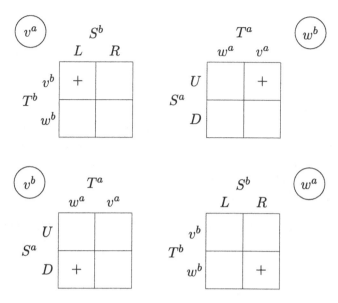

Figure 8.2.

9. Weakly Complete and Semi-Complete Models

In this section, we give two positive results. First, we introduce weakly complete models and use them to show that the notion "assumes" is an essential ingredient in our main impossibility result. Second, we introduce semi-complete models and use them to show that the existence problem for complete models is intrinsically multi-player, i.e., game-theoretic, in nature. We will get stronger positive results by building strategic models with strategy sets given in advance, rather than plain belief models.

To show that the notion "assumes" is essential, we will find strategic models which are complete in the weaker sense that every statement which is possible can be believed (instead of assumed) by the player. Thus, a paradox like the one in the Introduction does not arise in the ordinary modal logics which can express "believes" but cannot express "assumes".

Definition 9.1. Fix sets S^a and S^b. An (S^a, S^b)-based strategic model $\mathcal{M}$ will be called *weakly complete* if each nonempty set $Y \subseteq S^b \times T^b$ is believed by some $t^a \in T^a$, and each nonempty set $X \subseteq S^a \times T^a$ is believed by some $t^b \in T^b$.

Note that a weakly complete model $\mathcal{M}$ is "weakly complete for *every* language $\mathcal{L}$", even when extra relations are added to the vocabulary. That is, each nonempty set $Y \in \mathcal{L}^b$ is believed by some $t^a \in T^a$ and each nonempty set $X \in \mathcal{L}^a$ is believed by some $t^b \in T^b$.

Proposition 9.2. *Given sets S^a and S^b, there is a weakly complete (S^a, S^b)-based strategic model.*

Proof. Let T^a and T^b be sets of cardinality $\aleph_0 + |S^a| + |S^b|$. Then, $U^a = S^a \times T^a$ and $U^b = S^b \times T^b$ have the same cardinality as T^a and T^b. Let $f : T^a \to U^b$ and $g : T^b \to U^a$ be bijections. Let $P^a((s^a, t^a), y)$ hold if and only if $f(t^a) = y$, and define P^b in the analogous way from g. Then $\mathcal{M} = (S^a \times T^a, S^b \times T^b, P^a, P^b, ...)$ is weakly complete; if $y \in Y \subseteq U^b$ then $f^{-1}(y)$ believes Y, and similarly with a, b reversed. $\qquad\square$

We next show that the existence problem for complete models is intrinsically multi-player, i.e., game-theoretic, in nature. We do this by showing that there is no difficulty if the condition is that every nonempty set of Bob's states is assumed by one of Ann's states, or *vice versa*. The impossibility arises when we want both conditions simultaneously.

Definition 9.3. A belief model $\mathcal{M}$ is *semi-complete* (for player a) if every nonempty subset of U^b is assumed by some element of U^a. $\mathcal{M}$ is *first order semi-complete* (for a) if $\mathcal{M}$ is complete for the language $\mathcal{H}$ consisting of all first order formulas $\varphi(y)$ where y has sort U^b.

Note that a semi-complete model $\mathcal{M}$ is first order semi-complete, and remains first order semi-complete even when extra relations are added to the vocabulary. Moreover, semi-completeness for a depends only on the relation P^a, not on P^b.

Proposition 9.4. *Given sets S^a, S^b, T^b, there is an (S^a, S^b)-based strategic model $\mathcal{M}$ with the given type set T^b which is semi-complete for a. If the sets S^a, S^b, T^b are finite, then $\mathcal{M}$ may be taken to be finite.*

Proof. Let T^a be the set of all nonempty subsets of $S^b \times T^b$, and define P^a by $P^a(s^a, t^a, s^b, t^b)$ if and only if $(s^b, t^b) \in t^a$. Then for any P^b, the strategic model $\mathcal{M} = (S^a \times T^a, S^b \times T^b, P^a, P^b, ...)$ is semi-complete for a. $\quad\square$

We now show that if the sets S^a, S^b are at most countable, there is a countable (S^a, S^b)-based strategic model which is first order semi-complete.

Proposition 9.5. *Given finite or countable strategy sets S^a, S^b, there is a countable (S^a, S^b)-based strategic model $\mathcal{N}$ which is first order semi-complete.*

Proof. By Proposition 9.4 there is an infinite (S^a, S^b)-based strategic model $\mathcal{M}$ which is semi-complete for a. Then $\mathcal{M}$ is first order semi-complete for a. By the downward Lowenheim–Skolem–Tarski theorem, $\mathcal{M}$ has a count able elementary submodel $\mathcal{N}$. By Proposition 8.2, $\mathcal{N}$ is an (S^a, S^b)-based strategic model. It follows that $\mathcal{N}$ is first order semi-complete for a. $\quad\square$

The above argument actually gives the following more general fact.

Proposition 9.6. *Suppose M is an (S^a, S^b)-based strategic model which is complete for a language $\mathcal{L}$. Then any elementary submodel of $\mathcal{M}$ is an (S^a, S^b)-based strategic model which is complete for $\mathcal{L}$. If the vocabulary of $\mathcal{M}$ is countable, then $\mathcal{M}$ has a countable elementary submodel, giving a countable (S^a, S^b)-based strategic model which is complete for $\mathcal{L}$.*

Proof. By the downward Löwenheim–Skolem–Tarski theorem and Propositions 4.3 and 8.2. $\quad\square$

Our next example shows that Theorem 5.7, which pinpointed the location of the holes in a belief model, cannot be improved by shortening the list of formulas (i)–(v) to (i)–(iv). The example produces a belief model which is complete for U^a, U^b and formulas (i)–(iv), and as a bonus, is semi-complete for b.

Example 9.7. Let S^a, S^b be nonempty sets. There is an (S^a, S^b)-based strategic model which is semi-complete for b and is complete for a language containing U^a, U^b and the formulas (i)–(iv) from Theorem 5.7.

Moreover, if S^a, S^b are finite or countable, there is a countable (S^a, S^b)-based strategic model which is first order semi-complete for b and complete for the above language.

Proof. Let T^a be any set with at least 3 distinct elements c, d, e. Put $U^a = S^a \times T^a$, let T^b be the set of all nonempty subsets of U^a, and put $U^b = S^b \times T^b$. Let P^b be the relation $P^b((s^b, t^b), x)$ iff $x \in t^b$. Then choose any relation P^a such that c assumes U^b, d assumes $S^b \times \{U^a\}$, e assumes $S^b \times \{\{d\}\}$, and every other element of T^a assumes a subset of U^b which is not contained in $S^b \times \{U^a\}$.

As in Proposition 9.4, the resulting model $\mathcal{M}$ is a semi-complete strategic model for b. One can readily check that in $\mathcal{M}$, formula (i) defines the set $S^b \times \{U^a\}$ which is assumed by d, and formula (iii) defines the set $S^b \times \{\{d\}\}$ which is assumed by e. Since $\mathcal{M}$ is semi-complete for b, it is complete for every subset of U^a. Thus, $\mathcal{M}$ is complete for U^a, U^b, and the formulas (i)–(iv).

The moreover clause now follows from Proposition 9.6. $\quad\square$

10. Positively and Topologically Complete Models

We now give some more positive results on the existence of complete models. (We already had Example 4.4 in Section 4, and also the semi-complete models of the previous section.) We show that complete models exist for the fragment of first order logic which has the positive connectives $\wedge, \vee$, the quantifiers $\forall, \exists$, and the belief and assumption operators, but does not have the negation symbol. We will work with strategic models. As usual, it is understood that x is a variable of sort U^a and y is a variable of sort U^b.

Definition 10.1. Let $\mathcal{M}$ be a strategic model. A *positive formula* is a first order formula for $\mathcal{M}$ which is built according to the following rules:

- Every atomic formula is positive.
- If φ, ψ are positive formulas, then so are $\varphi \wedge \psi, \varphi \vee \psi, \forall x \varphi, \forall y \varphi,$ $\exists x \varphi, \exists y \varphi$.
- If $\varphi(y)$ is a positive formula, then so are $[x \text{ believes } \varphi]$ and $[x \text{ assumes } \varphi]$.
- If $\varphi(x)$ is a positive formula, then so are $[y \text{ believes } \varphi]$ and $[y \text{ assumes } \varphi]$.

The *positive language* for $\mathcal{M}$ is the set $\mathcal{P}$ of positive formulas in the first order language for $\mathcal{M}$. Thus, $\mathcal{P}^a$ is the set of all subsets of U^a which are definable by a positive formula $\varphi(x)$, and similarly for $\mathcal{P}^b$.

Theorem 10.2. *For every pair of finite or countable strategy sets S^a, S^b, there is a countable (S^a, S^b)-based strategic model $\mathcal{M}$ which is complete for its positive language.*

Before giving the proof, which uses topological methods, let us consider what our main impossibility theorem tells us about models which are complete for the positive language. Since $\mathcal{M}$ cannot be complete for its first order language, there must be a set which is definable in the first order language $\mathcal{L}$ but is not definable in the positive language $\mathcal{P}$. That is, if the players use the positive language, there must be a first order property that the players cannot express. Using Theorem 5.7, we can pinpoint exactly where this happens.

Proposition 10.3. *Let $\mathcal{M}$ be a strategic model which is complete for its positive language. Then the set D defined by the diagonal formula*

$$\forall y[P^a(x, y) \to \neg P^b(y, x)]$$

does not belong to $\mathcal{P}^a$, but its negation does belong to $\mathcal{P}^a$.

Proof. The sets U^a, U^b are definable by the positive formulas $x = x$ and $y = y$, and the formulas (i)–(iii) of Theorem 5.7 are positive formulas. Moreover, the complement $U^a \backslash D$ is defined by the positive formula $\exists y [P^a(x, y) \wedge P^b(y, x)]$.

Assume that D is definable in $\mathcal{M}$ by a positive formula. Since the positive formulas are closed under the belief and assumption operators, the set defined by the formula (v) is also defined by a positive formula. But then, since $\mathcal{M}$ is complete for its positive language, there cannot be a hole at any of the formulas (i)–(v), contradicting Theorem 5.7. We conclude that the set D does not belong to $\mathcal{P}^a$. $\qquad\square$

We now construct strategic models which are complete in a topological sense. We will then show that such models are also complete for the positive language. Given a topological space X, let $K(X)$ be the space of all nonempty compact subsets of X endowed with the Vietoris topology. If X is compact metrizable, then so is $K(X)$ (Kechris [1995, Theorem 4.26 and Exercise 4.20i]).[7]

Theorem 10.4. *Let S^a and S^b be compact metrizable spaces. There are compact metrizable spaces T^a, T^b and an (S^a, S^b)-based strategic model*

$$\mathcal{M} = (U^a, U^b, P^a, P^b, \tau^a, \tau^b, s^a, s^b, \ldots : s^a \in S^a, s^b \in S^b),$$

such that;

(a) Each $t^a \in T^a$ assumes a compact set $\kappa^a(t^a) \in K(U^b)$, and each $t^b \in T^b$ assumes a compact set $\kappa^b(t^b) \in K(U^a)$.

(b) The mappings $\kappa^a : T^a \to K(U^b)$ and $\kappa^b : T^b \to K(U^a)$ are continuous surjections.

Proof. Let $\mathcal{C}$ be the Cantor space, i.e., the space $\{0, 1\}^{\mathbb{N}}$. There is a continuous surjection from $\mathcal{C}$ to any compact metrizable space (Kechris [1995, Theorem 4.18]). Set $T^a = T^b = C$, $U^a = S^a \times T^a$, $U^b = S^b \times T^b$. The space $K(U^b)$ is compact metrizable, so there is a continuous surjection κ^a from T^a to $K(U^b)$, and similarly with a, b reversed. The model $\mathcal{M}$ is obtained by setting $P^a((s^a, t^a), y)$ if and only if $y \in \kappa^a(t^a)$, and similarly for P^b. Then conditions (a) and (b) hold. The relation P^a is a proper subset of $U^a \times U^b$ because the space T^b has compact nonempty proper subsets. Therefore, $\mathcal{M}$ is a strategic belief model. $\qquad\square$

---

[7] All topological spaces are understood to be nonempty.

Definition 10.5. We will call a strategic model with the properties (a)–(b) of Theorem 10.4 an (S^a, S^b)-based *topologically complete model.*

Note that a topologically complete model $\mathcal{M}$ is complete for any language $\mathcal{K}$ such that $\mathcal{K}^a \subseteq K(U^a)$ and $\mathcal{K}^b \subseteq K(U^b)$.

We now make the connection between topologically complete models and the positive language.

Lemma 10.6. *Let*

$$M = (U^a, U^b, P^a, P^b, \tau^a, \tau^b, s^a, s^b, \ldots : s^a \in S^a, s^b \in S^b)$$

be a topologically complete model such that each of the extra relations in the list is compact. Then every positive formula $\varphi(x)$ defines a compact set in U^a and similarly for $\varphi(y)$ and U^b. Hence, $\mathcal{M}$ is complete for its positive language.

Proof. It is clear that the unary relations are compact sets, that is, for each strategy $s^a \in S^a$ the set of states x for Ann with strategy s^a is compact, and similarly for S^b. It is also clear that the relations τ^a, τ^b, which hold for pairs of states with the same type, are compact. We next show that the relation P^a is compact. By Kechris (1995, Exercise 4.29.i), the relation $C = \{(y, Y) : y \in Y\}$ is compact in $U^b \times K(U^b)$. Since the function $f^a : (s^a, t^a) \longmapsto \kappa^a(t^a)$ is continuous from U^a to $K(U^b)$, the set $P^a = \{(x, y) : (y, f^a(x)) \in C\}$ is compact. Similarly, the relation P^b is compact. Thus, every atomic formula defines a compact set. Since the spaces are Hausdorff, compact sets are closed, finite unions and intersections of compact sets are compact, and projections of compact relations by universal or existential quantifiers are compact.

To complete the proof, it suffices to show that for each compact set $Y \subseteq U^b$, the sets

$$A = \{x \in U^a : x \text{ assumes } Y\}, \quad B = \{x \in U^a : x \text{ believes } Y\}$$

are compact (and similarly with a and b reversed). We have $A = (f^a)^{-1}(\{Y\})$. Since the finite set $\{Y\}$ is closed in $K(U^b)$, A is compact in U^a. Similarly, $B = (f^a)^{-1}(\{Z : Z \subseteq Y\})$ and the set $\{Z : Z \subseteq Y\}$ is closed in $K(U^b)$, so B is compact in U^a. $\qquad\square$

We remark that the positive language $\mathcal{P}$ depends only on the relations of the model $\mathcal{M}$, while the sets $K(U^a)$ and $K(U^b)$ depend on the topology of $\mathcal{M}$. If the vocabulary of $\mathcal{M}$ is countable, then the positive language $\mathcal{P}$ will be countable, and since $K(U^a)$ is uncountable, $\mathcal{P}^a$ will be a proper subset of $K(U^a)$. In fact, the sets U^a and U^b are uncountable and each finite set is compact, so $K(U^a)\backslash\mathcal{P}$ will even contain finite sets. At the other extreme, in the above lemma we can take the vocabulary of $\mathcal{M}$ to be uncountable and even to contain all compact relations, and in this case we will have $\mathcal{P}^a = K(U^a)$ and $\mathcal{P}^b = K(U^b)$.

Proof of Theorem 10.2. Since S^a and S^b are finite or countable, there exist compact metrizable topologies on S^a and S^b. By Theorem 10.4, there is an (S^a, S^b)-based topologically complete model

$$\mathcal{M} = (U^a, U^b, P^a, P^b, \tau^a, \tau^b, s^a, s^b, \ldots : s^a \in S^a, s^b \in S^b).$$

It is clear from the definition that $\mathcal{M}$ without the extra relations indicated by the three dots is still topologically complete, so we may take $\mathcal{M}$ to have no extra relations. Then $\mathcal{M}$ has a countable vocabulary, and $\mathcal{M}$ is complete for its positive language by Lemma 10.6. In general, $\mathcal{M}$ is uncountable, but it follows from Proposition 9.6 that there is a countable (S^a, S^b)-based strategic model $\mathcal{N}$ (a countable elementary submodel of $\mathcal{M}$) which is complete for $\mathcal{P}$. $\qquad\square$

11. Other Models in Game Theory

The purpose of this section is to explain briefly how complete belief models are related to other models in the game theory literature. Here is an attempt to classify models of all possible beliefs that have been considered.

(i) *Universal models*: Start with a space of underlying uncertainty. (This could be the players' strategy sets, or sets of possible payoff functions for the players, etc.) The players then form beliefs over this space (their zeroth-order beliefs), beliefs over this space and the spaces of zeroth-order beliefs for the other players (their first-order beliefs), and so on inductively through the ordinals. The question is whether this process 'ends' at some level? More precisely, do the beliefs at some ordinal level α determine the beliefs at all subsequent levels? If so, we get a universal model. If not, i.e., if the set of α-level beliefs increases through all ordinals α, we get a nonexistence result.

There are many papers on universal models. Existence results are given by Armbruster and Böge (1979), Böge and Eisele (1979), Mertens and Zamir (1985), Brandenburger and Dekel (1993), Heifetz (1993), Epstein and Wang (1996), Battigalli and Siniscalchi (1999, 2002), Mariotti, Meier, and Piccione (2005), and Pinter (2005), among others. Fagin, Halpern, and Vardi (1991), Fagin (1994), Fagin *et al.* (1999), and Heifetz and

Samet (1999) give nonexistence results. The positive results are obtained by making various topological or measure-theoretic hypotheses (as we did in the previous section). We should also note that many of these papers formalize belief as probability and not possibility as we have done. But — again in line with our results in Section 10 — the key point is the topological assumptions. (Epstein–Wang consider preferences, with a topological structure.) Aumann (1999) treats knowledge rather than belief, and uses S5 logic to get a positive result. (On this last result, see also Heifetz [1999]).

(ii) *Complete models*: These are defined as in our Definition 4.2 or Definition 10.5 — i.e., in terms of 'two-way surjectivity.' One way to obtain complete models is to construct a universal model. For example, Mariotti, Meier, and Piccione (2005) get a complete model for compact Hausdorff spaces, as a corollary of the existence of a universal model. Our proof of Theorem 10.4 gives a simple direct construction of a complete model for compact metrizable spaces. Battigalli and Siniscalchi (1999, 2002) get a complete model for spaces of conditional probability systems, again as a corollary of the existence of a universal model. Brandenburger, Friedenberg, and Keisler (2006) has a direct construction of a complete model for spaces of lexicographic probability systems. Salonen (1999) gives a variety of existence results on completeness, under a variety of assumptions.

(iii) *Terminal models*: Given a category **C** of models of beliefs, call a model $\mathcal{M}$ in **C** terminal if for any other model $\mathcal{N}$ in **C**, there is a unique belief-preserving morphism from $\mathcal{N}$ to $\mathcal{M}$.[8] Heifetz and Samet (1998a) show existence of a terminal model for probabilities, without topological assumptions. Meier (2006) shows existence of a terminal model with finitely additive measures and κ-measurability (for a fixed regular cardinal κ) and nonexistence if all subsets are required to be measurable. Heifetz and Samet (1998b) show nonexistence for the case of knowledge models, and Meier (2005) extends this nonexistence result to Kripke frames.

We end by noting that, to the best of our knowledge, no general treatment exists of the relationship between universal, complete, and terminal models (absent specific structure). Such a treatment would be very useful.

---

[8] Meier (2006) proposes the same terminology.

References

Armbruster, W and W Böge (1970). Bayesian game theory. In Moeschlin, O and D Pallaschke (Eds.), *Game Theory and Related Topics*. Amsterdam: North-Holland.

Aumann, R (1999). Interactive epistemology I: Knowledge. *International Journal of Game Theory*, 28, 263–300.

Battigalli, P and M Siniscalchi (1999). Hierarchies of conditional beliefs and interactive epistemology in dynamic games. *Journal of Economic Theory*, 88, 188–230.

Battigalli, P and M Siniscalchi (2002). Strong belief and forward-induction reasoning. *Journal of Economic Theory*, 106, 356–391.

Böge, W and Th Eisele (1979). On solutions of Bayesian games. *International Journal of Game Theory*, 8, 193–215.

Bonanno, G (2005). A simple modal logic for belief revision. *Synthese (Knowledge, Rationality and Action)*, 147, 193–228.

Boolos, G (1993). *The Logic of Provability*. Cambridge, UK: Cambridge University Press.

Brandenburger, A (2003). On the existence of a 'complete' possibility structure. In Basili, M, N Dimitri and I Gilboa (Eds.), *Cognitive Processes and Economic Behavior*, pp. 30–34. London: Routledge.

Brandenburger, A and E Dekel (1993). Hierarchies of beliefs and common knowledge. *Journal of Economic Theory*, 59, 189–198.

Brandenburger, A, A Friedenberg and H J Keisler (2006). Admissibility in games. *Econometrica*, 76, 307–352.

Epstein, L and T Wang (1996). Beliefs about beliefs without probabilities. *Econometrica*, 64, 1343–1373.

Fagin, R (1994). A quantitative analysis of modal logic. *Journal of Symbolic Logic*, 59, 209–252.

Fagin, R, J Geanakoplos, J Halpern and M Vardi (1999). The hierarchical approach to modeling knowledge and common knowledge. *International Journal of Game Theory*, 28, 331–365.

Fagin, R, J Halpern and M Vardi (1991). A model-theoretic analysis of knowledge. *Journal of the Association of Computing Machinery*, 38, 382–428.

Heifetz, A (1993). The bayesian formulation of incomplete information — The non-compact case. *International Journal of Game Theory*, 21, 329–338.

Heifetz, A (1999). How canonical is the canonical model? A comment on Aumann's interactive epistemology. *International Journal of Game Theory*, 28, 435–442.

Heifetz, A and D Samet (1998a). Topology-free typology of beliefs. *Journal of Economic Theory*, 82, 324–381.

Heifetz, A and D Samet (1998b). Knowledge spaces with arbitrarily high rank. *Games and Economic Behavior*, 22, 260–273.

Heifetz, A and D Samet (1999). Coherent beliefs are not always types. *Journal of Mathematical Economics*, 32, 475–488.

Humberstone, I (1987). The modal logic of all and only. *Notre Dame Journal of Formal Logic,* 28, 177–188.

Huynh, HL and B Szentes (1999). Believing the unbelievable: The dilemma of self-belief. Available at http://home.uchicago.edu/~szentes

Kechris, A (1999). *Classical Descriptive Set Theory.* New York, NY: Springer-Verlag.

Lomuscio, A (1999). Knowledge sharing among ideal agents. Doctoral dissertation, University of Birmingham.

Mariotti, T, M Meier and M Piccione (2005). Hierarchies of beliefs for compact possibility models. *Journal of Mathematical Economics,* 41, 303–324.

Meier, M (2005). On the nonexistence of universal information structures. *Journal of Economic Theory,* 122, 132–139.

Meier, M (2006). Finitely additive beliefs and universal type spaces. *The Annals of Probability* 34, 386–422.

Mertens, J-F and S Zamir (1985). Formulation of Bayesian analysis for games with incomplete information. *International Journal of Game Theory,* 14, 1–29.

Pacuit, E (2007). Understanding the Brandenburger–Keisler paradox. *Studia Logica* 86, 435–454.

Pearce, D (1984). Rational strategic behavior and the problem of perfection. *Econometrica,* 52, 1029–1050.

Pintér, M (2005). Type space on a purely measurable parameter space. *Economic Theory,* 26, 129–139.

Salonen, H (1999). Beliefs, filters, and measurability. University of Turku.

Thomason, R (1980). A note on syntactical treatments of modality. *Synthese,* 44, 391–395.

Yanofsky, N (2003). A universal approach to self-referential paradoxes, incompleteness and fixed points. *The Bulletin of Symbolic Logic,* 9, 362–386.

Chapter 2

Hierarchies of Beliefs and Common Knowledge

Adam Brandenburger and Eddie Dekel

Game-theoretic analysis often leads to consideration of an infinite hierarchy of beliefs for each player. Harsanyi suggested that such a hierarchy of beliefs could be summarized in a single entity, called the player's *type*. This chapter provides an elementary construction, complementary to the construction already given in the paper by Mertens and Zamir (1985), of Harsanyi's notion of a type. It is shown that if a player's type is *coherent* then it induces a belief over the types of the other players. Imposing common knowledge of coherency closes the model of beliefs. We go on to discuss the question that often arises as to the sense in which the structure of a game-theoretic model is, or can be assumed to be, common knowledge.

1. Introduction

Hierarchies of beliefs arise in an essential way in many problems in decision and game theory. For example, the analysis of a game, even of one with complete information, leads to consideration of an "infinite regress" in beliefs. Thus, supposing for simplicity that there are just two players i and j, the choice of strategy by i will depend on what i believes j's choice will be, which in turn will depend on what i believes j believes i's choice will be, and so on. An infinite regress of this kind underlies the idea of "rationalizable"

Originally published in *Journal of Economic Theory*, 59, 189–198.
Financial support: Harkness Fellowship, Harvard Business School Division of Research, Sloan Dissertation Fellowship, and NSF Grant SES 8808133.
Acknowledgments: We thank Jerry Green, David Kreps, Andrew Mas-Colell, and a referee for helpful comments.

strategies, introduced by Bernheim (1984) and Pearce (1984). For games with complete information, this regress in beliefs has traditionally been "cut through" by the imposition of an equilibrium concept such as Nash equilibrium. It was in the context of games with incomplete information, in which some parameters of the game are not common knowledge among the players, that the problem of an infinite regress in beliefs was first tackled by Harsanyi (1967–68). Harsanyi's solution was to summarize the entire stream of beliefs of a player in a single entity, called the player's *type*, such that each type induces a belief over the types of the other players. Harsanyi's formulation of a game with incomplete information has become an indispensable tool in many areas of economics, but it is only relatively recently that rigorous arguments have been given in support of Harsanyi's notion of a type by Armbruster and Böge (1979), Böge and Eisele (1979), and Mertens and Zamir (1985). This chapter provides an alternative construction of types, which is similar to that in Mertens and Zamir (1985), but which relies on more elementary mathematics and is more explicit about what is assumed to be common knowledge.

Our construction of types has two stages. First, we show that if an individual's type is *coherent* then it induces a belief over the types of the other individuals. (Coherency is a requirement that the various levels of beliefs of an individual do not contradict one another — see Definition 1.) This result (Proposition 1) is essentially just a statement of Kolmogorov's Existence Theorem from the theory of stochastic processes (see, for example, Chung [1974, p. 60]). Second, the model of beliefs is closed by imposing, via a simple inductive definition, the requirement that each type knows (in the probabilistic sense of assigning probability 1) that the other individuals' types are coherent, that each type knows that the other types know this, and so on. That is, the model is closed by imposing common knowledge of coherency. (What is meant here by "closed" is elucidated in the paragraph preceding Definition 1.)

At a technical level, we replace the assumption in the paper by Mertens and Zamir (1985) that the underlying state space is compact with the assumption that it is complete separable metric. (Remark 2.18 in Mertens and Zamir [1985] suggests that such a replacement is possible.) Recently, Heifetz (1990) has provided a general construction of types, assuming only that the underlying state space is Hausdorff.

Having completed our construction of types in Section 2, we go on in Section 3 to discuss the question that often arises as to the sense in which the structure of a game-theoretic model is, or can be assumed to be, common knowledge. This question has been discussed by Aumann (1976, 1987) and others. (See Bacharach [1987], Gilboa [1988], Kaneko [1987], Samet [1990], Shin [1986], and Tan and Werlang [1985].) Aumann has argued that if the structure is not a common knowledge, then the description of the states of the world is incomplete and so the state space should be expanded. The construction of types shows what the expanded state space should be, namely the product of the underlying state space and the individuals' type spaces. Moreover, on the expanded state space, common knowledge of the structure is captured by the assumption of common knowledge of coherency.

2. Construction of Types

In this section, hierarchies of beliefs are constructed. The notions of type and coherency are defined and it is shown that a coherent type induces a belief over other individuals' types. We go on to prove that common knowledge of coherency closes the model of beliefs in the sense that all beliefs are then completely specified.

There are two individuals i and j who face some common (underlying) space of uncertainty S.[1] The space S is assumed to be complete separable metric (Polish). For any metric space Z, let $\Delta(Z)$ denote the space of probability measures on the Borel field of Z, endowed with the weak topology. According to Bayesian decision theory, each individual must have a belief over the space S; the individuals' first-order beliefs are then elements of $\Delta(S)$. Since each individual may not know the belief of the other, each must have a second-order belief. That is, i's second-order belief is a joint belief over S and the space of j's first-order beliefs; i's second-order belief is thus an element of $\Delta(S \times \Delta(S))$. Similarly for j. Formally, define spaces

$$X_0 = S$$
$$X_1 = X_0 \times \Delta(X_0)$$
$$\vdots$$
$$X_n = X_{n-1} \times \Delta(X_{n-1})$$

A *type* t^i of i is just a hierarchy of beliefs $t^i = (\delta_1^i, \delta_2^i, \ldots) \in \times_{n=0}^{\infty} \Delta(X_n)$. Similarly for j. Let $T_0 = \times_{n=0}^{\infty} \Delta(X_n)$ denote the space of all possible types of i or j.

---

[1]All our arguments generalize immediately to the case of more than two individuals.

Of course, i only knows his own type and not the type of j. (Likewise for j.) So it seems that a "second level" hierarchy of beliefs is required, wherein i has a belief over j's type, over j's belief over i's type, and so on. Thus, in the absence of further assumptions, a model which specifies only the hierarchy of beliefs $(\delta_1^i, \delta_2^i, \ldots) \in \times_{n=0}^{\infty} \Delta(X_n)$ for i, and likewise for j, is *not* closed. The condition under which the specification of i's type already determines his belief over j's type is defined next.

Definition 1. A type $t = (\delta_1, \delta_2, \ldots) \in T_0$ is *coherent* if for every $n \geq 2$, $\mathrm{marg}_{X_{n-2}} \delta_n = \delta_{n-1}$, where $\mathrm{marg}_{X_{n-2}}$ denotes the marginal on the space X_{n-2}.

Coherency says that the different levels of beliefs of an individual do not contradict one another.[2] Let T_1 denote the set of all coherent types. The following proposition shows that a coherent type induces a belief over S and the space of types of the other individual.

Proposition 1. *There is a homeomorphism $f : T_1 \to \Delta(S \times T_0)$.*

Proposition 1 will be an easy consequence of the following lemma, which itself is essentially a statement of Kolmogorov's Existence Theorem.

Lemma 1. *Suppose $\{Z_n\}_{n=0}^{\infty}$ is a collection of Polish spaces, and let*

$$D = \{(\delta_1, \delta_2, \ldots) : \ \delta_n \in \Delta(Z_0 \times \cdots \times Z_{n-1}) \quad \forall n \geq 1,$$
$$\mathrm{marg}_{Z_0 \times \cdots \times Z_{n-2}} \delta_n = \delta_{n-1} \quad \forall n \geq 2\}.$$

Then, there is a homeomorphism $f : D \to \Delta(\times_{n=0}^{\infty} Z_n)$.

Proof. Consider any element $(\delta_1, \delta_2, \ldots) \in D$. By a version of Kolmogorov's Existence Theorem (Dellacherie and Meyer [1978, p. 68]) there is a unique measure $\delta \in \Delta(\times_{n=0}^{\infty} Z_n)$ such that $\mathrm{marg}_{Z_0 \times \cdots \times Z_{n-1}} \delta = \delta_n$ for all $n \geq 1$. Let f map $(\delta_1, \delta_2, \ldots)$ into this δ. The map f is 1-1 since the value of δ on the cylinders is given by the δ_n's; f is onto since given any $\delta \in \Delta(\times_{n=0}^{\infty} Z_n)$, $f(\mathrm{marg}_{Z_0} \delta, \mathrm{marg}_{Z_0 \times Z_1} \delta, \ldots) = \delta$. Note that $f^{-1}(\delta) = (\mathrm{marg}_{Z_0} \delta, \mathrm{marg}_{Z_0 \times Z_1} \delta, \ldots)$ so f^{-1} is continuous since the maps $\delta \mapsto \mathrm{marg}_{Z_0 \times \cdots \times Z_{n-1}} \delta$, $n \geq 1$, are all continuous. To see that f is continuous, consider a sequence $(\delta_1^r, \delta_2^r, \ldots) \to (\delta_1, \delta_2, \ldots)$ in D, i.e., δ_n^r converges weakly to δ_n for all $n \geq 1$. Let $\delta^r = f(\delta_1^r, \delta_2^r, \ldots)$,

---

[2]What is here called coherency is usually called consistency in the theory of stochastic processes. The term coherency is used to avoid confusion with Harsanyi's use of the term consistency, which means something different.

$\delta = f(\delta_1, \delta_2, \ldots)$. We have to show that δ^r converges weakly to δ. But this follows from the fact that the cylinders form a convergence-determining class and the values of δ^r, δ on the cylinders are given by the δ_n^r's, δ_n's, respectively. □

Proof of Proposition 1. In Lemma 1, set $Z_0 = X_0$, $Z_n = \Delta(X_{n-1})$ for $n \geq 1$. So $Z_0 \times \cdots \times Z_n = X_n$ and $\times_{n=0}^\infty Z_n = S \times T_0$. If S is a Polish space then so is $\Delta(S)$ (Dellacherie and Meyer [1978, p. 73]), hence the Z_n's will be Polish spaces provided S is. The set of coherent types T_1 is exactly D. So Lemma 1 implies that there is a homeomorphism $f : T_1 \to \Delta(S \times T_0)$. □

An obvious question to ask is why the particular homeomorphism f, just constructed, is "natural." The reason is the following property of f: the marginal probability assigned by $f(\delta_1, \delta_2, \ldots)$ to a given event in X_{n-1} is equal to the probability that δ_n assigns to that same event. That is, in deriving probabilities on the product space $S \times T_0 = X_0 \times \Delta(X_0) \times \Delta(X_1) \times \cdots$ from $(\delta_1, \delta_2, \ldots)$, the function f preserves the probabilities specified by each δ_n on each X_{n-1}.

Coherency implies that i's type determines i's belief over j's type. But i's type does not necessarily determine i's belief over j's belief over i's type — in particular, this is so if i believes it possible that j's type is not coherent. For a type to determine *all* beliefs (including beliefs over beliefs over types), common knowledge of coherency must be imposed. To do so, define a sequence of sets T_k, $k \geq 2$, by

$$T_k = \{t \in T_1 : f(t)(S \times T_{k-1}) = 1\}.$$

(It is straightforward to show inductively that T_{k-1} is a Borel set, so T_k is indeed well defined.) Let $T = \cap_{k=1}^\infty T_k$. The set $T \times T$ is the subset of $T_1 \times T_1$ obtained by requiring the following statements to hold: (1) i knows j's type is coherent; (2) j knows i's type is coherent; (3) i knows j knows i's type is coherent; and so on. That is, $T \times T$ is the set of types which satisfy common knowledge of coherency. The following proposition shows that the space T closes the model, and corresponds to the "universal type space" of Theorem 2.9 in Mertens and Zamir (1985).

Proposition 2. *There is a homeomorphism* $g : T \to \Delta(S \times T)$.

Proof. It is easy to check that $T = \{t \in T_1 : f(t)(S \times T) = 1\}$, so $f(T) = \{\delta \in \Delta(S \times T_0): \delta(S \times T) = 1\}$ since f is onto. But $f(T)$ is homeomorphic to

T and $\{\delta \in \Delta(S \times T_0): \delta(S \times T) = 1\}$ is homeomorphic to $\Delta(S \times T)$ (for any metric space Z and measurable subset W of Z, $\{\delta \in \Delta(Z): \delta(W) = 1\}$ is homeomorphic to $\Delta(W)$). So T is homeomorphic to $\Delta(S \times T)$. $\square$

Once again an immediate question arises as to why the homeomorphism g is "natural." The answer is that g preserves the beliefs of each individual in exactly the same way as the function f of Proposition 1 preserves beliefs. (See the discussion following the proof of Proposition 1.) Moreover, the development in Section 3 — where we show how the model of hierarchies of beliefs can be transformed into a standard model of differential informa- tion — uses the homeomorphism g, and, in particular, Proposition 3 relies essentially on the specific homeomorphism g.

A technical aspect of the construction worth noting is that closure of the model of hierarchies of beliefs is not a purely measure-theoretic result. Recall in fact that we assumed S to be a Polish space. This is because (cf. Halmos [1974, pp. 211–212]) Kolmogorov's Existence Theorem is itself not purely measure-theoretic, and relies on topological assumptions.

3. Relationship to the Standard Model of Differential Information

The standard formulation of a model of differential information as com- monly used in game theory and economics is a collection $\langle \Omega, H^i, H^j, p^i, p^j \rangle$.[3] The set Ω is the space of states of the world, H^i is i's information partition (if $\omega \in \Omega$ is the true state, i is informed of the cell of H^i that contains ω), p^i is i's prior probability measure on Ω, and H^j and p^j are the analogous objects for j. In this section, we discuss the relationship between the standard formulation and the types model constructed in Section 2. First, we use the types model to shed some light on the interpretational question mentioned in the Introduction, namely the sense in which the structure of the standard model is, or can be assumed to be, common knowledge. Second, we describe a transformation of the types model into a stantard model and demonstrate, by way of example, that the transformation is meaningful.

An interpretational question that often arises in discussions of the standard model of differential information is whether the information structure (consisting of partitions and priors) is "common knowledge"

---

[3]We maintain the simplifying assumption of only two individuals i and j.

in an informal sense. (We say in an informal sense because the information structure is not an event in Ω and hence the formal definition of common knowledge does not apply. In what follows, we will use quotation marks when we wish to signify informal usage.) The issue of "common knowledge" of the information structure arises in the following manner. Given an event A in Ω one can define, using i's information structure H^i and p^i, the event that i knows A (see, e.g., Aumann [1976]), to be denoted $K^i(A)$. Similarly for j. Now suppose in fact that $A = K^j(B)$ for some event B in Ω. Then $K^i(A) = K^i(K^j(B))$ is certainly interpretable as the event that i knows $K^j(B)$. But in practice we interpret $K^i(K^j(B))$ as the event that i knows j knows B — and this latter interpretation relies on an implicit assumption that i "knows" j's information structure. That is, it is assumed that i "knows" H^j and p^j. Applying the same argument to more complex events such as $K^i(K^j(K^i(C)))$ and the like shows that in fact "common knowledge" of the information structure is needed.

The nature of this "common knowledge" has been much discussed. (See Aumann [1976, 1987], Bacharach [1987], Gilboa [1988], Kaneko [1987], Samet [1990], Shin [1986], Tan and Werlang [1985], and others.[4]) Aumann has argued that "common knowledge" of the information structure is without loss of generality since the description of a state in Ω should include a description of the manner in which information is imparted to the individuals (the partitions) and a description of the players' beliefs (the priors). If this is not the case, Aumann argues that the description of the states is incomplete and so the state space should be expanded. The observation we wish to make is that the appropriate expanded state space is the product of the underlying state space S and the type spaces T. More precisely, the expanded state space is $S \times T \times T$ where the first copy of T is the type space of individual i and second copy of T is the type space of individual j.[5] The point is that "common knowledge" of the information structure on the expanded state space is captured by the assumption of common knowledge of coherency that we made in Section 2. To see this, consider, for example, the set $T_2 = \{t \in T_1 : f(t)(S \times T_1) = 1\}$ as defined

---

[4]There are also relevant literatures in computer science, artificial intelligence, linguistics, and philosophy; see Fagin, Halpern, and Vardi (1991), Halpern (1986), Vardi (1988), and the references therein.

[5]Thus, $\Omega = S \times T \times T$. A formal treatment of the information structure on Ω is given below.

there. The set T_2 is the set of types of i, say, which know that j's type is
coherent. So T_2 is the set of types of i which can calculate beliefs over j's
beliefs over i's type, or in other words the set of types of i which "know"
j's information structure. Similarly, T_3 is the set of types of i which can
calculate beliefs over j's beliefs over i's beliefs over j's type, or in other
words, the set of types of i which "know" that j "knows" i's information
structure. And so on. The upshot is that since common knowledge of
coherency is a natural rationality assumption (it merely states that it is
common knowledge that the various levels of beliefs of an individual do not
contradict one another), "common knowledge" of the information structure
(on the expanded state space) is indeed without loss of generality.

To summarize, the same model, namely the model of hierarchies of
beliefs, validates Harsanyi's notion of a type and Aumann's notion of a
space of completely specified states of the world.

We now show how to transform the types model into a standard model,
thus demonstrating that the standard model is in fact no less general than
the types model. From this, it follows that the standard model, which is,
of course, a simpler construct, can be employed whenever doing so is more
convenient.

Starting with an underlying space of uncertainty S and induced type
spaces T, we can construct a standard model as follows. The set Ω of states
of the world is the product space $S \times T \times T$, where, as before, the first copy of
T is the type space of individual i and the second copy of T is the type space
of individual j. Note that even if S is finite, $S \times T \times T$ is an uncountable
space and the information structure on $S \times T \times T$ must be specified in
terms of σ-fields rather than partitions. Let $\mathscr{H}$ denote the Borel field of
$S \times T \times T$. Since the information that i possesses is exactly knowledge of
his own type, the natural sub σ-field of $\mathscr{H}$ for i is $\{S \times B \times T : B$ is a Borel
subset of $T\}$. The homeomorphism g of Proposition 2 determines i's beliefs:
the natural conditional probability for i to access to an event $A \in \mathscr{H}$ at a
state (s, t^i, t^j) is $g(t^i)(A_{t^i})$, where $A_{t^i} = \{(s, t^j) : (s, t^i, t^j) \in A\}$. Individual
j's sub σ-field and beliefs are specified in analogous fashion. In sum, we
have shown, with one proviso about to be discussed, how the types model
can be transformed into a standard model. The proviso is that we have
specified i and j's system of *conditional* probability measures rather than
their *prior* probability measures. In fact, there would be no difficulty in
constructing for i and j (different) prior probability measures on $S \times T \times T$
with the indicated conditionals. (The technical conditions allowing this are
readily verified.) But since it is the conditionals, and not the priors, that

are of decision-theoretic significance, we refrain from going into details on constructing the priors.[6]

So far, we have shown how the types model can, formally speaking, be transformed into a standard model. That this is a sensible way of viewing the types model is best seen by working through an example. Suppose we wish to write down the statement that an event is *common knowledge* between i and j. There is a natural way, which we give in a moment, of doing this in the context of the types model. There is also the well-known definition, due to Aumann (1976), of common knowledge in the context of a standard model of differential information. What we are going to do is to show that the "types" definition of common knowledge is equivalent to the "standard" definition of common knowledge when the latter is applied to the standard model derived from the types model in the manner described in the preceding paragraph.

We start with the "types" definition. Given an event E in S, let

$$V_1(E) = \{t \in T : g(t)(E \times T) = 1\}$$

and then define a sequence of sets $V_k(E)$, $k \geq 2$, by

$$V_k(E) = \{t \in T : g(t)(S \times V_{k-1}(E)) = 1\}.$$

(It is straightforward to show inductively that $V_{k-1}(E)$ is a Borel set, so $V_k(E)$ is indeed well defined.) Let $V(E) = \cap_{k=1}^{\infty} V_k(E)$. Then we say that E is common knowledge between i and j according to the "types" definition if $(t^i, t^j) \in V(E) \times V(E)$. This definition simply states that i is of a type that assigns probability 1 to E, j is of a type that assigns probability 1 to E, i is of a type that assigns probability 1 to j being of a type that assigns probability 1 to E, and so on.

We now turn to the "standard" definition of common knowledge. Aumann's original definition was couched in terms of partitions. However, as was pointed out above, the set $\Omega = S \times T \times T$ is uncountable and σ-fields rather than partitions must be employed. A generalization of Aumann's definition to cover this case was proposed in Brandenburger and Dekel (1987) and we follow this approach here. The event that i knows an event $A \in \mathscr{H}$, to be denoted by $K^i(A)$, is given by

$$K^i(A) = \{(s, t^i, t^j) : g(t^i)(A_{t^i}) = 1\}.$$

---

[6] Also worth noting is that the constructed conditionals are regular and proper (the latter in the sense of Blackwell and Dubins [1975]).

The event that j knows A, to be denoted by $K^j(A)$, is defined in analogous fashion. Thus, $K(A) = K^i(A) \cap K^j(A)$ is the event that everyone knows A. We say that A is common knowledge at a state (s, t^i, t^j) according to the "standard" definition if $(s, t^i, t^j) \in K_\infty(A)$, where K_∞ denotes the infinite application of the K operator.

It remains to establish that the event E in S is common knowledge according to the "types" definition if and only if the event $E \times T \times T$ in Ω is common knowledge according to the "standard" definition.[7] The equivalence is stated formally in Proposition 3.

Proposition 3. $S \times V(E) \times V(E) = K_\infty(E \times T \times T)$.

Proof. The proof follows immediately from the definitions. Observe that

$$K^i(E \times T \times T) = \{(s, t^i, t^j) : g(t^i)(E \times T) = 1\} = S \times V_1(E) \times T.$$

Similarly, $K^j(E \times T \times T) = S \times T \times V_1(E)$ and hence $K(E \times T \times T) = S \times V_1(E) \times V_1(E)$. Continuing in this fashion establishes that $K_\infty(E \times T \times T) = S \times V(E) \times V(E)$. $\qquad\square$

Proposition 3 confirms that our transformation of the types model into a standard model makes sense although, strictly speaking, this has been shown to be true only insofar as common knowledge of events is concerned. Nevertheless, it should be clear that in fact *any* calculation involving the individuals' beliefs is preserved under the transformation.

The reserve transformation has been considered by Tan and Werlang (1985). They show how, starting from the standard formulation of a model of differential information, to calculate the induced hierarchies of beliefs and hence how to construct an associated types model. They also demonstrate that their transformation is meaningful by showing that it preserves the notion of common knowledge.

References

Armbruster, W and W Böge (1979). Bayesian game theory. In Moeschlin, O and D Pallaschke (Eds.), *Game Theory and Related Topics*. Amsterdam: North-Holland.

Aumann, R (1976). Agreeing to disagree. *Annals of Statistics*, 4, 1236–1239.

---

[7]Note that the equivalence relates E in S to $E \times T \times T$ in Ω. This is because the set E is not an event in Ω, but is naturally identified with the event $E \times T \times T$.

Aumann, R (1987). Correlated equilibrium as an expression of Bayesian rationality. *Econometrica*, 55, 1–18.

Bacharach, M (1987). When do we have information partitions? Unpublished, Christchurch College, Oxford.

Bernheim, D (1984). Rationalizable strategic behavior. *Econometrica*, 52, 1007–1028.

Blackwell, D and L Dubins (1975). On existence and non-existence of proper, regular, conditional distributions. *Annals of Probability*, 3, 741–752.

Böge, W and Th Eisele (1979). On solutions of Bayesian games. *International Journey of Game Theory*, 8, 193–215.

Brandenburger, A and E Dekel (1987). Common knowledge with probability 1. *Journal of Mathematical Economics*, 16, 237–245.

Chung, KL (1974). *A Course in Probability Theory*, 2nd Edition. New York, NY: Academic Press.

Dellacherie, C and P-A Meyer (1978). *Probabilities and Potential*. Mathematics Studies, 29. Amsterdam: North-Holland.

Fagin, R, J Halpern and M Vardi (1991). A model-theoretic analysis of knowledge. *Journal of the Association for Computer Machinery*, 38, 382–428.

Gilboa, I (1988). Information and meta-information. In Vardi, M (Ed.), *Proceedings of the Second Conference on Theoretical Aspects of Reasoning about Knowledge*. Los Altos: Kaufmann.

Halmos, P (1976). *Measure Theory*. New York, NY: Springer-Verlag.

Halpern, J (Ed.) (1986). *Theoretical Aspects of Reasoning about Knowledge: Proceedings of the 1986 Conference*. Los Altos: Kaufmann.

Harsanyi, J (1967–68). Games with incomplete information played by "Bayesian" players, I–III. *Management Science*, 14, 159–182, 320–334, 486–502.

Heifetz, A (1990). The Bayesian formulation of incomplete information — The noncompact case. Unpublished, School of Mathematical Sciences, Tel Aviv University, Tel Aviv.

Kaneko, M (1987). Structural common knowledge and factual common knowledge. RUEE Working Paper No. 87–27, Department of Economics, Hitotsubashi University, 74–85.

Mertens, J-F and S Zamir (1985). Formulation of Bayesian analysis for games with incomplete information. *International Journal of Game Theory*, 14, 1–29.

Pearce, D (1984). Rationalizable strategic behavior and the problem of perfection. *Econometrica*, 52, 1029–1050.

Samet, D (1990). Ignoring ignorance and agreeing to disagree. *Journal of Economic Theory*, 52, 190–207.

Shin, H (1986). Logical structure of common knowledge. Unpublished, Nuffield College, Oxford.

Tan, T and S Werlang (1985). On Aumann's notion of common knowledge — An alternative approach. Unpublished, Department of Economics, Princeton University, Princeton, NJ.

Vardi, M (Ed.) (1988). *Proceedings of the Second Conference on Theoretical Aspects of Reasoning about Knowledge*. Los Altos: Kaufmann.

Chapter 3

Rationalizability and Correlated Equilibria

Adam Brandenburger and Eddie Dekel

We discuss the unity between the two standard approaches to noncooperative solution concepts for games. The decision-theoretic approach starts from the assumption that the rationality of the players is common knowledge. This leads to the notion of correlated rationalizability. It is shown that correlated rationalizability is equivalent to a posteriori equilibrium — a refinement of subjective correlated equilibrium. Hence a decision-theoretic justification for the equilibrium approach to game theory is provided. An analogous equivalence result is proved between independent rationalizability, which is the appropriate concept if each player believes that the others act independently, and conditionally independent a posteriori equilibrium. A characterization of Nash equilibrium is also provided.

1. Introduction

The fundamental solution concept for noncooperative games is that of a Nash equilibrium (Nash [1951]). Many justifications for Nash equilibrium have been provided in the literature. Probably the most common view of Nash equilibrium is as a self-enforcing agreement. A game is envisaged as being preceded by a more or less explicit period of communication between the players. It is argued that if the players agree on a certain profile of strategies, then these must constitute a Nash equilibrium. Otherwise some

Originally published in *Econometrica*, 55, 1391–1402.

Keywords: Rationalizability; correlated equilibrium; subjective and common priors; independence; Nash equilibrium.

Financial support: Harkness Fellowship, Sloan Dissertation Fellowship, and Miller Institute for Basic Research in Science.

Acknowledgments: We wish to thank Jerry Green, David Kreps, and two anonymous referees for helpful comments, and Robert Aumann for many enlightening discussions.

player will have an incentive to deviate from the agreement. Aumann (1974) proposed the ideas of objective and subjective correlated equilibrium as extensions of Nash equilibrium to allow for correlation between the players' randomizations and for subjectivity in the players' probability assessments.

The Nash equilibrium solution concept has been criticized from two opposing directions. On the one hand, the literature on refinements of Nash equilibrium (Selten [1965, 1975], Myerson [1978], Kreps and Wilson [1982], Kohlberg and Mertens [1986], and others) starts from the contention that not every Nash equilibrium can be viewed as a plausible agreed-upon way to play the game. On the other hand, Bernheim (1984) and Pearce (1984) have argued that Nash equilibrium is too restrictive in that it rules out behavior that does not contradict the rationality of the players. Bernheim and Pearce propose instead the concept of rationalizability as the logical consequence of assuming that the structure of the game and the rationality of the players (and nothing more) is common knowledge.

This chapter starts with the solution concept of rationalizability, since this is what is implied by the basic decision-theoretic analysis of a game. However, it is shown that rationalizability is more closely related to an equilibrium approach than one might at first think. The main result we prove in this chapter is an equivalence between rationalizability and a posteriori equilibrium — a refinement of subjective correlated equilibrium. So in fact a certain kind of equilibrium arises from assuming no more than common knowledge of rationality of the players in a game. This chapter therefore provides a formal decision-theoretic justification for using equilibrium concepts in game theory.

The solution concepts of rationalizability and a posteriori equilibrium will now be briefly described. Call a strategy of player i justifiable if it is optimal given some belief (probability measure) over the possible strategies of player j. (For simplicity suppose that there are only two players.) Define a justifiable strategy of j similarly. A strategy of i is rationalizable if it is justifiable using a belief which assigns positive probability only to strategies of j which are justifiable, if these latter strategies are justified using beliefs which assign positive probability only to justifiable strategies of i, and so on. In this way, the notion of rationalizability captures the idea that a player should only choose a strategy which respects common knowledge of rationality. Tan and Werlang (1984) and Bernheim (1985) provide formal proofs of the equivalence between rationalizability and common knowledge of rationality.

Aumann (1974) introduced various notions of objective and subjective correlated equilibrium, including the notion of a posteriori equilibrium which refines subjective correlated equilibrium in a way we now discuss. Objective and subjective correlated equilibrium differ in that the first requires the players' priors to be the same while the second allows them to be different. In both cases the equilibrium requirement is that each player i's strategy should be ex ante optimal, that is, should maximize i's expected utility before any private information is observed. Of course this requirement is equivalent to having i's strategy maximize conditional expected utility on every information cell which is assigned positive prior probability. A possible strengthening of the definition of equilibrium is to require optimality even on null information cells. In the case of objective equilibrium this strengthening makes no difference, but it is significant for subjective equilibrium (see the example in Figure 1 of Section 2). A posteriori equilibrium is exactly this strengthening of subjective correlated equilibrium.

The equivalence result in this chapter comes in two parts depending on whether one starts with "correlated" or "independent" rationalizability. The difference is that the second requires a player to believe that the other players choose their strategies independently, while the first does not. (Of course, the two versions of rationalizability coincide for two-person games.) Independent rationalizability is the concept originally defined by Bernheim (1984) and Pearce (1984). It is appropriate if one thinks of the players in a "laboratory" situation: any correlating devices are explicitly modelled, the players are placed in separate rooms, and then informed of the game they are to play. Correlated rationalizability seems more appropriate when the players are able to coordinate their actions via a large collection of correlating devices (such as sunspots) which are not explicitly modelled in the game but are taken into account by allowing for correlated beliefs.

Our starting point is that rationalizability is the solution concept implied by common knowledge of rationality of the players in a game. It is then shown that there is an equivalence between rationalizability and a posteriori equilibrium. Of course, most applications of game theory in economics assume that the players have a common prior, that is, most applications use either the Nash or objective correlated equilibrium concepts. Section 4 of the chapter discusses characterizations of these solution concepts.

In a related paper, Aumann (1987) adopts a somewhat different notion of Bayesian rationality from that in this chapter. Bayesian rationality is formalized using a standard model of differential information with the additional feature that the state space includes the actions of the players. Under an assumption of common knowledge of rationality together with an assumption of common priors (the Common Prior Assumption) one is again led to objective correlated equilibrium. For the details of this characterization and a discussion of the Common Prior Assumption the reader should consult Aumann (1987). Alternative characterizations of objective solution concepts can also be found in Tan and Werlang (1984) and Bernheim (1985).

The organization of the rest of the chapter is as follows. Section 2 provides formal definitions of correlated rationalizability and a posteriori equilibrium, and proves the equivalence result between these two concepts. In Section 3, an analogous equivalence result is proved between independent rationalizability and conditionally independent a posteriori equilibrium. Section 4 discusses characterizations of objective correlated equilibrium and Nash equilibrium.

2. Correlated Rationalizability and A Posteriori Equilibria

This section starts by defining the sets of correlated rationalizable strategies and payoffs in a game. The approach is based on that in Pearce (1984). However, unlike Pearce's paper, players are not allowed to select mixed strategies — allowing them to do so would not expand the set of rationalizable payoffs. Also, a player's beliefs over the actions of the other players may be correlated (cf. Pearce [1984, p. 1035]). The next section examines the case in which these beliefs are independent.

Consider an n-person game $\Gamma = \langle A^1, \ldots, A^n; u^1, \ldots, u^n \rangle$ where, for each $i = 1, \ldots, n, A^i$ is a finite set of pure strategies (henceforth actions) of player i and $u^i : \prod_{j=1}^n A^j \to \mathscr{R}$ is i's payoff function. For any finite set Y, let $\Delta(Y)$ denote the set of probability measures on Y. Given sets $Y^1, \ldots, Y^n, Y^{-i}$ denotes the set $Y^1 \times \cdots \times Y^{i-1} \times Y^{i+1} \times \cdots \times Y^n$, and $y^{-i} = (y^1, \ldots, y^{i-1}, y^{i+1}, \ldots, y^n)$ is a typical element of Y^{-i}.

Definition 2.1. A subset $B^1 \times \cdots \times B^n$ of $A^1 \times \cdots \times A^n$ is a *best reply set* if for every i and each $a^i \in B^i$ there is a $\sigma \in \Delta(B^{-i})$ to which a^i is a best reply.

The set of correlated rationalizable actions $R^1 \times \cdots \times R^n$ is the (finite) component-by-component union $(\cup_\alpha B_\alpha^1) \times \cdots \times (\cup_\alpha B_\alpha^n)$ of all best reply sets $B_\alpha^1 \times \cdots \times B_\alpha^n$. It is easy to check that $R^1 \times \cdots \times R^n$ is itself a best reply set. This fact will be used below. There are two equivalent definitions of the set $R^1 \times \cdots \times R^n$. One is in terms of the systems of justifiable actions discussed in the Introduction. The other is in terms of iterated deletion of strongly dominated actions. (Proofs of the equivalence of the three definitions are easily adapted from arguments in Bernheim [1984] and Pearce [1984].) i's maximal expected payoff against a $\sigma \in \Delta(R^{-i})$ is a correlated rationalizable payoff to i. Let Π^i denote the set of all possible correlated rationalizable payoffs to i.

We now want to define an a posteriori equilibrium (Aumann [1974, Section 8]) of the game Γ. First, the definition of a subjective correlated equilibrium of Γ is reviewed, and then an a posteriori equilibrium is defined as a special kind of subjective correlated equilibrium. To define a subjective correlated equilibrium of Γ, one must add to the basic description of the game a finite space Ω. The finiteness of Ω entails no loss of generality. Each player i has a prior P^i — a probability measure on Ω — and a partition $\mathscr{H}^i$ of Ω. A strategy of player i is an $\mathscr{H}^i$-measurable map $f^i : \Omega \to A^i$. An n-tuple of strategies $(f^1, \ldots, f^n)$ is a subjective correlated equilibrium if for every i

$$\sum_{\omega \in \Omega} P^i(\{\omega\}) u^i[f^i(\omega), f^{-i}(\omega)] \geq \sum_{\omega \in \Omega} P^i(\{\omega\}) u^i[\tilde{f}^i(\omega), f^{-i}(\omega)]$$

for every strategy $\tilde{f}^i$ of i. The definition of subjective correlated equilibrium is more general than that of objective correlated equilibrium in that it allows the players' priors P^i to be different. If the P^i's are required to be equal then one gets objective correlated equilibrium.

In a subjective correlated equilibrium the players' strategies are only required to be ex ante optimal. In an a posteriori equilibrium the players' strategies must be optimal even after they have learned their private information. The following simple example motivates this distinction. Refer to Figure 1. The set Ω consists of two points ω_1, ω_2. Row is informed of the true state, Column has no private information. Row assigns (prior) probability 1 to ω_1, Column assigns probability $\frac{1}{2}$ to $\omega_1, \frac{1}{2}$ to ω_2. The following strategies form a subjective correlated equilibrium: Row plays U if informed that ω_1 happens, D if ω_2 happens; Column plays L. Notice that this equilibrium relies on Row playing a strongly dominated action if ω_2 happens. As in the literature on refinements of Nash equilibrium, it

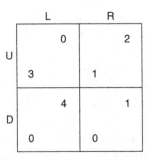

Figure 1.

seems natural to rule out such situations by requiring optimal behavior even on null events — in this case after a move by Nature which is assigned prior probability zero. The definition of an a posteriori equilibrium (Definition 2.2) is designed to deal with this issue. The unique a posteriori equilibrium of the game in Figure 1 has Row playing U and hence Column playing R for sure.

To define an a posteriori equilibrium formally, start again with the game Γ. As for a subjective correlated equilibrium, add to Γ a finite space Ω and for each player i a probability measure P^i on Ω and a partition $\mathcal{H}^i$ of Ω. Furthermore, in order to deal with the kind of difficulty raised in the example above, the players' posterior beliefs at every $\omega \in \Omega$ must be specified. So for each player i choose a version of conditional probability which is regular and proper. That is, for *every* $H^i \in \mathcal{H}^i$, $P^i(\cdot|H^i)$ is required to be a probability measure on Ω and to satisfy $P^i(H^i|H^i) = 1$. (This last requirement is properness in the sense of Blackwell and Dubins [1975].) Of course, if $P^i(H^i) > 0$, then by Bayes' rule $P^i(\cdot|H^i)$ automatically satisfies both requirements, but the point is that $P^i(\cdot|H^i)$ is required to satisfy them even if $P^i(H^i) = 0$. For each i, let $\mathcal{H}^i(\omega)$ denote the cell of i's partition that contains ω.

Definition 2.2. An n-tuple of strategies $(f^1, \ldots, f^n)$ is an *a posteriori equilibrium of* Γ if for each i

$$\forall \omega \in \Omega \quad \sum_{\omega' \in \Omega} P^i[\{\omega'\}|\mathcal{H}^i(\omega)]u^i[f^i(\omega), f^{-i}(\omega')]$$

$$\geq \sum_{\omega' \in \Omega} P^i[\{\omega'\}|\mathcal{H}^i(\omega)]u^i[a^i, f^{-i}(\omega')] \quad \forall a^i \in A^i.$$

Notice that by a change of variables, i's optimality condition can be rewritten as:

$$\forall \omega \in \Omega \qquad \sum_{a^{-i} \in A^{-i}} P^i[\{\omega' : f^{-i}(\omega') = a^{-i}\} | \mathcal{H}^i(\omega)] u^i[f^i(\omega), a^{-i}]$$

$$\geq \sum_{a^{-i} \in A^{-i}} P^i[\{\omega' : f^{-i}(\omega') = a^{-i}\} | \mathcal{H}^i(\omega)] u^i(a^i, a^{-i})$$

$$\forall a^i \in A^i.$$

From the point of view of the players there are two stages to the game: the ex ante and the interim stages corresponding to before and after they receive their private information. It will be helpful to distinguish between a player's payoffs at these two stages.

Definition 2.3. Given an a posteriori equilibrium $(f^1, \ldots, f^n)$ of Γ, i's *interim payoff* at ω is

$$\sum_{a^{-i} \in A^{-i}} P^i[\{\omega' : f^{-i}(\omega') = a^{-i}\} | \mathcal{H}^i(\omega)] u^i[f^i(\omega), a^{-i}];$$

i's *ex ante payoff* is

$$\sum_{\omega \in \Omega} P^i(\{\omega\}) \sum_{a^{-i} \in A^{-i}} P^i[\{\omega' : f^{-i}(\omega') = a^{-i}\} | \mathcal{H}^i(\omega)] u^i[f^i(\omega), a^{-i}].$$

The basic equivalence result in this section (Proposition 2.1) is between correlated rationalizable payoffs and interim payoffs from a posteriori equilibria. The idea behind rationalizability is that (according to Bayesian decision theory) player i has a certain given belief over the actions of the other players, and this determines i's (maximal) expected payoff. On the other hand, at the ex ante stage in an a posteriori equilibrium i does not yet know what his/her belief over the other players' actions will be. This belief will be equal to i's conditional probability which is determined by i's information, i.e., it is i's belief at the interim stage. This is why the basic equivalence result is stated in terms of interim payoffs. In fact, because of convexity of the set of correlated rationalizable payoffs to i (Lemma 2.1) one can also prove an equivalence between correlated rationalizable payoffs and ex ante payoffs from a posteriori equilibria — see Proposition 2.2.

Proposition 2.1. $(\pi^1, \ldots, \pi^n) \in \Pi^1 \times \cdots \times \Pi^n$ *if and only if there is an a posteriori equilibrium of Γ in which $(\pi^1, \ldots, \pi^n)$ is a vector of interim payoffs.*

Proof. *Only if.* Given a vector $(\pi^1, \ldots, \pi^n) \in \Pi^1 \times \cdots \times \Pi^n$, we have to show that there is an a posteriori equilibrium of Γ in which $(\pi^1, \ldots, \pi^n)$ is a vector of interim payoffs. To see this, consider a mediator (cf. Myerson [1985]) who randomly selects a joint action $(a^1, \ldots, a^n) \in R^1 \times \cdots \times R^n$ and recommends to each player i to play a^i. Since π^i is a correlated rationalizable payoff to i, there is an $\tilde{a}^i \in R^i$ and a $\tilde{\sigma} \in \Delta(R^{-i})$ such that $\tilde{a}^i$ is a best reply to $\tilde{\sigma}$ and π^i is i's expected payoff from playing $\tilde{a}^i$ against $\tilde{\sigma}$. If i is recommended to play $\tilde{a}^i$, then the conditional probability with which i believes the mediator chooses actions in R^{-i} is $\tilde{\sigma}$. For any other $a^i \in R^i$ choose $\sigma \in \Delta(R^{-i})$ to which a^i is a best reply. If i is recommended to play a^i, then the conditional probability with which i believes the mediator chooses actions in R^{-i} is σ. With these conditional probabilities i will be willing to follow the mediator's recommendations, and when informed of $\tilde{a}^i$, i's conditional expected payoff from this a posteriori equilibrium is π^i.

If. We have to show that a vector of interim payoffs from an a posteriori equilibrium $(f^1, \ldots, f^n)$ of Γ is an element of $\Pi^1 \times \cdots \times \Pi^n$. For each i let $A^i_+ = \{a^i \in A^i : a^i = f^i(\omega) \text{ for some } \omega \in \Omega\}$. The set $A^1_+ \times \cdots \times A^n_+$ is a best reply set. To prove this, it has to be shown that for every i and each $a^i \in A^i_+$ there is a $\sigma \in \Delta(A^{-i}_+)$ to which a^i is a best reply. Given an $a^i \in A^i_+$, choose an ω such that $f^i(\omega) = a^i$. Since $(f^1, \ldots, f^n)$ is an a posteriori equilibrium, and only strategies $a^{-i} \in A^{-i}_+$ "enter" into the equilibrium, i's optimality condition at ω can be written as

$$\sum_{a^{-i} \in A^{-i}_+} P^i[\{\omega' : f^{-i}(\omega') = a^{-i}\} | \mathscr{H}^i(\omega)] u^i(a^i, a^{-i})$$

$$\geq \sum_{a^{-i} \in A^{-i}_+} P^i[\{\omega' : f^{-i}(\omega') = a^{-i}\} | \mathscr{H}^i(\omega)] u^i(\tilde{a}^i, a^{-i}) \quad \forall \tilde{a}^i \in A^i.$$

This says that a^i is a best reply to the strategy σ which assigns probability $P^i[\{\omega' : f^{-i}(\omega') = a^{-i}\} | \mathscr{H}^i(\omega)]$ to a^{-i} for $a^{-i} \in A^{-i}_+$. It follows that i's conditional expected payoff on $\mathscr{H}^i(\omega)$ is a correlated rationalizable payoff to i. $\qquad \square$

Proposition 2.1 also implies that for each player i the set of actions in A^i which are played in some a posteriori equilibrium of Γ is equal to R^i, i.e., the set of correlated rationalizable actions of i. Hence, the equivalence result in Proposition 2.1 could be stated in terms of actions as well as payoffs. The same remark applies to Propositions 3.1 and 4.1 later in the chapter.

Suppose that in the first half of the proof of Proposition 2.1, player i assigns (prior) probability 1 to being recommended to play $\tilde{a}^i$. Then, i's ex ante payoff is also π^i. It was shown in the second half of the proof that i's conditional expected payoff on $\mathscr{H}^i(\omega)$ is a correlated rationalizable payoff to i. So i's ex ante payoff from the a posteriori equilibrium $(f^1, \ldots, f^n)$ is a convex combination of correlated rationalizable payoffs to i. Lemma 2.1 says that the set of correlated rationalizable payoffs to i is convex. Putting these observations together shows that the term "interim" in Proposition 2.1 can be replaced with "ex ante."

Proposition 2.2. *The set of ex ante payoff vectors from the a posteriori equilibria of* Γ *is equal to* $\Pi^1 \times \cdots \times \Pi^n$.

As argued above, Proposition 2.2 will be implied by the following Lemma.

Lemma 2.1. Π^i *is convex for every* i.

Proof. For any $\sigma \in \Delta(R^{-i})$, let $v^i(a^i, \sigma)$ be i's expected payoff from playing a^i against σ. Then, $\Pi^i = \{\max_{a^i \in A^i} v^i(a^i, \sigma) : \sigma \in \Delta(R^{-i})\}$. But $\{\max_{a^i \in A^i} v^i(a^i, \sigma) : \sigma \in \Delta(R^{-i})\}$ is the image of the continuous map $\sigma \mapsto \max_{a^i \in A^i} v^i(a^i, \sigma)$, and is therefore a closed interval since the domain $\Delta(R^{-i})$ is compact and connected. $\square$

3. Independent Rationalizability and Conditionally Independent A Posteriori Equilibria

The previous section established an equivalence between correlated rationalizability and a posteriori equilibrium. In this section, analogous results are obtained starting from independent rather than correlated rationalizability. Independent rationalizability is the concept originally defined by Bernheim (1984) and Pearce (1984). The set of independent rationalizable payoffs is the subset of the set of correlated rationalizable payoffs obtained by restricting each player's beliefs over the actions of the other players to be independent. Clearly these sets of payoffs are the same in two-person games. To see that the set of independent rationalizable payoffs is a strict subset of the set of correlated rationalizable payoffs in games with three or more players, consider the example in Figure 2. Player 1 chooses the row, 2 the column, 3 the matrix. 0.7 is a correlated rationalizable payoff to 3 as follows. Player 3 believes 1, 2 play (U, L) with probability $\frac{1}{2}$, (D, R) with probability $\frac{1}{2}$ (to which B is the best reply). Player 1 believes 2 plays L

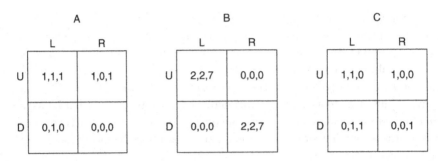

Figure 2.

with probability $\frac{1}{2}$, R with probability $\frac{1}{2}$, and 3 plays B (to which U and D are best replies). Player 2 believes 1 plays U with probability $\frac{1}{2}$, D with probability $\frac{1}{2}$, and 3 plays B (to which L and R are best replies). On the other hand, 1 is the unique independent rationalizable payoff to 3. To see this, first note that B is not a best reply to any pair of mixed strategies of 1, 2. Hence, 1, 2 must assign probability 0 to 3 playing B. But then $U, L,$ strongly dominate D, R for 1, 2, respectively.

Definition 3.1. A subset $\hat{B}^1 \times \cdots \times \hat{B}^n$ of $A^1 \times \cdots \times A^n$ is an *independent best reply set* if for every i and each $a^i \in \hat{B}^i$ there is $\sigma^{-i} \in \prod_{j \neq i} \Delta(\hat{B}^j)$ to which a^i is a best reply.

The set of independent rationalizable actions $\hat{R}^1 \times \cdots \times \hat{R}^n$ is the (finite) component-by-component union $(\cup_\alpha \hat{B}^1_\alpha) \times \cdots \times (\cup_\alpha \hat{B}^n_\alpha)$ of all independent best reply sets $\hat{B}^1_\alpha \times \cdots \times \hat{B}^n_\alpha$. Player i's maximal expected payoff against a $\sigma^{-i} \in \prod_{j \neq i} \Delta(\hat{R}^j)$ is then an independent rationalizable payoff to i. Let $\hat{\Pi}^i$ denote the set of all possible independent rationalizable payoffs to i.

The results of the previous section would suggest an equivalence between independent rationalizable payoffs and interim payoffs from "mixed" a posteriori equilibria (and if the set of independent rationalizable payoffs is convex, that the equivalence holds for ex ante payoffs as well). This intuition is correct; however, "mixed" should not be taken to mean independence of the players' partitions of Ω in terms of their priors (which is the usual approach). What is needed is a form of conditional independence, which is not implied by a definition in terms of priors.

Definition 3.2. $\mathcal{H}^1, \ldots, \mathcal{H}^n$ are P^i-*prior independent* if $P^i(\cap_{j=1}^n H^j) = \prod_{j=1}^n P^i(H^j)$ for every $H^j \in \mathcal{H}^j, j = 1, \ldots, n.$ $\mathcal{H}^1, \ldots, \mathcal{H}^{i-1}, \mathcal{H}^{i+1},$

$\dots, \mathcal{H}^n$ are P^i-*conditionally independent* given $\mathcal{H}^i$ if for every $H^i \in \mathcal{H}^i$, $P^i(\cap_{j \neq i} H^j | H^i) = \prod_{j \neq i} P^i(H^j | H^i)$ for every $H^j \in \mathcal{H}^j, j \neq i$.

Prior independence is the standard definition of independent σ-fields (Chung [1974, p. 61]). It is also the notion of independence used in Aumann (1974) to define mixed strategies. Our definition of conditional independence is a strengthening of the standard definition of conditionally independent σ-fields (Chung [1974, p. 306]) from an almost everywhere to an everywhere requirement. Conditional independence says that *whatever* information i receives, i believes that the other players choose their actions independently. Prior independence implies that if $P^i(H^i) > 0$, then $P^i(\cap_{j \neq i} H^j | H^i) = \prod_{j \neq i} P^i(H^j)$. So prior independence only implies that i believes with P^i-probability 1 that the others play independently. Prior independence says nothing about $P^i(\cap_{j \neq i} H^j | H^i)$ if H^i is P^i-null, so prior independence does not imply conditional independence. Nor does conditional independence imply prior independence. (Let $\Omega = \{\omega_1, \omega_2\}$ and suppose all the players have the finest partition. If $P^i(\{\omega_1\}) = \frac{1}{2}$ then conditional independence is satisfied but prior independence is not.)

Definition 3.3. A *conditionally independent a posteriori equilibrium of* Γ is an a posteriori equilibrium of Γ in which for every i, $\mathcal{H}^1, \dots, \mathcal{H}^{i-1}$, $\mathcal{H}^{i+1}, \dots, \mathcal{H}^n$ are P^i-conditionally independent given $\mathcal{H}^i$.

Proposition 3.1. *The sets of interim and ex ante payoff vectors from the conditionally independent a posteriori equilibria of* Γ *are both equal to* $\hat{\Pi}^1 \times \dots \times \hat{\Pi}^n$.

Proof. *Only if.* Given a vector $(\pi^1, \dots, \pi^n) \in \hat{\Pi}^1 \times \dots \times \hat{\Pi}^n$, we have to show that there is a conditionally independent a posteriori equilibrium of Γ in which $(\pi^1, \dots, \pi^n)$ is a vector of interim and ex ante payoffs. The proof is like the first half of the proof of Proposition 2.1. A mediator randomly selects a joint action $(a^1, \dots, a^n) \in \hat{R}^1 \times \dots \times \hat{R}^n$ and recommends to each player i to play a^i. Since π^i is an independent rationalizable payoff to i, there is an $\tilde{a}^i \in \hat{R}^i$ and a $\tilde{\sigma}^{-i} \in \prod_{j \neq i} \Delta(\hat{R}^j)$ such that $\tilde{a}^i$ is a best reply to $\tilde{\sigma}^{-i}$ and π^i is i's expected payoff from playing $\tilde{a}^i$ against $\tilde{\sigma}^{-i}$. If i is recommended to play $\tilde{a}^i$ then the conditional probability with which i believes the mediator chooses actions in $\hat{R}^{-i}$ is $\tilde{\sigma}^{-i}$. Notice that $\tilde{\sigma}^{-i}$ is a product measure on $\hat{R}^{-i}$. Continuing in this way, after any recommendation i's conditional probability on $\hat{R}^{-i}$ is a product measure. So the a posteriori equilibrium which is constructed is conditionally independent, and π^i is an interim payoff to i. By letting i assign (prior) probability 1 to the mediator recommending $\tilde{a}^i$, this is also the ex ante payoff to i.

If. We have to show that a vector of interim or ex ante payoffs from a conditionally independent a posteriori equilibrium $(f^1, \ldots, f^n)$ of Γ is an element of $\hat{\Pi}^1 \times \cdots \times \hat{\Pi}^n$. The proof is essentially the same as the second half of the proof of Proposition 2.1. For each i let $A^i_+ = \{a^i \in A^i : a^i = f^i(\omega)$ for some $\omega \in \Omega\}$. The set $A^1_+ \times \cdots \times A^n_+$ is an independent best reply set. This follows from the same argument as before, noting that

$$P^i[\{\omega' : f^{-i}(\omega') = a^{-i}\}|\mathscr{H}^i(\omega)] = \prod_{j \neq i} P^i[\{\omega' : f^j(\omega') = a^j\}|\mathscr{H}^i(\omega)]$$

because of conditional independence. It follows that i's conditional expected payoff on $\mathscr{H}^i(\omega)$ is an independent rationalizable payoff to i. Player i's ex ante payoff will then be a convex combination of independent rationalizable payoffs, and a trivial modification of Lemma 2.1 shows that each $\hat{\Pi}^i$ is convex. $\square$

To prove an equivalence with independent rationalizability it was necessary to use conditionally independent a posteriori equilibrium. Recall that conditional independence does not in general imply prior independence. Nevertheless, when considering conditionally independent a posteriori equilibria, prior independence can be assumed without loss of generality. More precisely, the sets of interim and ex ante payoffs from the conditionally independent a posteriori equilibria which also satisfy prior independence are again both equal to $\hat{\Pi}^1 \times \cdots \times \hat{\Pi}^n$. To see this, first note that by definition these sets must be contained in $\hat{\Pi}^1 \times \cdots \times \hat{\Pi}^n$. Second note that the a posteriori equilibrium constructed in the first half of the proof of Proposition 3.1 satisfies prior independence (when each player i assigns probability 1 to the mediator recommending $\tilde{a}^i$).

4. Objective Solution Concepts

The starting point of this chapter is that rationalizability (either correlated or independent) is the solution concept implied by common knowledge of rationality of the players in a game. The previous two sections established equivalences between rationalizability (correlated and independent) and equilibrium concepts (a posteriori and conditionally independent a posteriori). It follows that common knowledge of rationality alone implies equilibrium behavior on the part of the players. Notice that the players may have subjective, i.e., different, priors. However, as discussed in the Introduction, it is usual in applications to assume that the players have the same prior. If an assumption of common priors is adopted then, as Aumann (1987) has shown, one is led to objective correlated equilibrium rather than a posteriori equilibrium. In order to go further and characterize Nash equilibrium, additional assumptions must be made.

One way to characterize Nash equilibrium is to adopt, in addition to a common prior, the assumption of prior independence (Definition 3.2). In the alternative characterization provided below the assumption of common priors is weakened to the requirement that any two players share the same beliefs about a third player's choices of action. This requirement is met by assuming "concordant" priors. Technically this assumption differs only slightly from common priors, in that under concordant priors player i's belief over events in $\mathscr{H}^i$ need not be the same as the (common) beliefs of the other players. However, this is perhaps more natural, since i's beliefs over events in $\mathscr{H}^i$ have no decision theoretic significance for the play of the game.

Definition 4.1. $P^1, \ldots, P^n$ are *concordant* if for each i and every $j, k \neq i$, $P^j(H^i) = P^k(H^i)$ for every $H^i \in \mathscr{H}^i$.

To compensate for weakening the assumption of common priors, prior independence must be strengthened to hold "everywhere." Recall that prior independence says that with probability 1: (i) i's beliefs over the other players' choices of action is a product measure; and (ii) i will not update his/her prior. Conditional independence is designed to strengthen (i) to an everywhere condition. It remains to strengthen (ii) to an everywhere condition. This is achieved by assuming "informational independence." (We are grateful to a referee for suggesting the following definition.) Notice that both concordant priors and conditional independence are automatically satisfied for two-person games (whereas, the assumption that players' beliefs do not vary with their private information *is* needed in two-person games).

Definition 4.2. $\mathscr{H}^1, \ldots, \mathscr{H}^{i-1}, \mathscr{H}^{i+1}, \ldots, \mathscr{H}^n$ are P^i-*informationally independent* of $\mathscr{H}^i$ if for every H^i and $\tilde{H}^i \in \mathscr{H}^i$, $P^i(\cap_{j \neq i} H^j | H^i) = P^i(\cap_{j \neq i} H^j | \tilde{H}^i)$ for every $H^j \in \mathscr{H}^j, j \neq i$.

Proposition 4.1. *Consider the a posteriori equilibria of* Γ *which have concordant priors and in which for every* i, $\mathscr{H}^1, \ldots, \mathscr{H}^{i-1}, \mathscr{H}^{i+1}, \ldots, \mathscr{H}^n$ *are* P^i-*conditionally independent given* $\mathscr{H}^i$ *and* P^i-*informationally independent of* $\mathscr{H}^i$. *The sets of interim and ex ante payoffs from these equilibria* are both equal to the set of expected payoff vectors from the Nash equilibria of Γ.

Proof. Consider an a posteriori equilibrium $(f^1, \ldots, f^n)$ of Γ and for each i let $A^i_+ = \{a^i \in A^i : a^i = f^i(\omega)$ for some $\omega \in \Omega\}$. i's conditional expected payoff on $H^i \in \mathscr{H}^i$ from playing a^i is

$$\sum_{a^{-i} \in A^{-i}_+} P^i[\{\omega : f^{-i}(\omega) = a^{-i}\} | H^i] u^i(a^i, a^{-i})$$

which is equal to

$$\sum_{a^{-i} \in A^{-i}_+} \prod_{j \neq i} P^i[\{\omega : f^j(\omega) = a^j\} | H^i] u^i(a^i, a^{-i})$$

by conditional independence, which in turn equals

$$\sum_{a^{-i} \in A^{-i}_+} \prod_{j \neq i} P^i[\{\omega : f^j(\omega) = a^j\}] u^i(a^i, a^{-i})$$

by informational independence. Write $P^i[\{\omega : f^j(\omega) = a^j\}] = \sigma^j(a^j)$ and let $\sigma^j \in \Delta(A^j_+)$ be the mixed strategy which assigns probability $\sigma^j(a^j)$ to each $a^j \in A^j_+$. Note that σ^j does not depend on i by the assumption of concordant priors. In other words, i's conditional expected payoff from playing a^i is the expected payoff from playing a^i against the $(n-1)$-tuple of mixed strategies σ^{-i}. Let $BR(\sigma^{-i})$ denote the set of i's best replies to σ^{-i}. Then, $A^i_+ \subset BR(\sigma^{-i})$. So there are sets $A^1_+ \subset A^1, \ldots,$ $A^n_+ \subset A^n$ and mixed strategies $\sigma^1 \in \Delta(A^1_+), \ldots, \sigma^n \in \Delta(A^n_+)$ such that $A^1_+ \subset BR(\sigma^{-1}), \ldots, A^n_+ \subset BR(\sigma^{-n})$. That is, $(\sigma^1, \ldots, \sigma^n)$ is a Nash equilibrium. So i's conditional expected payoff on any H^i — and hence i's ex ante payoff — is equal to i's expected payoff from a Nash equilibrium. The converse direction is immediate. $\square$

References

Aumann, R (1974). Subjectivity and correlation in randomized strategies. *Journal of Mathematical Economics*, 1, 67–96.

Aumann, R (1987). Correlated equilibrium as an expression of Bayesian rationality. *Econometrica*, 55, 1–18.

Bernheim, D (1984). Rationalizable strategic behavior. *Econometrica*, 52, 1007–1028.

Bernheim, D (1985). Axiomatic characterizations of rational choice in strategic environments. Unpublished, Department of Economics, Stanford University.

Blackwell, D and L Dubins (1975). On existence and non-existence of proper, regular, conditional distributions. *The Annals of Probability*, 3, 741–752.

Chung, KL (1974). *A Course in Probability Theory*, 2nd Edition. New York, NY: Academic Press.

Kohlberg, E and J-F Mertens (1986). On the strategic stability of equilibria. *Econometrica*, 54, 1003–1037.

Kreps, D and R Wilson (1982). Sequential equilibria. *Econometrica*, 50, 863–894.

Myerson, R (1978). Refinements of the Nash equilibrium concept. *International Journal of Game Theory*, 7, 73–80.

Myerson, R (1985). Bayesian equilibrium and incentive-compatibility: An introduction. In Hurwicz, L, D Schmeidler, and H Sonnenschein (Eds.), *Social Goals and Social Organization*. New York, NY: Cambridge University Press.

Nash, J (1951). Non-cooperative games. *Annals of Mathematics*, 54, 286–295.

Pearce, D (1984). Rationalizable strategic behavior and the problem of perfection. *Econometrica*, 52, 1029–1050.

Selten, R (1965). Spieltheoretische Behandlung eines Oligopolmodells mit Nachfrageträgheit. *Zeitschrift für die gesamte Staatswissenschaft*, 121, 301–324, 667–689.

Selten, R (1975). Reexamination of the perfectness concept for equilibrium points in extensive games. *International Journal of Game Theory*, 4, 25–55.

Tan, T and S Werlang (1984). The Bayesian foundations of rationalizable strategic behavior and Nash equilibrium behavior. Unpublished, Department of Economics, Princeton University.

Chapter 4

Intrinsic Correlation in Games

Adam Brandenburger and Amanda Friedenberg

Correlations arise naturally in noncooperative games, e.g., in the equivalence between undominated and optimal strategies in games with more than two players. But the noncooperative assumption is that players do not coordinate their strategy choices, so where do these correlations come from? The epistemic view of games gives an answer. Under this view, the players' hierarchies of beliefs (beliefs, beliefs about beliefs, etc.) about the strategies played in the game are part of the description of a game. This gives a source of correlation: A player believes other players' strategy choices are correlated, because he believes their hierarchies of beliefs are correlated. We refer to this kind of correlation as "intrinsic," since it comes from variables — viz., the hierarchies of beliefs — that are part of the game. We compare the intrinsic route with the "extrinsic" route taken by Aumann (1974), which adds signals to the original game.

Originally published in *Journal of Economic Theory*, 141, 28–67.

Keywords: Correlation; epistemic game theory; intrinsic correlation; conditional independence; correlated equilibrium; rationalizability.

Financial support: Stern School of Business and Olin School of Business.

Acknowledgments: We are indebted to thank Pierpaolo Battigalli, Ethan Bueno de Mesquita, Yossi Feinberg, Jerry Keisler, and Gus Stuart for important imput. Geir Asheim, Philip Dawid, Konrad Grabiszewski, Jerry Green, Qingmin Liu, Paul Milgrom, Stephen Morris, John Nachbar, Andres Perea, John Pratt, Dov Samet, Michael Schwarz, Jeroen Swinkels, Daniel Yamins, and audiences at the 2005 Econometric Society World Congress, 2005 SAET Conference, University College London Conference in Honor of Ken Binmore (August 2005), 7th Augustus de Morgan Workshop (November 2005), Berkeley, CUNY, Davis, University of Leipzig, Notre Dame, Pittsburgh, Stanford, UCLA, Washington University, and Yale gave valuable comments. An associate editor and two referees made very helpful observations and suggestions.

1. Introduction

Correlation is basic to game theory. For example, consider the equivalence between undominated strategies — strategies that are not strongly dominated — and strategies that are optimal under some measure on the strategy profiles of the other players. As is well known, for this to hold in games with more than two players, the measure may need to be dependent (i.e., correlated).

In Figure 1, Ann chooses the row, Bob chooses the column, Charlie chooses the matrix, and the payoffs are written in the order Ann then Bob then Charlie. The strategy Y is optimal for Charlie, under a measure that puts probability $\frac{1}{2}$ on (U, L) and probability $\frac{1}{2}$ on (D, R). It is therefore undominated. But there is no product measure under which Y is optimal.[1]

Where does a correlated assessment — such as probability $\frac{1}{2}$ on (U, L) and probability $\frac{1}{2}$ on (D, R) — come from? After all, the noncooperative assumption is that players do not coordinate their strategy choices. Alternatively put, there is no physical correlation across players.

The epistemic approach to game theory suggests an answer. Under the epistemic approach, a complete description of a game includes specifying not only the players' strategy sets and payoff functions, but also their hierarchies of beliefs (beliefs, beliefs about beliefs, etc.) about the strategies played in the game. This gives a source of correlation. A player can think that other players' strategy choices are correlated, because he thinks what they believe about the game is correlated. For example, in Figure 1, Charlie might assign: (i) probability $\frac{1}{2}$ to "Ann's having hierarchy of beliefs h^a and playing U, and Bob's having hierarchy of beliefs h^b and playing L,"; and (ii) probability $\frac{1}{2}$ to "Ann's having hierarchy of beliefs $\tilde{h}^a$ and playing D, and Bob's having hierarchy of beliefs $\tilde{h}^b$ and playing R."

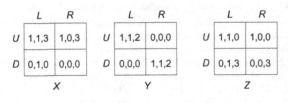

Figure 1.

---

[1]Let p be the probability on U, and q be the probability on L. It is straightforward to check that $\max\{3p, 3(1-p)\} > 2pq + 2(1-p)(1-q)$.

This chapter formalizes this line of argument. There are two require-ments. (Here, and throughout, we will use the shorthand "hierarchy of beliefs" for "hierarchy of beliefs about the strategies played in the game." Figure 4 gives a numerical example of such hierarchies.)

(a) We impose a *conditional independence (CI)* requirement on hierarchies of beliefs. Conditional on Ann's and Bob's hierarchies of beliefs, Charlie assesses Ann's and Bob's strategies are independent. (But this need not hold unconditionally!)

(b) We impose a *sufficiency (SUFF)* requirement on hierarchies of beliefs. Conditional on Ann's hierarchy of beliefs, Charlie's assessment about Ann's strategy does not change if she (i.e., Charlie)[2] learns Bob's hierarchy of beliefs.

Both requirements are immediately satisfied in the example above. We show later that under CI and SUFF, if a player has an independent assessment about the other players' hierarchies of beliefs, he must have an independent assessment about their strategy choices (Proposition 9.1). Equivalently, if a player has a correlated assessment about the other players' strategy choices, he must have a correlated assessment about their hierarchies of beliefs. This is exactly our view of correlation: The correlation in strategies is "nonphysical" as per noncooperative game theory, and comes from correlation in what the players believe about the play of the game.

What strategies can be played under our view of correlation? Our restrictions limit how Charlie (for example) thinks Ann's and Bob's strategies and hierarchies of beliefs are related. For these restrictions to matter, we need to be in a setting where there is a connection between strategies and hierarchies. Rationality is the natural such connection — a strategy may be optimal under some (first-order) beliefs but not under others. This leads to a third requirement:

(c) We impose *rationality and common belief of rationality (RCBR)*. That is, each player is rational, assigns probability 1 to the event the other players are rational, and so on. This is the usual "baseline" epistemic condition on a game.

In sum, this chapter asks: What strategies can be played in a game under the requirements of CI, SUFF, and RCBR?

---

[2]In this chapter, Charlie is female.

At a broad level, our analysis of correlation in games is analogous to the Bayesian or subjective view of coin tossing. To rule out physical correlation across tosses, one subscribes to an assessment that is conditionally independent given an extra variable not included in the original description — in this case, the parameter or "bias" of the coin. For the game setting, there are also extra variables — in this chapter, they are the players' hierarchies of beliefs about the strategies played in the game. We ask for CI relative to these variables. We also impose a SUFF condition to ensure that we "attach" the correct extra variable to the correct player.

2. Intrinsic vs. Extrinsic Correlation

We will refer to correlation of the kind we study in this chapter as *intrinsic correlation*, because the correlations come from variables — viz., hierarchies of beliefs — that are part of the description of the game. (At least, they are part of the game under the epistemic approach to game theory.)

The existing route, initiated by Aumann (1974), can be termed *extrinsic correlation*. This is because it adds to the given game payoff-irrelevant moves by Nature (often called "signals," "sunspots," or similar). Figure 2 is a typical scenario. A coin is tossed. Ann and Bob observe the outcome and choose strategies as shown. Charlie does not observe the outcome, and assigns probabilities $\frac{1}{2}$ to *Heads* and $\frac{1}{2}$ to *Tails*. Thus, Charlie has a correlated assessment about Ann's and Bob's strategies, because Ann and Bob get a correlated signal and choose strategies according to the realization of the signal.

What is the relationship between intrinsic and extrinsic correlation? We will answer at both the formal and the conceptual level.

Aumann (1974, 1987) gave two definitions: objective and subjective correlated equilibrium, depending on whether there is or is not a common

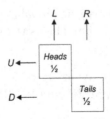

Figure 2.

prior. The idea of a common prior is irrelevant to our project, so we will focus on the subjective case.

The following equivalence gives the relationship to intrinsic correlation. Fix a game G. The set of strategy profiles that can be played in some subjective correlated equilibrium of G is the same as the set of correlated rationalizable profiles in G. (See Brandenburger and Dekel [1987].) The *correlated rationalizable* strategies are those that survive iterated elimination of strategies that are never best replies — or, equivalently, are dominated.) It is well known that the condition of RCBR in a game is characterized by correlated rationalizability. (This result goes back to Brandenburger and Dekel [1987] and Tan and Werlang [1988]. Proposition 10.1 is a statement in the set-up of this chapter.) It follows that intrinsic correlation is a weakly stronger theory than extrinsic correlation: Any strategy profile allowed under intrinsic correlation is certainly allowed under extrinsic correlation.

But — surprisingly, in our view — the inclusion can be strict. The main result of the chapter (Theorem 11.1) exhibits a game G in which there is a correlated rationalizable strategy that cannot be played under intrinsic correlation. The two theories of correlation are different.

We also record the relationship between intrinsic correlation and the Bernheim (1984) and Pearce (1984) concept of independent rationalizability. (The *independent rationalizable* strategies are those that survive iterated elimination of strategies that are never best replies under product measures.) Proposition G.1 says that any independent rationalizable profile is allowed under intrinsic correlation. This is not quite obvious. Under independent rationalizability, each player assesses the other players' strategy choices as independent. But, as is well known in probability theory, independence does not imply CI. So, demonstrating this inclusion takes some work. The inclusion is strict: The game of Figure 1 is an example. Strategy Y can be played under intrinsic correlation (see Footnote 5 for the details), but it is easy to check that only X is independent rationalizable. (So, the expected payoffs to Charlie are also different.)

Intrinsic correlation is a distinct theory of how games are played. Refer to Figure 3.

3. Comparison

We now ask about the conceptual meaning of intrinsic correlation vs. the other routes in Figure 3.

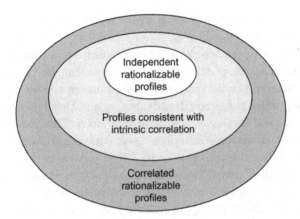

Figure 3.

Bernheim (1984, p. 1014) argued for independent rationalizability, on the grounds that "in a noncooperative framework ... the choices of any two agents are by definition independent events; they cannot affect each other." Likewise, Pearce (1984, p. 1035) distinguishes independent rationalizability from the situation where "a player's opponents could *coordinate* their randomized strategic actions."

The epistemic view of games reaches a different conclusion. Even if players choose strategies independently, correlation is possible because some aspect of who the players are (their hierarchies of beliefs) may be correlated. This line of reasoning — the intrinsic route — is really just an adaptation to game theory of the usual idea of common-cause correlation.

The Bernheim–Pearce analysis now appears as a special case. Suppose each player thinks that the other players' hierarchies of beliefs are uncorrelated. Then, under CI and SUFF, our Proposition 9.1 implies that each player thinks the other players' strategy choices are uncorrelated. If we add RCBR, we get exactly independent rationalizability.[3]

Why do we prefer the case of correlated hierarchies of beliefs? To us, such correlation is in line with Savage's Small-Worlds idea in decision theory

---

[3]So, as a by-product, we also get new epistemic conditions for independent rationalizability. Existing treatments (e.g., Tan and Werlang [1988]) directly assume that each player assesses the other players' choices are independent. Our results show that this can be deduced from more basic conditions on the hierarchies of beliefs. See Corollary G.1 for a formal statement.

(Savage [1954, pp. 82–91]). The given game is only a part of a larger whole. In particular, there is a "context" or "history" to the game. Who the players are then becomes a shorthand for their prior experiences. That is, the players' prior experiences (partially) determine their characteristics — including their beliefs, beliefs about beliefs, etc. It seems natural that these experiences could be (though do not have to be) correlated, in which case the players' hierarchies of beliefs could be correlated, too.

Next, the distinction between the intrinsic and extrinsic (Aumann) routes. We establish they are formally different. (This is our main result, Theorem 11.1.) But how do they differ conceptually?

The answer is that Aumann takes the next step and actually changes the game to include its "context." The signals in his analysis are what the players see before they play the given game — and these signals are added to the game. (Recall that the correlated equilibria of a game are the Nash equilibria of the extended game.) So, the starting point for Aumann is also the Savage Small-Worlds idea.[4] But Aumann's and our analysis then differ in whether or not the "context" of the game is made part of the game.

We should point out a subtlety: Our approach certainly allows for signals beyond the game as given. Signals might give a story for why the players have the beliefs they have. But the signals would remain in the background and are not brought into the game itself.

Summarizing, both the intrinsic and the extrinsic routes to correlation in games get correlation by recognizing the larger context in which a game is played. The difference is whether we analyze the game as originally given or the larger game.

Logically, the intrinsic route comes before the extrinsic route (per Figure 3). Historically, the extrinsic route came first. There is a simple 'mechanical' reason for this. Correlation requires additional variables. Nowadays, with the epistemic view of games well established, these variables are immediately at hand — they are the players' hierarchies of beliefs. But Aumann, writing prior to the epistemic program (Aumann [1974]), had to look beyond the game to find extra variables. This is why the outer part of Figure 3 (extrinsic correlation) came before the middle part (intrinsic correlation). In a literal way, we see this chapter as filling in the existing picture of correlation in games.

---

[4]We thank a referee for clarifying the connection to Savage (1954).

4. Organization of the Chapter

In Sections 5–7, we give a heuristic treatment of our theory of correlation. The formal presentation and proofs are in Sections 8–11 and accompanying appendices. (The formal treatment turns out to be quite involved and lengthy.) Section 12 concludes. The heuristic treatment can be read either before or in parallel with the formal treatment.

5. Type Structures

Return to the game in Figure 1. Even if the matrix itself is "transparent" to the players, each player may be uncertain about the strategies chosen by the other players, about what the other players believe about the strategies chosen, and so on. A key feature of the epistemic approach is incorporating this kind of uncertainty into the description of a game situation.

Figure 4 is an example of a *type structure* which, when added to a game like Figure 1, gives what can be called an epistemic game. The type structure shown describes the players' possible hierarchies of beliefs about the strategies chosen, as follows. There are two *types* for Ann, viz. t^a and u^a, two types for Bob, viz. t^b and u^b, and one type for Charlie, viz. t^c. Each type is associated with a probability measure on the strategies and types of other players. In the usual way, each type induces a *hierarchy of beliefs* about the strategies in the game.

Thus, type t^c assigns probability $\frac{1}{2}$ to (U, t^a, L, t^b) and probability $\frac{1}{2}$ to (D, u^a, R, u^b), and so has a first-order belief that assigns: (i) probability $\frac{1}{2}$ to "Ann's playing U and Bob's playing L,"; and (ii) probability $\frac{1}{2}$ to "Ann's playing D and Bob's playing R." Type t^a has a first-order belief that assigns probability 1 to (L, Y), and type u^a has a first-order belief that assigns probability 1 to (R, Y). Type t^b has a first-order belief that assigns probability 1 to (U, Y), and type u^b has a first-order belief that assigns probability 1 to (D, Y). So, type t^c has a second-order belief that assigns: (i) probability $\frac{1}{2}$ to "Ann's playing U and assigning probability 1 to (L, Y)," and "Bob's playing L and assigning probability 1 to (U, Y),"; and (ii) probability $\frac{1}{2}$ to "Ann's playing D and assigning probability 1 to (R, Y)," and "Bob's playing R and assigning probability 1 to (D, Y)." And so on.

Notice that Charlie's type t^c has a correlated assessment about the strategies Ann and Bob choose. Charlie also has a correlated assessment about Ann's and Bob's hierarchies of beliefs. To see this, note that Ann's types t^a and u^a (resp. Bob's types t^b and u^b) are associated with distinct

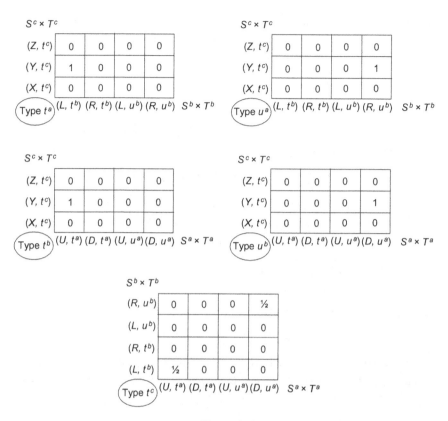

Figure 4.

hierarchies. (Associating different hierarchies with different types keeps the example simple. We do not impose this condition in our formal treatment.)

Next, we want to capture our idea that Charlie's correlated assessment about Ann's and Bob's strategies comes from the fact that she is uncertain about their hierarchies of beliefs. Indeed, if this is the source of the correlation, then, if Charlie's uncertainty about Ann's and Bob's hierarchies is resolved, she should have an independent assessment about their strategies. This leads to the CI condition from Section 1.

But CI alone does not fully capture the meaning of correlation via hierarchies of beliefs. We need to take account of the information in the game. Ann's information is her own hierarchy, not Bob's. So, Bob's hierarchy should provide information about Ann's choice of strategy only insofar as it provides information about her hierarchy of beliefs. It follows that if Charlie's uncertainty about Ann's hierarchy is resolved, she should not change her assessment about Ann's strategy if she learns Bob's hierarchy. (And vice versa with Ann and Bob interchanged.) This leads to the SUFF condition from Section 1.

Let us state CI and SUFF in the context of the example:

(a) *Conditional Independence (CI)*: The measure associated with each type assesses the other players' strategy choices as independent, conditional on their hierarchies of beliefs. Consider, for example, Charlie's type t^c and the associated measure — which we will denote $\lambda^c(t^c)$. We have

$$\lambda^c(t^c)(U, L|t^a, t^b) = \lambda^c(t^c)(U|t^a, t^b) \times \lambda^c(t^c)(L|t^a, t^b).$$

(Here and below, the conditioning is to be thought of as on the hierarchies. This again uses the fact that for Ann and Bob, different types induce different hierarchies.) The corresponding equality holds when conditioning on (the hierarchies associated with) u^a and u^b. So, Charlie's type t^c assesses Ann's and Bob's strategies as independent conditional on their hierarchies.

(b) *Sufficiency (SUFF)*: We have

$$\lambda^c(t^c)(U|t^a, t^b) = \lambda^c(t^c)(U|t^a),$$

and similarly for D. If Charlie knows Ann's hierarchy of beliefs, and comes to learn Bob's hierarchy of beliefs too, this will not change her (Charlie's) assessment of Ann's choice. Likewise,

$$\lambda^c(t^c)(L|t^a, t^b) = \lambda^c(t^c)(L|t^b),$$

and similarly for R. Ann's hierarchy of beliefs is sufficient for the assessment Charlie's type t^c makes about her choice of strategy. Bob's hierarchy of beliefs is sufficient in the same way.

We repeat that we show later (Proposition 9.1): *Under CI and SUFF, if a player has an independent assessment about the other players' hierarchies of beliefs, he must have an independent assessment about their strategy choices.* (In Appendix A, we show that the CI and SUFF conditions are independent, and neither can be dropped in this result.) CI and SUFF give us correlation in strategies from correlation in hierarchies of beliefs (about the strategies). This is our theory of intrinsic correlation.[5]

---

[5]The type structure of Figure 4 proves the claim in Section 2 that Y can be played under intrinsic correlation. All types satisfy CI and SUFF. (We checked for Charlie's type t^c above. The checks for Ann's types t^a, u^a and Bob's types t^b, u^b are immediate, since

6. The Main Result

What is the prediction of intrinsic correlation? Figure 3 gave the relationship to independent and correlated rationalizability. Here, we give a heuristic treatment of our main finding — that the second inclusion is strict. There are correlated rationalizable strategies that cannot be played under intrinsic correlation. In fact, we show:

(i) There is a game G and a correlated rationalizable strategy in G that cannot be played under RCBR and CI.

(ii) There is a game G' and a correlated rationalizable strategy in G' that cannot be played under RCBR and SUFF.

We now give a sketch of the proofs. (The formal proofs are quite involved and are given in Section 11.)

For statement (i), consider the game in Figure 5. Here, all strategies are correlated rationalizable. Yet we will argue that if Y is consistent with RCBR then Charlie's type cannot satisfy CI.

Fix an associated type structure. First note the following three facts:

(a) If strategy-type pairs (U, t^a) and (M, u^a) are rational, then t^a and u^a must each assign probability 1 to the strategies (L, Y).

	L	C	R
U	0,0,2	0,0,2	0,1,2
M	0,0,0	0,0,0	0,1,0
D	1,0,2	1,0,2	1,1,2

X

	L	C	R
U	1,1,1	0,1,0	0,1,0
M	1,0,0	0,0,1	0,1,0
D	1,0,0	1,0,0	1,1,0

Y

	L	C	R
U	0,0,0	0,0,0	0,1,0
M	0,0,2	0,0,2	0,1,2
D	1,0,2	1,0,2	1,1,2

Z

Figure 5.

each of these types is associated with a degenerate measure.) For Ann, the strategy-type pairs (U, t^a) and (D, u^a) are rational: Strategy U (resp. D) maximizes her expected payoff under the marginal on $S^b \times S^c$ of the measure associated with t^a (resp. u^a). Similarly, (L, t^b) and (R, u^b) are rational for Bob; and (Y, t^c) is rational for Charlie. Also, each type for each player assigns positive probability only to rational strategy-type pairs for the other players. That is, each player believes the other players are rational. By induction, each of these strategy-type pairs is therefore consistent with RCBR. In particular, Charlie can play Y.

(b) If (L, t^b) and (C, u^b) are rational, then t^b and u^b must each assign probability 1 to the strategies (U, Y).

(c) If (Y, t^c) and (Y, u^c) are rational, then t^c and u^c must each assign probability $\frac{1}{2}$ to (U, L) and probability $\frac{1}{2}$ to (M, C).

In words, U and M are each optimal only if Ann assigns probability 1 to (L, Y). Likewise, L and C are each optimal only if Bob assigns probability 1 to (U, Y). Finally, Y is optimal only if Charlie assigns probability $\frac{1}{2}$ to (U, L) and probability to (M, C).

It follows that if (U, t^a) and (M, u^a) are rational, then the hierarchies of beliefs associated with t^a and u^a must agree at the first level. Likewise, for (L, t^b) and (C, u^b), and for (Y, t^c) and (Y, u^c).

Next, suppose that (U, t^a) and (M, u^a) are rational and t^a and u^a believe the other players are rational. We already know that t^a and u^a must each assign probability 1 to (L, Y). Given this, they must also assign probability 1 to: (i) Bob's assigning probability 1 to (U, Y); and (ii) Charlie's assigning probability $\frac{1}{2}$ to (U, L) and probability $\frac{1}{2}$ to (M, C). This follows from (b) and (c). Thus, the hierarchies of beliefs associated with t^a and u^a must agree up to the second level. And so on, inductively.

This gives:

(a′) If (U, t^a) and (M, u^a) are consistent with RCBR, then t^a and u^a must induce the same hierarchy of beliefs.

(b′) If (L, t^b) and (C, u^b) are consistent with RCBR, then t^b and u^b must induce the same hierarchy of beliefs.

Let h^a (resp. h^b) be this hierarchy of beliefs for Ann (resp. Bob). Now refer to Figure 6. We know that for Y to be consistent with even rationality (a fortiori, RCBR), Charlie must assign probability $\frac{1}{2}$ to each of (U, L) and (M, C). Also, Charlie must assign probability 1 to the event "RCBR with respect to Ann and Bob." (This uses a conjunction property of belief.) Taken together with (a′) and (b′), this means that Charlie must assign probability $\frac{1}{2}$ to each of the two indicated points on the (h^a, h^b)-plane. But CI requires Charlie's conditional measure, conditioned on any such horizontal plane, to be a product measure. So we have a contradiction, completing the argument for statement (i).

The argument for statement (ii) is very similar. This time, consider the game in Figure 7. (This is the same as Figure 5, except for swapping Bob's payoffs in (U, C, Y) and (M, C, Y).) Again, all strategies are correlated

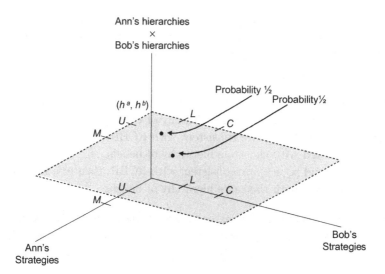

Figure 6.

	L	C	R
U	0,0,2	0,0,2	0,1,2
M	0,0,0	0,0,0	0,1,0
D	1,0,2	1,0,2	1,1,2

X

	L	C	R
U	1,1,1	0,0,0	0,1,0
M	1,0,0	0,1,1	0,1,0
D	1,0,0	1,0,0	1,1,0

Y

	L	C	R
U	0,0,0	0,0,0	0,1,0
M	0,0,2	0,0,2	0,1,2
D	1,0,2	1,0,2	1,1,2

Z

Figure 7.

rationalizable. We will see that if Y is consistent with RCBR then Charlie's type cannot satisfy SUFF.

Fix an associated type structure, and note the following four facts:

(a) If (U, t^a) and (M, u^a) are rational, then t^a and u^a must each assign probability 1 to the strategies (L, Y).

(b_L) If (L, t^b) and (L, u^b) are rational, then t^b and u^b must each assign probability 1 to the strategies (U, Y).

(b_C) If (C, v^b) and (C, w^b) are rational, then v^b and w^b must each assign probability 1 to the strategies (M, Y).

(c) If (Y, t^c) and (Y, u^c) are rational, then t^c and u^c must each assign probability $\frac{1}{2}$ to (U, L) and probability $\frac{1}{2}$ to (M, C).

It follows that if (U, t^a) and (M, u^a) are rational, then the hierarchies of beliefs associated with t^a and u^a must agree at the first level. Likewise, for (L, t^b) and (L, u^b), for (C, v^b) and (C, w^b), and for (Y, t^c) and (Y, u^c).

Using a similar argument:

(a′) If (U, t^a) and (M, u^a) are consistent with RCBR, then t^a and u^a must be associated with the same hierarchy of beliefs.

(b′$_L$) If (L, t^b) and (L, u^b) are consistent with RCBR, then t^b and u^b must be associated with the same hierarchy of beliefs.

(b′$_C$) If (C, v^b) and (C, w^b) are consistent with RCBR, then v^b and w^b must be associated with the same hierarchy of beliefs.

Write h^a for the hierarchy associated with requirement (a′). Write h^b_L (resp. h^b_C) for the hierarchy associated with requirement (b′$_L$) (resp. (b′$_C$)). Note, by (b_L)–(b_C), the hierarchies h^b_L and h^b_C are necessarily distinct.

Now refer to Figure 8. Charlie must assign probability $\frac{1}{2}$ to each of (U, L) and (M, C), and must assign probability 1 to the event "RCBR with respect to Ann and Bob." Together with (a′)–(b′$_C$), this means that

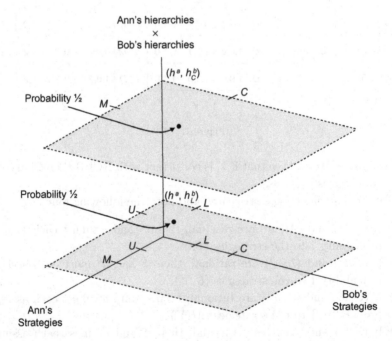

Figure 8.

Charlie must assign probability $\frac{1}{2}$ to the point (U, L) on the (h^a, h^b_L)-plane and $\frac{1}{2}$ probability to the point (M, C) on the (h^a, h^b_C)-plane. Notice, Charlie assigns: (i) probability $\frac{1}{2}$ to U, conditional on Ann's hierarchy h^a; and (ii) probability 1 to U, conditional on Ann's hierarchy h^a and Bob's hierarchy h^b_L. This contradicts SUFF, completing the argument for statement (ii).

We have shown that there is a gap between the middle and outer sets in Figure 3. This raises the question: How "big" is this gap? In Appendix H, we give a sufficient condition on a game under which the two sets coincide, and also discuss genericity properties.[6]

Finally, we note an open question. Fix a game G. We say a strategy in G can be played under intrinsic correlation if it can be played under RCBR, CI, and SUFF. But these conditions refer to some type structure. Obviously, it would be desirable to have a (type-free) characterization of the strategies that can be played. Unfortunately, we do not have one — and leave this to future work.

7. Comparison Contd.

Let us return to the comparison of intrinsic and extrinsic correlation — now in the specific case of the key game of Figure 5. We give an explicit construction — in the style of Aumann (1974) — under which Y can be played under RCBR. In Figure 9, which adds a coin toss to the original game, Ann, Bob, Charlie, and Nature move simultaneously. Figure 10 is an associated type structure, where each player's type is now associated with a measure on the strategies and types for the other players — and the outcome *Heads* vs. *Tails* of the coin toss.

There are two types for Ann, t^a and u^a, two types for Bob, t^b and u^b, and one type for Charlie, t^c. Ann's type t^a assigns probability 1 to $(L, t^b, Y, t^c, Heads)$, while her type u^a assigns probability 1 to $(L, t^b, Y, t^c, Tails)$; and similarly for Bob's types t^b and u^b, as shown. Charlie's (unique) type t^c assigns probability $\frac{1}{2}$ to $(U, t^a, L, t^b, Heads)$ and probability $\frac{1}{2}$ to $(D, u^a, R, u^b, Tails)$. Notice that each type assigns positive probability only to rational strategy-type pairs for the other players. So, by induction, RCBR holds at the state $(U, t^a, L, t^b, Y, t^c, Heads)$, for example.

---

[6] It would be equally interesting to find conditions under which the inner and middle sets in Figure 3 coincide. We do not have any results on this.

		L	C	R
	U	0,0,2	0,0,2	0,1,2
Heads	M	0,0,0	0,0,0	0,1,0
	D	1,0,2	1,0,2	1,1,2
			X	

	L	C	R
U	1,1,1	0,1,0	0,1,0
M	1,0,0	0,0,1	0,1,0
D	1,0,0	1,0,0	1,1,0
		Y	

	L	C	R
U	0,0,0	0,0,0	0,1,0
M	0,0,2	0,0,2	0,1,2
D	1,0,2	1,0,2	1,1,2
		Z	

		L	C	R
	U	0,0,2	0,0,2	0,1,2
Tails	M	0,0,0	0,0,0	0,1,0
	D	1,0,2	1,0,2	1,1,2
			X	

	L	C	R
U	1,1,1	0,1,0	0,1,0
M	1,0,0	0,0,1	0,1,0
D	1,0,0	1,0,0	1,1,0
		Y	

	L	C	R
U	0,0,0	0,0,0	0,1,0
M	0,0,2	0,0,2	0,1,2
D	1,0,2	1,0,2	1,1,2
		Z	

Figure 9.

The construction also satisfies CI — defined for the game with a coin toss. Ann's types t^a and u^a induce different hierarchies of beliefs over what she is uncertain about — viz., the strategies played and the outcome of the coin toss. Likewise for Bob's types t^b and u^b. So, when she conditions on Ann's and Bob's hierarchies of beliefs (now about strategies and the coin toss), Charlie gets a degenerate (marginal) measure on their strategies — probability 1 on (U, L) or probability 1 on (M, C). CI is immediately satisfied. It is easy to check that SUFF — defined for the game with a coin toss — is likewise satisfied.

The conclusion is that in the game with a coin toss, Charlie can now play Y under the conditions of RCBR, CI, and SUFF — unlike what we found in Section 6. This construction is general: *Starting with any game, if we add a (nondegenerate) coin toss to the game, then we can construct a type structure where*: (i) *each type satisfies CI and SUFF; and* (ii) *for each correlated rationalizable profile there is a state at which there is RCBR and the profile is played.*[7]

---

[7]See Appendix I for the proof, and also a discussion of other sources of extrinsic correlation.

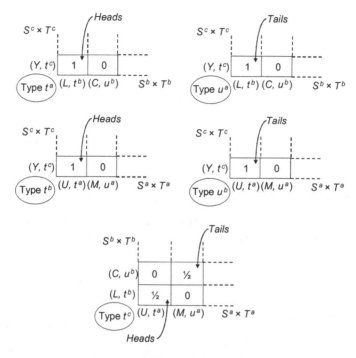

Figure 10.

Of course, the reason for the different conclusion is that when we add a coin toss (or any other external signal), we change the game. We are no longer analyzing the original game. Intrinsic correlation does not allow this.

The distinction between variables internal vs. external to the game also arises in the literature on "interim rationalizability." (See Battigalli and Siniscalchi [2003], Ely and Peski [2006], Dekel, Fridenberg, and Morris [2007].) An important issue in this literature is the presence or absence of redundant types — i.e., two or more types for a player that induce the same hierarchy of beliefs (Mertens and Zamir [1985, Definition 2.4]). Conceptually, redundant types arise if the analysis is done on a partial state space — in this case, we are to think that there are signals external to the game as given. (Liu [2004]) gives a formal treatment of signals for the case of Bayesian equilibrium.)

We define and study the intrinsic route. For us, then, signals affect the analysis via the players' hierarchies of beliefs about the strategies chosen. Go back to the statements of CI and SUFF in Section 5 and note that the conditioning is on hierarchies and not types.

8. Formal Presentation

We now begin the formal treatment. Given a Polish space Ω, write $\mathcal{B}(\Omega)$ for the Borel σ-algebra on Ω. Also, write $\mathcal{M}(\Omega)$ for the space of all Borel probability measures on Ω, where $\mathcal{M}(\Omega)$ is endowed with the topology of weak convergence (and so is again Polish).

Given sets $X^1, \ldots, X^n$, write $X = \prod_{i=1}^{n} X^i, X^{-i} = \prod_{j\neq i} X^j$, and $X^{-i-j} = \prod_{k\neq i,j} X^k$. (Throughout, we adopt the convention that if X is a product set and $X^j = \emptyset$ then, $X^i = \emptyset$ for all i.) An n-*player strategic-form game* is given by $G = \langle S^1, \ldots, S^n; \pi^1, \ldots, \pi^n \rangle$, where S^i is player i's finite strategy set and $\pi^i : S \to \mathbb{R}$ is i's payoff function. Extend π^i to $\mathcal{M}(S)$ in the usual way.

Definition 8.1. Fix an n-player strategic-form game G. An $(S^1, \ldots, S^n)$-*based type structure* is a structure

$$\Phi = \langle S^1, \ldots, S^n; T^1, \ldots, T^n; \lambda^1, \ldots, \lambda^n \rangle,$$

where each T^i is a Polish space and each $\lambda^i : T^i \to \mathcal{M}(S^{-i} \times T^{-i})$ is continuous. Members of T^i are called *types* of player i. Members of $S \times T$ are called *states*.

Associated with each type t^i for each player i in a type structure Φ is a hierarchy of beliefs about the strategies played.[8] To see this, inductively define sets Y_m^i, by setting $Y_1^i = S^{-i}$ and

$$Y_{m+1}^i = Y_m^i \times \prod_{j\neq i} \mathcal{M}(Y_m^j).$$

Now define continuous maps $\rho_m^i : S^{-i} \times T^{-i} \to Y_m^i$ inductively by

$$\rho_1^i(s^{-i}, t^{-i}) = s^{-i},$$

$$\rho_{m+1}^i(s^{-i}, t^{-i}) = (\rho_m^i(s^{-i}, t^{-i}), (\delta_m^j(t^j))_{j\neq i}),$$

where $\delta_m^j = \underline{\rho}_m^j \circ \lambda^j$ and, for each $\mu \in \mathcal{M}(S^{-j} \times T^{-j}), \underline{\rho}_m^j(\mu)$ is the image measure under ρ_m^j. (Appendix B shows that these maps are indeed continuous and so are well defined.) Define a continuous map $\delta^i : T^i \to \prod_{m=1}^{\infty} \mathcal{M}(Y_m^i)$ by $\delta^i(t^i) = (\delta_1^i(t^i), \delta_2^i(t^i), \ldots)$. (Again, see Appendix B.) In words, $\delta^i(t^i)$ is simply the hierarchy of beliefs (about strategies) induced by type t^i.

---

[8]The formulation below closely follows Mertens–Zamir (1985, Section 2) and Battigalli–Siniscalchi (1999, Section 3).

For each player i, define a map $\delta^{-i} : T^{-i} \to \prod_{j \neq i} \prod_{m=1}^{\infty} \mathcal{M}(Y_m^j)$ by

$$\delta^{-i}(t^1, \ldots, t^{i-1}, t^{i+1}, \ldots, t^n)$$
$$= (\delta^1(t^1), \ldots, \delta^{i-1}(t^{i-1}), \delta^{i+1}(t^{i+1}), \ldots, \delta^n(t^n)).$$

Since each δ^j is continuous, δ^{-i} is continuous.

Later, we will use the following definitions:

Definition 8.2. The map δ^i is *bimeasurable* if, for each Borel subset E of T^i, the image $\delta^i(E)$ is a Borel subset of $\prod_{m=1}^{\infty} \mathcal{M}(Y_m^i)$. The type structure Φ is *bimeasurable* if, for each i, the map δ^i is bimeasurable.

Applying a theorem due to Purves (1966) (see also Mauldin [1981]), the map δ^i is bimeasurable if and only if the set

$$\left\{ y \in \prod_{m=1}^{\infty} \mathcal{M}(Y_m^i) : (\delta^i)^{-1}(y) \text{ is uncountable} \right\}$$

is countable.

Recall that types $t^i, u^i \in T^i$ are *redundant* (Mertens and Zamir [1985, Definition 2.4]) if they induce the same hierarchies of beliefs, i.e., $\delta^i(t^i) = \delta^i(u^i)$. So, a type structure is bimeasurable if and only if, for each player i, there are (at most) a countable number of uncountable redundancies. In particular, a nonredundant type structure, i.e., a structure where each type induces a distinct hierarchy of beliefs, is bimeasurable. Any belief-closed subset (Mertens and Zamir [1985, Definition 2.15]) of the universal type structure is bimeasurable.

9. CI and SUFF Formalized

Fix a type structure Φ, and a player $i = 1, \ldots, n$. For each $j \neq i$, define random variables $\vec{s}_i^j$ and $\vec{t}_i^j$ on $S^{-i} \times T^{-i}$ by $\vec{s}_i^j = \text{proj}_{S^j}$ and $\vec{t}_i^j = \text{proj}_{T^j}$. (Here, proj denotes the projection map.) Let $\vec{s}_i$ and $\vec{t}_i$ be random variables on $S^{-i} \times T^{-i}$ with $\vec{s}_i = \text{proj}_{S^{-i}}$ and $\vec{t}_i = \text{proj}_{T^{-i}}$. Also, define the composite maps $\eta_i^j = \delta^j \circ \vec{t}_i^j$ and $\eta^{-i} = \delta^{-i} \circ \vec{t}_i$.

Write $\sigma(\vec{s}_i^j)$ (resp. $\sigma(\vec{s}_i), \sigma(\eta_i^j), \sigma(\eta^{-i})$) for the σ-algebra on $S^{-i} \times T^{-i}$ generated by $\vec{s}_i^j$ (resp. $\vec{s}_i, \eta_i^j, \eta^{-i}$). Similarly, let $\sigma(\vec{s}_i^j : j \neq i)$ (resp. $\sigma(\eta_i^j : j \neq i)$) be the σ-algebra on $S^{-i} \times T^{-i}$ generated by the random variables $\vec{s}_i^1, \ldots, \vec{s}_i^{i-1}, \vec{s}_i^{i+1}, \ldots, \vec{s}_i^n$ (resp. $\eta_i^1, \ldots, \eta_i^{i-1}, \eta_i^{i+1}, \ldots, \eta_i^n$). Note that $\sigma(\vec{s}_i^j : j \neq i) = \sigma(\vec{s}_i)$ and $\sigma(\eta_i^j : j \neq i) = \sigma(\eta^{-i})$. (See, Dellacherie and Meyer [1978, p. 9]).

Fix a measure $\lambda^i(t^i) \in \mathcal{M}(S^{-i} \times T^{-i})$, an event $E \in \mathcal{B}(S^{-i} \times T^{-i})$, and a sub-$\sigma$-algebra $\mathcal{S}$ of $\mathcal{B}(S^{-i} \times T^{-i})$. Write $\lambda^i(t^i)(E\|\mathcal{S}) : S^{-i} \times T^{-i} \to \mathbb{R}$ for a version of conditional probability of E given $\mathcal{S}$.

Definition 9.1. The random variables $\vec{s}_i^1, \ldots, \vec{s}_i^{i-1}, \vec{s}_i^{i+1}, \ldots, \vec{s}_i^n$ are $\lambda^i(t^i)$-*conditionally independent given the random variable* η^{-i} if, for all $j \neq i$ and $E^j \in \sigma(\vec{s}_i^j)$,

$$\lambda^i(t^i)\left(\bigcap_{j \neq i} E^j \| \sigma(\eta^{-i})\right) = \prod_{j \neq i} \lambda^i(t^i)(E^j \| \sigma(\eta^{-i})) \quad \text{a.s.}$$

Say the type t^i satisfies *conditional independence (CI)* if $\vec{s}_i^1, \ldots, \vec{s}_i^{i-1}$, $\vec{s}_i^{i+1}, \ldots, \vec{s}_i^n$ are $\lambda^i(t^i)$-conditionally independent given η^{-i}.

Definition 9.2. The random variable η_i^j is $\lambda^i(t^i)$-*sufficient for the random variable* $\vec{s}_i^j$ if, for each $j \neq i$ and $E^j \in \sigma(\vec{s}_i^j)$,

$$\lambda^i(t^i)(E^j \| \sigma(\eta^{-i})) = \lambda^i(t^i)(E^j \| \sigma(\eta_i^j)) \quad \text{a.s.}$$

Say the type t^i satisfies *sufficiency (SUFF)* if, for each $j \neq i, \eta_i^j$ is $\lambda^i(t^i)$-sufficient for $\vec{s}_i^j$.

In words, Definition 9.1 says that a type t^i satisfies CI if, conditional on knowing the hierarchies of beliefs for the other players j, type t^i's assessment of their strategies is independent. For SUFF, suppose type t^i knows player j's hierarchy of beliefs, and comes to learn the hierarchies of beliefs for the players other than j. Type t^i satisfies SUFF if this new information does not change t^i's assessment of j's strategy.

The next result assumes the type structure is bimeasurable (Definition 8.2). It says that under CI and SUFF, if a type t^i for player i assesses other players' hierarchies as independent, then t^i assesses their strategies as independent.[9] Equivalently, if t^i assesses other players' strategies as correlated, then t^i must assess their hierarchies as correlated. So, taken together, CI and SUFF capture our concept of correlation via the players' hierarchies of beliefs.

Proposition 9.1. *Suppose the type structure* Φ *is bimeasurable, and consider a type* t^i *that satisfies CI and SUFF. If the random variables*

---

[9]See the appendices for proofs not given in the text.

$\eta_i^1, \ldots, \eta_i^{i-1}, \eta_i^{i+1}, \ldots, \eta_i^n$ *are* $\lambda^i(t^i)$-*independent, then the random variables* $\bar{s}_i^1, \ldots, \bar{s}_i^{i-1}, \bar{s}_i^{i+1}, \ldots, \bar{s}_i^n$ *are* $\lambda^i(t^i)$-*independent.*

10. RCBR Formalized

Definition 10.1. Say $(s^i, t^i) \in S^i \times T^i$ is *rational* if

$$\sum_{s^{-i} \in S^{-i}} \pi^i(s^i, s^{-i}) \mathrm{marg}_{S^{-i}} \lambda^i(t^i)(s^{-i})$$

$$\geq \sum_{s^{-i} \in S^{-i}} \pi^i(r^i, s^{-i}) \mathrm{marg}_{S^{-i}} \lambda^i(t^i)(s^{-i})$$

for every $r^i \in S^i$. Let R_1^i be the set of all rational pairs (s^i, t^i).

Definition 10.2. Say $E \subseteq S^{-i} \times T^{-i}$ is *believed under* $\lambda^i(t^i)$ if E is Borel and $\lambda^i(t^i)(E) = 1$. Let

$$B^i(E) = \{t^i \in T^i \colon E \text{ is believed under } \lambda^i(t^i)\}.$$

For $m \geq 1$, define R_m^i inductively by

$$R_{m+1}^i = R_m^i \cap [S^i \times B^i(R_m^{-i})].$$

Definition 10.3. If $(s^1, t^1, \ldots, s^n, t^n) \in R_{m+1}$, say there is *rationality and mth-order belief of rationality* (*RmBR*) at this state. If $(s^1, t^1, \ldots, s^n, t^n) \in \bigcap_{m=1}^{\infty} R_m$, say there is RCBR at this state.

Next, define sets S_m^i inductively by $S_0^i = S^i$, and

$$S_{m+1}^i = \{s^i \in S_m^i \colon \text{there exists } \mu \in \mathcal{M}(S^{-i}), \text{with } \mu(S_m^{-i}) = 1, \text{ such that}$$

$$\pi(s^i, \mu) \geq \pi^i(r^i, \mu) \quad \text{for every } r^i \in S_m^i\}.$$

Note there is an M such that $\bigcap_{m=0}^{\infty} S_m^i = S_M^i \neq \emptyset$, for all i. A strategy $s^i \in S_M^i$ (resp. strategy profile $s \in S_M$) is called *correlated rationalizable* (Brandenburger and Dekel [1987]).

We will make use of the following concept adapted from Pearce (1984):

Definition 10.4. Fix a game $G = \langle S^1, \ldots, S^n; \pi^1, \ldots, \pi^n \rangle$ and subsets $Q^i \subseteq S^i$, for $i = 1, \ldots, n$. The set $\prod_{i=1}^n Q^i$ is a *best-response set* (*BRS*) if, for every i and each $s^i \in Q^i$, there is a $\mu \in \mathcal{M}(S^{-i})$ with $\mu(Q^{-i}) = 1$, such that $\pi^i(s^i, \mu) \geq \pi^i(r^i, \mu)$ for every $r^i \in S^i$.

Standard facts about BRS's are: (i) the set S_M of correlated rationalizable profiles is a BRS and (ii) every BRS is contained in S_M.

Next is a statement of the result that RCBR is characterized by the correlated rationalizable strategies:[10]

Proposition 10.1. *Consider a game* $G = \langle S^1, \ldots, S^n; \pi^1, \ldots, \pi^n \rangle$.

(i) *Fix a type structure* $\langle S^1, \ldots, S^n; T^1, \ldots, T^n; \lambda^1, \ldots, \lambda^n \rangle$, *and suppose there is RCBR at the state* $(s^1, t^1, \ldots, s^n, t^n)$. *Then the strategy profile* $(s^1, \ldots, s^n)$ *is correlated rationalizable in* G.

(ii) *There is a type structure* $\langle S^1, \ldots, S^n; T^1, \ldots, T^n; \lambda^1, \ldots, \lambda^n \rangle$ *such that, for each correlated rationalizable strategy profile* $(s^1, \ldots, s^n)$, *there is a state* $(s^1, t^1, \ldots, s^n, t^n)$ *at which there is RCBR*.

11. Main Result Formalized

Here we give a formal treatment of statements (i) and (ii) in Section 6.

Theorem 11.1.

(i) *There is a game* G *and a correlated rationalizable strategy* s^i *of* G, *such that the following holds: For any type structure* Φ, *there does not exist a state at which each type satisfies CI, RCBR holds, and* s^i *is played*.

(ii) *There is a game* G' *and a correlated rationalizable strategy* s^i *of* G', *such that the following holds: For any type structure* Φ, *there does not exist a state at which each type satisfies SUFF, RCBR holds, and* s^i *is played*.

Begin with part (i). A suitable game G was given in Figure 5.

Lemma 11.1. *For the game* G:

(i) *the strategy* U *(resp.* M*) is optimal under* $\mu \in \mathcal{M}(S^b \times S^c)$ *if and only if* $\mu(L, Y) = 1$;

---

[10]Brandenburger–Dekel (1987) and Tan–Werlang (1988) show related results. Proposition 2.1 in Brandenburger and Dekel (1987) demonstrates an equivalence between common knowledge of rationality and correlated rationalizability. Theorem 5.1 in Tan and Werlang (1988) shows that (in a universal structure) $RmBR$ yields strategy profiles that survive $(m+1)$ rounds of correlated rationalizability. (Tan and Werlang [1988] also state a converse (Theorem 5.3) and reference the proof to the unpublished version of the paper.)

(ii) *the strategy L (resp. C) is optimal under $\mu \in \mathcal{M}(S^a \times S^c)$ if and only if $\mu(U, Y) = 1$;*

(iii) *the strategy Y is optimal under $\mu \in \mathcal{M}(S^a \times S^b)$ if and only if $\mu(U, L) = \mu(M, C) = \frac{1}{2}$ (moreover, this measure is not independent).*

Proof. Parts (i) and (ii) are immediate.

For part (iii), note that Y is optimal under $\mu \in \mathcal{M}(S^a \times S^b)$ if and only if

$$\mu(U, L) + \mu(M, C) \geq \max\{2(1 - \mu(M)), 2(1 - \mu(U))\},$$

where we write M for the set $\{M\} \times S^b$ and U for the set $\{U\} \times S^b$. Since $1 \geq \mu(U, L) + \mu(M, C)$, it follows that

$$1 \geq \max\{2(1 - \mu(M)), 2(1 - \mu(U))\},$$

or $\mu(M) \geq \frac{1}{2}$ and $\mu(U) \geq \frac{1}{2}$. Since M and U are disjoint, we get $\mu(M) = \frac{1}{2}$ and $\mu(U) = \frac{1}{2}$. From this, $\mu(U, L) + \mu(M, C) = 1$. But $\mu(U) \geq (U, L)$ and $\mu(M) \geq \mu(M, C)$, and so $\mu(U, L) = \mu(M, C) = \frac{1}{2}$.

Finally, notice that μ is not independent, since $\frac{1}{2} = \mu(U, L) \neq \mu(U) \times \mu(L) = \frac{1}{2} \times \frac{1}{2}$. $\square$

Corollary 11.1. *The correlated rationalizable set in G is $\{U, M, D\} \times \{L, C, R\} \times \{X, Y, Z\}$.*

Proof. We have just established the optimality of U, M, L, C, and Y. The optimality of D, R, X, and Z is clear. So, $S_1 = S$. Then, by induction, $S_m = S$ for all m. $\square$

In particular then, correlated rationalizability allows Charlie to play Y, and to have an expected payoff of 1. By Proposition 10.1(ii), there is a type structure Φ and a state at which there is RCBR and Charlie plays Y. However, we will see below (Corollary 11.5) that this cannot happen when we add CI. Moreover, Charlie must then have an (expected) payoff of 2 not 1.

We introduce some notation. Fix a player i. For $j \neq i$, we write $[s^j]$ for the subset $\{s^j\} \times S^{-i-j} \times T^{-i}$ of $S^{-i} \times T^{-i}$. We also write $[t^j]$ for the subset $S^{-i} \times \{u^j \in T^j : \delta^j(u^j) = \delta^j(t^j)\} \times T^{-i-j}$ of $S^{-i} \times T^{-i}$. Note, $[t^j] = (\eta_i^j)^{-1}(\delta^j(t^j))$ and is measurable.

Corollary 11.2. *Fix a type structure Φ for G, with $(Y, t^c) \in R_1^c$. Then,*

$$\lambda^c(t^c)(E) = \lambda^c(t^c)([U] \cap [L] \cap E) + \lambda^c(t^c)([M] \cap [C] \cap E)$$

for any event E in $S^a \times T^a \times S^b \times T^b$. Moreover, if $\lambda^c(t^c)(E) = 1$, then,

$$\lambda^c(t^c)([U] \cap [L] \cap E) = \lambda^c(t^c)([M] \cap [C] \cap E) = \frac{1}{2}.$$

Proof.   The first part is immediate from $\lambda^c(t^c)([U] \cap [L]) = \lambda^c(t^c)([M] \cap [C]) = \frac{1}{2}$ (Lemma 11.1), and the fact that $[U] \cap [M] = \emptyset$. The second part follows from $[U] \cap [L] \cap E \subseteq [U] \cap [L]$ and $[M] \cap [C] \cap E \subseteq [M] \cap [C]$.   □

Lemma 11.2. *Fix a type structure Φ for G. Suppose $(Y, t^c) \in \cap_m R_m^c$ where t^c satisfies CI. Then there are $(t^a, t^b), (u^a, u^b) \in T^a \times T^b$, with $(U, t^a, L, t^b), (M, u^a, C, u^b) \in \cap_m (R_m^a \times R_m^b)$ and either $\delta^a(t^a) \neq \delta^a(u^a)$ or $\delta^b(t^b) \neq \delta^b(u^b)$ (or both).*

The idea of the proof was given in Section 6 (see Figure 6 and the surrounding discussion).

Proof of Lemma 11.2. Fix Φ with $(Y, t^c) \in \cap_m R_m^c$. Then, $\lambda^c(t^c)(R_m^a \times R_m^b) = 1$ for all m, so that $\lambda^c(t^c)(\cap_m(R_m^a \times R_m^b)) = 1$. Corollary 11.2 then gives

$$\lambda^c(t^c) \left([U] \cap [L] \cap \bigcap_m (R_m^a \times R_m^b) \right)$$

$$= \lambda^c(t^c) \left([M] \cap [C] \cap \bigcap_m (R_m^a \times R_m^b) \right) = \frac{1}{2}.$$

Suppose, for any $(U, t^a, L, t^b), (M, u^a, C, u^b) \in \cap_m(R_m^a \times R_m^b), \delta^a(t^a) = \delta^a(u^a)$ and $\delta^b(t^b) = \delta^b(u^b)$. Then,

$$[U] \cap [L] \cap \bigcap_m (R_m^a \times R_m^b) \subseteq [U] \cap [L] \cap [t^a] \cap [t^b],$$

$$[M] \cap [C] \cap \bigcap_m (R_m^a \times R_m^b) \subseteq [M] \cap [C] \cap [t^a] \cap [t^b].$$

So, $[U] \cap [M] = \emptyset$ implies $\lambda^c(t^c)([t^a] \cap [t^b]) = 1$. From this, Corollary 11.2 implies

$$\lambda^c(t^c)([U] \cap [L]) = \lambda^c(t^c)([U]) = \lambda^c(t^c)([L]) = \frac{1}{2}.$$

From this, the fact that $\lambda^c(t^c)([t^a] \cap [t^b]) = 1$, and Corollary E.2 in Appendix E,

$$\lambda^c(t^c)([U] \cap [L] \| \sigma(\eta^{-c})) = \frac{1}{2} \quad \text{a.s.,}$$

$$\lambda^c(t^c)([U] \| \sigma(\eta^{-c})) = \frac{1}{2} \quad \text{a.s.,}$$

$$\lambda^c(t^c)([L] \| \sigma(\eta^{-c})) = \frac{1}{2} \quad \text{a.s.}$$

Thus, in particular,

$$\lambda^c(t^c)([U] \| \sigma(\eta^{-c})) \times \lambda^c(t^c)([L] \| \sigma(\eta^{-c})) = \frac{1}{4} \quad \text{a.s.}$$

so that t^c does not satisfy CI.   □

Proposition 11.1. *Fix a game* $\langle S^1, \ldots, S^n; \pi^1, \ldots, \pi^n \rangle$ *and a BRS* $\prod_{i=1}^n Q^i$ *satisfying: For every* i *and each* $s^i \in Q^i$*, there is a unique* $\mu(s^i) \in \mathcal{M}(S^{-i})$ *under which* s^i *is optimal. Fix also a type structure* Φ*. Then for every* i *and all* m *the following hold:*

(i) *If* $(s^{-i}, t^{-i}), (s^{-i}, u^{-i}) \in R_m^{-i} \cap (Q^{-i} \times T^{-i})$ *then* $\rho_m^i(s^{-i}, t^{-i}) = \rho_m^i(s^{-i}, u^{-i})$.
(ii) *If* $(s^i, t^i), (r^i, u^i) \in R_m^i \cap (Q^i \times T^i)$ *and* $\mu(s^i) = \mu(r^i)$ *then* $\delta_n^i(t^i) = \delta_n^i(u^i)$ *for all* $n \leq m$.

Again, the idea of the proof was given in Section 6.

Proof of Proposition 11.1. By induction on m.

Begin with $m = 1$: Part (i) is immediate from the fact that $\rho_1^i(s^{-i}, t^{-i}) = \rho_1^i(s^{-i}, u^{-i}) = s^{-i}$. For part (ii), fix $(s^i, t^i), (r^i, u^i) \in R_1^i \cap (Q^i \times T^i)$ with $\mu(s^i) = \mu(r^i)$. By definition, $\rho_{-1}^i(\lambda^i(t^i)) = \text{marg}_{S^{-i}} \lambda^i(t^i)$ and $\rho_{-1}^i(\lambda^i(u^i)) = \text{marg}_{S^{-i}} \lambda^i(u^i)$. Since $(s^i, t^i), (r^i, u^i) \in R_1^i$,

$$\text{marg}_{S^{-i}} \lambda^i(t^i) = \mu(s^i) = \mu(r^i) = \text{marg}_{S^{-i}} \lambda^i(u^i).$$

Now assume the lemma is true for m. Begin with part (i). Suppose $(s^{-i}, t^{-i}), (s^{-i}, u^{-i}) \in R_{m+1}^{-i} \cap (Q^{-i} \times T^{-i})$. The induction hypothesis applied to part (i) gives $\rho_m^i(s^{-i}, t^{-i}) = \rho_m^i(s^{-i}, u^{-i})$. Also, the induction hypothesis applied to part (ii) gives $\delta_m^j(t^j) = \delta_m^j(u^j)$ for each $j \neq i$. With this, $\rho_{m+1}^i(s^{-i}, t^{-i}) = \rho_{m+1}^i(s^{-i}, u^{-i})$, establishing part (i) for $(m+1)$.

Turn to part (ii). Suppose $(s^i, t^i), (r^i, u^i) \in R_{m+1}^i \cap (Q^i \times T^i)$ and $\mu(s^i) = \mu(r^i)$. Then $(s^i, t^i), (r^i, u^i) \in R_m^i \cap (Q^i \times T^i)$, and so the induction hypothesis applied to part (ii) gives $\delta_n^i(t^i) = \delta_n^i(u^i)$ for all $n \le m$. As such, it suffices to show $\delta_{m+1}^i(t^i) = \delta_{m+1}^i(u^i)$.

Fix an event E in Y_{m+1}^i, and a point $(s^{-i}, t^{-i}) \in (\rho_{m+1}^i)^{-1}(E) \cap$ Supp $\lambda^i(t^i)$. Then, for each $(s^{-i}, u^{-i}) \in$ Supp $\lambda^i(t^i) \cup$ Supp $\lambda^i(u^i)$, it must be that $(s^{-i}, u^{-i}) \in (\rho_{m+1}^i)^{-1}(E)$. To see this, first notice that, by Corollary D.1 in Appendix D, Supp $\lambda^i(t^i) \cup$ Supp $\lambda^i(u^i) \subseteq R_m^{-i}$. Also note that since $(s^i, t^i), (r^i, u^i) \in R_1^i$,

$$\text{marg}_{S^{-i}} \lambda^i(t^i) = \mu(s^i) = \mu(r^i) = \text{marg}_{S^{-i}} \lambda^i(u^i).$$

Since $\mu(s^i)(Q^{-i}) = 1$, it follows that Supp $\lambda^i(t^i) \cup$ Supp $\lambda^i(u^i) \subseteq Q^{-i} \times T^{-i}$. So $(s^{-i}, t^{-i}), (s^{-i}, u^{-i}) \in R_m^{-i} \cap (Q^{-i} \times T^{-i})$. By part (i) of the induction hypothesis, $\rho_m^i(s^{-i}, t^{-i}) = \rho_m^i(s^{-i}, u^{-i})$. By part (ii) of the induction hypothesis, for each $j \ne i, \delta_m^j(t^j) = \delta_m^j(u^j)$. So, $\rho_{m+1}^i(s^{-i}, t^{-i}) = \rho_{m+1}^i (s^{-i}, u^{-i})$. From this it follows that $(s^{-i}, u^{-i}) \in (\rho_{m+1}^i)^{-1}(E)$, as required.

Using this, we can now write

$$\lambda^i(t^i)((\rho_{m+1}^i)^{-1}(E))$$

$$= \lambda^i(t^i)((\rho_{m+1}^i)^{-1}(E) \cap \text{Supp } \lambda^i(t^i))$$

$$= \sum_{s^{-i} \in \text{proj}_{S^{-i}}(\rho_{m+1}^i)^{-1}(E)} \lambda^i(t^i)(\{s^{-i}\}$$

$$\times \{t^{-i} : (s^{-i}, t^{-i}) \in (\rho_{m+1}^i)^{-1}(E) \cap \text{Supp } \lambda^i(t^i)\})$$

$$= \sum_{s^{-i} \in \text{proj}_{S^{-i}}(\rho_{m+1}^i)^{-1}(E)} \lambda^i(t^i)(\{s^{-i}\} \times \{t^{-i} : (s^{-i}, t^{-i}) \in \text{Supp } \lambda^i(t^i)\})$$

$$= \sum_{s^{-i} \in \text{proj}_{S^{-i}}(\rho_{m+1}^i)^{-1}(E)} \text{marg}_{S^{-i}} \lambda^i(t^i)(s^{-i}).$$

A corresponding argument shows that

$$\lambda^i(u^i)((\rho_{m+1}^i)^{-1}(E)) = \sum_{s^{-i} \in \text{proj}_{S^{-i}}(\rho_{m+1}^i)^{-1}(E)} \text{marg}_{S^{-i}} \lambda^i(u^i)(s^{-i}).$$

Now note

$$\text{marg}_{S^{-i}} \lambda^i(t^i) = \mu(s^i) = \mu(r^i) = \text{marg}_{S^{-i}} \lambda^i(u^i),$$

establishing $\delta_{m+1}^i(t^i) = \delta_{m+1}^i(u^i)$, as required. $\qquad\square$

Corollary 11.3. *Fix a game* $\langle S^1, \ldots, S^n; \pi^1 \ldots, \pi^n \rangle$ *and a BRS* $\prod_{i=1}^n$ $Q^i \subseteq S$ *satisfying: For every i and each $s^i \in Q^i$, there is a unique $\mu(s^i) \in \mathcal{M}(S^{-i})$ under which s^i is optimal. Fix also a type structure Φ. If (s^i, t^i), $(r^i, u^i) \in \bigcap_m R_m^i \cap (Q^i \times T^i)$ and $\mu(s^i) = \mu(r^i)$, then $\delta^i(t^i) = \delta^i(u^i)$.*

Proof. Suppose instead that $\delta^i(t^i) \neq \delta^i(u^i)$. Then there exists m such that $\delta_m^i(t^i) \neq \delta_m^i(u^i)$. Since $(s^i, t^i), (r^i, u^i) \in R_m^i$ this contradicts Proposition 11.1. $\square$

Corollary 11.4. *Let $Q^a = \{U, M\}, Q^b = \{L, C\}, Q^c = \{Y\}$ in the game G. Fix a type structure Φ. For each i, if $(s^i, t^i), (r^i, u^i) \in \bigcap_m R_m^i \cap (Q^i \times T^i)$, then $\delta^i(t^i) = \delta^i(u^i)$.*

Proof. Immediate from Corollaries 11.1 and 11.3. $\square$

Proof of Theorem 11.1(i). Fix a type structure Φ for G. Corollary 11.4 implies that if $(U, t^a), (M, u^a) \in \bigcap_m R_m^a$, then $\delta^a(t^a) = \delta^a(u^a)$. Likewise, if $(L, t^b), (C, u^b) \in \bigcap_m R_m^b$, then $\delta^b(t^b) = \delta^b(u^b)$. With this, Lemma 11.2 implies that if $(Y, t^c) \in \bigcap_m R_m^c$, then t^c does not satisfy CI. But Y is a correlated rationalizable strategy, by Lemma 11.1. Setting $s^i = Y$ establishes the theorem. $\square$

Corollary 11.5. *Fix a type structure Φ for G, and a state $(s^a, t^a, s^b, t^b, s^c, t^c)$ at which there is RCBR. If t^c satisfies CI, then $s^a = D, s^b = R$, and $s^c = X$ or Z.*

To prove part (ii) of Theorem 11.1, we use the game G' in Figure 7.

Corollary 11.6. *Let $Q^a = \{U, M\}, Q^b = \{L, C\}, Q^c = \{Y\}$ in the game G'. For $i = a, c$, if $(s^i, t^i), (r^i, u^i) \in \bigcap_m R_m^i \cap (Q^i \times T^i)$, then $\delta^i(t^i) = \delta^i(u^i)$. For $i = b$, if $(s^i, t^i), (r^i, u^i) \in \bigcap_m R_m^i \cap (Q^i \times T^i)$, then $\delta^i(t^i) = \delta^i(u^i)$ only if $s^i = r^i$.*

Proof. Same as for Corollary 11.4, except that while L is optimal only under the measure that assigns probability one to (U, Y), now C is optimal only under the measure that assigns probability one to (M, Y). So if $(L, t^b), (C, u^b) \in \bigcap_m R_m^b$, then $t^b \neq u^b$. $\square$

Proof of Theorem 11.1(ii). Fix a type structure Φ for G'. Suppose $(Y, t^c) \in \bigcap_m R_m^c$. Then, $\lambda^c(t^c)(\bigcap_m (R_m^a \times R_m^b)) = 1$, so that Corollary 11.2 gives

$$\lambda^c(t^c) \left([U] \cap [L] \cap \bigcap_m (R_m^a \times R_m^b) \right)$$

$$= \lambda^c(t^c) \left([M] \cap [C] \cap \bigcap_m (R_m^a \times R_m^b) \right) = \frac{1}{2}.$$

Using Corollary 11.6, there are t^a, t^b, u^b, with $\delta^b(t^b) \neq \delta^b(u^b)$, such that

$$[U] \cap [L] \cap \bigcap_m (R_m^a \times R_m^b) \subseteq [U] \cap [L] \cap [t^a] \cap [t^b],$$

$$[M] \cap [C] \cap \bigcap_m (R_m^a \times R_m^b) \subseteq [M] \cap [C] \cap [t^a] \cap [u^b].$$

Paralleling the argument in the proof of Lemma 11.2, we then have

$$\lambda^c(t^c)([U] \cap [L] \cap [t^a] \cap [t^b]) = \frac{1}{2},$$

$$\lambda^c(t^c)([M] \cap [C] \cap [t^a] \cap [u^b]) = \frac{1}{2},$$

since $[U] \cap [M] = \emptyset$. Paralleling Corollary 11.2, we get that for any event E,

$$\lambda^c(t^c)(E) = \lambda^c(t^c)([U] \cap [L] \cap [t^a] \cap [t^b] \cap E)$$

$$+ \lambda^c(t^c)([M] \cap [C] \cap [t^a] \cap [u^b] \cap E).$$

Setting $E = [t^a], [U] \cap [t^a], [U] \cap [t^a] \cap [t^b]$, and $[t^a] \cap [t^b]$, yields respectively

$$\lambda^c(t^c)([t^a]) = 1,$$

$$\lambda^c(t^c)([U] \cap [t^a]) = \frac{1}{2},$$

$$\lambda^c(t^c)([U] \cap (t^a) \cap [t^b]) = \frac{1}{2},$$

$$\lambda^c(t^c)([t^a] \cap [t^b]) = \frac{1}{2}.$$

Corollary E.1 in Appendix E gives, for any $(s^a, v^a, s^b, v^b) \in [t^a]$,

$$\lambda^c(t^c)([U] \| \sigma(\eta_c^a))(s^a, v^a, s^b, v^b) = \frac{\lambda^c(t^c)([U] \cap [t^a])}{\lambda^c(t^c)([t^a])} = \frac{1}{2}.$$

The same corollary yields for any $(s^a, v^a, s^b, v^b) \in [t^a] \cap [t^b]$,

$$\lambda^c(t^c)([U] \| \sigma(\eta^{-c}))(s^a, v^a, s^b, v^b) = \frac{\lambda^c(t^c)([U] \cap [t^a] \cap [t^b])}{\lambda^c(t^c)([t^a] \cap [t^b])} = 1.$$

Since, $[t^a] \cap [t^b] \subseteq [t^a]$ and $\lambda^c(t^c)([t^a] \cap [t^b]) > 0$, this says t^c does not satisfy SUFF. $\qquad \square$

Appendix F considers an immediate implication of the results of this section, corresponding to the case where players reason only up to some finite number of levels.[11]

12. Conclusion

We have looked at three routes to correlations in games — independent rationalizability (no correlation!), intrinsic correlation (our route), and correlated rationalizability (extrinsic correlation). But there is also another route — related to physical correlation.

Return to the game of Figure 5. Suppose Charlie is sitting in her own "cubicle" (the term is from Kohlberg–Mertens [1986, p. 1005]). Charlie may nonetheless play Y because she thinks that Ann and Bob are coordinating their strategy choices — they jointly choose (U, L) or jointly choose (M, C). While Charlie makes her decision in her own cubicle, she does not think that Ann and Bob do the same.

From the perspective of the analyst, this justification of Y is asymmetric. The analyst thinks of each player as a separate decision maker sitting in his own cubicle. But, at the same time, the analyst allows each player to think others coordinate.

The intrinsic route avoids this asymmetry. Under this route, we can do noncooperative game theory without physical correlation, by specifying the players' hierarchies of beliefs (about strategies) and doing the analysis relative to them. As we noted earlier, this leads naturally to the question of the characterization of the strategies that can be played under this analysis. We leave this as open.

---

[11]We thank Yossi Feinberg for prompting us to investigate this case.

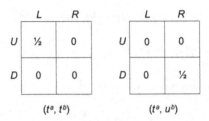

Figure A1.

Appendix A. CI and SUFF Contd.

The following two examples illustrate our CI and SUFF conditions.[12] In particular, they show that the conditions are independent, and neither can be dropped in Proposition 9.1.

Example A.1. Here, Charlie's type t^c satisfies CI and assesses Ann's and Bob's hierarchies as independent, but does not assess their strategies as independent. Let the type spaces be $T^a = \{t^a\}, T^b = \{t^b, u^b\}, T^c = \{t^c\}$. Suppose that t^b and u^b induce different hierarchies of beliefs for Bob (we do not need to specify the hierarchies). Figure A1 depicts the measure associated with Charlie's type t^c. We have,

$$\lambda^c(t^c)([U] \cap [L] | [t^a] \cap [t^b])$$
$$= 1 = 1 \times 1 = \lambda^c(t^c)([U] | [t^a] \cap [t^b]) \times \lambda^c(t^c)([L] | [t^a] \cap [t^b]).$$

Corresponding equalities hold for each of $\lambda^c(t^c)([D] \cap [L] | [t^a] \cap [t^b]), \lambda^c(t^c)([U] \cap [R] | [t^a] \cap [t^b])$, and $\lambda^c(t^c)([D] \cap [R] | [t^a] \cap [t^b])$, and also for $\lambda^c(t^c)([\cdot] \cap [\cdot] | [t^a] \cap [u^b])$, so that CI holds.

Independence over hierarchies is immediate since $\lambda^c(t^c)([t^a]) = 1$, so that

$$\lambda^c(t^c)([t^a] \cap [t^b]) = \lambda^c(t^c)([t^a]) \times \lambda^c(t^c)([t^b]),$$
$$\lambda^c(t^c)([t^a] \cap [u^b]) = \lambda^c(t^c)([t^a]) \times \lambda^c(t^c)([u^b]).$$

Yet we also have

$$\lambda^c(t^c)([U] \cap [L]) = \frac{1}{2} \neq \frac{1}{2} \times \frac{1}{2} = \lambda^c(t^c)([U]) \times \lambda^c(t^c)([L]),$$

---

[12]We are grateful to Pierpaolo Battigalli for simplifying the previous versions of these examples.

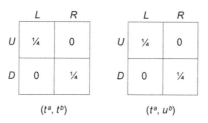

Figure A2.

so that independence over strategies is violated. (It is easy to check that SUFF is also violated, as it must be.)

Example A.2. Here, Charlie's type t^c satisfies SUFF and again assesses Ann's and Bob's hierarchies as independent, but again does not assess their strategies as independent. Let $T^a = \{t^a\}, T^b = \{t^b, u^b\}, T^c = \{t^c\}$. Suppose again that t^b and u^b induce different hierarchies of beliefs for Bob. Figure A2 depicts the measure associated with Charlie's type t^c.

Notice that

$$\lambda^c(t^c)([U]|[t^a] \cap [t^b]) = \lambda^c(t^c)([U]|[t^a] \cap [u^b]) = \frac{1}{2} = \lambda^c(t^c)([U]|[t^a]),$$

$$\lambda^c(t^c)([D]|[t^a] \cap [t^b]) = \lambda^c(t^c)([D]|[t^a] \cap [u^b]) = \frac{1}{2} = \lambda^c(t^c)([D]|[t^a]).$$

The corresponding equalities with respect to Bob are immediate, since $\lambda^c(t^c)([t^a]) = 1$. This establishes SUFF.

Independence over hierarchies is immediate, as in Example A.1, since again $\lambda^c(t^c)([t^a]) = 1$. Yet we also have

$$\lambda^c(t^c)([U] \cap [L]) = \frac{1}{2} \neq \frac{1}{2} \times \frac{1}{2} = \lambda^c(t^c)([U]) \times \lambda^c(t^c)([L]),$$

so that independence over strategies is violated. (It is readily checked that CI is also violated.)

Appendix B. Proofs for Section 8

Lemma B.1. *Fix Polish spaces A, B and a continuous map $f : A \to B$. Let $g : \mathcal{M}(A) \to \mathcal{M}(B)$ be given by $g(\mu) = \mu \circ f^{-1}$ for each $\mu \in \mathcal{M}(A)$. Then, g is continuous.*

Proof. We need to show that the inverse image of every closed set in
$\mathcal{M}(B)$ is closed in $\mathcal{M}(A)$. Let E be a closed set in $\mathcal{M}(B)$, then we want
that: Fix a sequence of measures μ_n in $g^{-1}(E)$, where μ_n converges weakly
to μ (in $\mathcal{M}(A)$). Then, $\mu \in g^{-1}(E)$. To show this, it suffices to show that
$g(\mu_n)$ converges weakly to $g(\mu)$ (in $\mathcal{M}(B)$). If so, $g(\mu) \in E$ and so $\mu \in$
$g^{-1}(E)$.

So: Fix an open set U in B. Then, $f^{-1}(U)$ is open in A. By the
Portmanteau theorem, $\liminf \mu_n(f^{-1}(U)) \geq \mu(f^{-1}(U))$. But this says that
$\liminf g(\mu_n)(U) \geq g(\mu)(U)$, and so, by the Portmanteau Theorem again,
$g(\mu_n)$ converges weakly to $g(\mu)$.   □

Proposition B.1. *The maps $\rho_m^i : S^{-i} \times T^{-i} \to Y_m^i$ and $\delta_m^i : T^i \to \mathcal{M}(Y_m^i)$
are continuous.*

Proof. First note that $\rho_1^i = \text{proj}_{S^{-i}}$ and so is certainly continuous. So $\underline{\rho}_1^i$
is continuous, by Lemma B.1, and thus, δ_1^i is continuous.

Assume ρ_m^i and δ_m^i are continuous, for all i. Fix a rectangular open set
$U \times \prod_{j \neq i} V^j \subseteq Y_{m+1}^i = Y_m^i \times \prod_{j \neq i} \mathcal{M}(Y_m^j)$. Notice

$$(\rho_{m+1}^i)^{-1} \left(U \times \prod_{j \neq i} V^j \right) = (\rho_m^i)^{-1}(U) \cap \bigcap_{j \neq 1} [S^{-i} \times (\delta_m^j)^{-1}(V^j)].$$

Thus, $(\rho_{m+1}^i)^{-1}(U \times \prod_{j \neq 1} V^j)$ is open since each set on the right-hand side
is open. Since the rectangular sets form a basis, this shows that ρ_{m+1}^i is
continuous. Again, by Lemma B.1, for each $i, \underline{\rho}_{m+1}^i$ is then continuous, and
so each δ_{m+1}^i is continuous.   □

Since each δ_m^i is continuous, it follows that the map δ^i is continuous.
(See, e.g., Munkres [1975, Theorem 8.5].)

Appendix C. Proofs for Section 9

Let $X^1, \ldots, X^m$ be finite sets, $Y^1, \ldots, Y^m, Z^1, \ldots, Z^m$ be Polish spaces,
and set

$$\Omega = \prod_j X^j \times \prod_j Y^j.$$

For each j, define a measurable maps $f^j : \Omega \to X^j$ and $g^j : \Omega \to Y^j$
by $f^j = \text{proj}_{X^j}$ and $g^j = \text{proj}_{Y^j}$. Also, for each j, let $h^j : Y^j \to Z^j$ be a

measurable map. Define the product maps $g : \Omega \to \prod_j Y^j$ and $h : \prod_j Y^j \to \prod_j Z^j$ by $g(\omega) = (g^1\omega), \ldots, g^m(\omega))$ and $h(y) = (h^1(y^1), \ldots, h^m(y^m))$. Note that these maps are measurable.

Fix a probability measure μ on Ω, an event E in Ω, and a sub σ-algebra $\mathcal{S}$ on Ω. Write $\mu(\mathrm{E}\|\mathcal{S}) : \Omega \to \mathbb{R}$ for (a version of) the conditional probability of E given $\mathcal{S}$.

Say the random variables $f^1, \ldots, f^m$ are *μ-conditionally independent given $h \circ g$* if, for all $j = 1, \ldots, m$ and $E^j \in \sigma(f^j)$,

$$\mu\left(\bigcap_j E^j \| \sigma(h \circ g)\right) = \prod_j \mu(E^j \| \sigma(h \circ g)) \quad \text{a.s.} \tag{C.1}$$

Say the random variable $h^j \circ g^j$ is *μ-sufficient for f^j* if, for each $E^j \in \sigma(f^j)$,

$$\mu(E^j \| \sigma(h \circ g)) = \mu(E^j \| \sigma(h^j \circ g^j)) \quad \text{a.s.} \tag{C.2}$$

Proposition C.1. *Fix bimeasurable maps $h^1, \ldots, h^m$. Suppose the random variables $f^1, \ldots, f^m$ are μ-conditionally independent given $h \circ g$, and that, for each $j = 1, \ldots, m$, the random variable $h^j \circ g^j$ is μ-sufficient for f^j. If $h^1 \circ g^1, \ldots, h^m \circ g^m$ are μ-independent, then $f^1, \ldots, f^m$ are μ-independent.*

For the proof, we will let $v = \mu \circ (h \circ g)^{-1}$ (resp. $v^j = \mu \circ (h^j \circ g^j)^{-1}$) be the image measure of μ under $h \circ g$ (resp. $h^j \circ g^j$).

Lemma C.1. *Fix events F^j in Z_j, for all j. Then,*

$$(h \circ g)^{-1}\left(\prod_j F^j\right) = \bigcap_j (h^j \circ g^j)^{-1}(F^j).$$

Proof. If $\omega \in (h \circ g)^{-1}(\prod_j F^j)$, then, $h(g(\omega)) \in \prod_j F^j$, i.e., $h_j(g^j(\omega)) \in F^j$ for all j. Thus, $\omega \in \bigcap_j (h^j \circ g^j)^{-1}(F^j)$. Conversely, if $\omega \in \bigcap_j (h^j \circ g^j)^{-1}(F^j)$, then, $h^j(g^j(\omega)) \in F^j$ for all j, i.e., $h(g(\omega)) \in \prod_j F^j$. Thus, $\omega \in (h \circ g)^{-1}(\prod_j F_j)$. $\square$

Lemma C.2. *If $h^1 \circ g^1, \ldots, h^m \circ g^m$ are μ-independent, then v is a product measure.*

Proof. Fix events F^j in Z^j, for all j. Then

$$v\left(\prod_j F^j\right) = \mu\left((h \circ g)^{-1}\left(\prod_j F^j\right)\right)$$

$$= \mu\left(\bigcap_j (h^j \circ g^j)^{-1}(F^j)\right)$$

$$= \prod_j \mu((h^j \circ g^j)^{-1}(F^j))$$

$$= \prod_j \mu\left((h \circ g)^{-1}\left(F^j \times \prod_{k \neq j} Z^k\right)\right)$$

$$= \prod_j v\left(F^j \times \prod_{k \neq j} Z^k\right).$$

where the first and fifth lines are from the definition of v, the second and fourth lines come from Lemma C.1, and the third line follows from the fact that $h^1 \circ g^1, \ldots, h^m \circ g^m$ are μ-independent. $\square$

Lemma C.3. $v^j = \mathrm{marg}_{Z^j} v$.

Proof. Fix an event F^j in Z^j. Then,

$$v^j(F^j) = \mu((h^j \circ g^j)^{-1}(F^j))$$

$$= \mu\left((h \circ g)^{-1}\left(F^j \times \prod_{k \neq j} Z^k\right)\right) = v\left(F^j \times \prod_{k \neq j} Z^k\right),$$

where the second line follows from Lemma C.1. $\square$

Below, we will sometimes write $\phi^j = \mu(E\|\sigma(h^j \circ g^j))$.

Lemma C.4. *Fix an event E in Ω, and $\omega, \tilde{\omega} \in \Omega$. If* $\mathrm{proj}_{Y^j}\omega = \mathrm{proj}_{Y^j}\tilde{\omega}$, *then*

$$\mu(E\|\sigma(h^j \circ g^j))_\omega = \mu(E\|\sigma(h^j \circ g^j))_{\tilde{\omega}}.$$

Proof. Since ϕ^j is $\sigma(h^j \circ g^j)$-measurable, for any number r we have $(\phi^j)^{-1}(\{r\})$ is contained in $\sigma(h^j \circ g^j)$. So, there exists some event G

in Z^j with $(h^j \circ g^j)^{-1}(G) = (\phi^j)^{-1}(\{r\})$ (Aliprantis and Border [1999, Lemma 4.22]). By construction, $(h^j \circ g^j)^{-1}(G) = \prod_k X^k \times F^j \times \prod_{k \neq j} Y^k$ for some event F^j in Y^j. □

Lemma C.5. *Fix* $j = 1, \ldots, m$ *and some* $E^j \in \sigma(f^j)$. *If* h^j *is a bimeasurable map then there is a measurable map* $\psi^j : Z^j \to \mathbb{R}$ *with* $\psi^j \circ h^j \circ g^j = \mu(E^j \| \sigma(h^j \circ g^j))$.

Proof. By Kechris (1995, Theorem 12.2), it suffices to show that there is a measurable map $\psi^j : h^j(Y^j) \to \mathbb{R}$ with $\psi^j \circ h^j \circ g^j = \phi^j$. By Lemma C.4, such a map is well defined. We show it is measurable.

Fix an event G in $\mathbb{R}$. Then, $(\phi^j)^{-1}(G)$ is measurable in $\prod_k X^k \times \prod_k Y^k$. Following the argument in the proof of Lemma C.4, $(\phi^j)^{-1}(G)$ must take the form $\prod_k X^k \times F^j \times \prod_{k \neq j} Y^k$ for some event F^j in Y^j. Then, $g^j((\phi^j)^{-1}(G)) = F^j$ is measurable. Now $h^j(F^j)$ is measurable in Z^j, since h^j is bimeasurable. Since $h^j(F^j) \subseteq h^j(Y^j), h^j(F^j)$ is Borel in $h^j(Y^j)$ (Aliprantis and Border [1999, Lemma 4.19]). Now note that $h^j(F^j) = (\psi^j)^{-1}(G^j)$, so that $(\psi^j)^{-1}(G^j)$ is indeed Borel in $h^j(Y^j)$. □

Proof of Proposition C.1. Assume $h^1 \circ g^1, \ldots, h^m \circ g^m$ are μ-independent. By Lemma C.2, v is a product measure. We will use this fact below.

For each $j = 1, \ldots, m$, fix $E^j \in \sigma(f^j)$. Then, by definition of conditional probability,

$$\mu\left(\bigcap_j E^j\right) = \int_\Omega \mu\left(\bigcap_j E^j \| \sigma(h \circ g)\right)_\omega \, d\mu(\omega).$$

Using CI (Equation (C.1)) and SUFF (Equation (C.2)),

$$\mu\left(\bigcap_j E^j\right) = \int_\Omega \prod_j \mu(E^j \| \sigma(h^j \circ g^j))_\omega \, d\mu(\omega). \tag{C.3}$$

For each $j = 1, \ldots, m$, use Lemma C.5 to define a measurable map $\psi^j : Z^j \to \mathbb{R}$ with $\psi^j \circ h^j \circ g^j = \mu(E^j \| \sigma(h^j \circ g^j))$. Also define $\psi : \prod_j Z^j \to \mathbb{R}$ by $\psi(z^1, \ldots, z^m) = \prod_j \psi^j(z^j)$, which is again measurable. Note, $\psi \circ h \circ g = \prod_j \phi^j$.

Using the properties above,

$$\mu\left(\bigcap_j E^j\right) = \int_\Omega \prod_j \mu(E^j \| \sigma(g^j))_\omega \, d\mu(\omega)$$

$$= \int_{Z^1 \times \cdots \times Z^m} \prod_j \psi^j(z^j) \, d\nu(z^1, \ldots, z^m),$$

where the first line is Equation (C.3) and the second line is a change of variables (Aliprantis and Border [1999, Theorem 12.46]). Now use the fact that ν is a product measure, and Fubini's Theorem, to get

$$\mu\left(\bigcap_j E^j\right) = \int_{Z^1 \times \cdots \times Z^m} \prod_j \psi^j(z^j) d\nu(z^1, \ldots, z^m)$$

$$= \prod_j \int_{Z^j} \psi^j(z^j) d\,\mathrm{marg}_{Z^j}\nu(z^1, \ldots, z^m).$$

Using this and Lemma C.3,

$$\mu\left(\bigcap_j E^j\right) = \prod_j \int_{Z^j} \psi^j(z^j) \, d\nu^j(z^j). \tag{C.4}$$

Now note that, by definition of conditional probability, we also have that, for each $j = 1, \ldots, m$,

$$\mu(E^j) = \int_\Omega \mu(E^j \| \sigma(h^j \circ g^j))_\omega \, d\mu(\omega).$$

Using this fact and another change of variables,

$$\mu(E^j) = \int_\Omega \mu(E^j \| \sigma(h^j \circ g^j))_\omega \, d\mu(\omega) = \int_{Z^j} \psi^j(z^j) \, d\nu^j(z^j). \tag{C.5}$$

So, by Equations (C.4) and (C.5),

$$\mu\left(\bigcap_j E^j\right) = \prod_j \mu(E^j),$$

as required. $\qquad\square$

Proposition 9.1 is an immediate corollary of Proposition C.1. Set $X^j = S^j$, $Y^j = T^j$, and $Z^j = \prod_{m=1}^\infty \mathcal{M}(Y_m^i)$, $f^j = \bar{s}_i^j$, $g^j = \bar{t}_i^j$, and $h^j = \delta^j$.

Appendix D. Proofs for Section 10

Lemma D.1. *Let E be a closed subset of a Polish space X, and $\mathcal{M}(X; E)$ be the set of $\mu \in \mathcal{M}(X)$ with $\mu(E) = 1$. Then $\mathcal{M}(X; E)$ is closed.*

Proof. Take a sequence μ_n of measures in $\mathcal{M}(X; E)$, with $\mu_n \to \mu$. It follows from the Portmanteau Theorem that $\limsup \mu_n(E) \leq \mu(E)$. Since $\limsup \mu_n(E) = 1$ for all n, $\mu(E) = 1$ and so $\mu \in \mathcal{M}(X; E)$ as desired. $\square$

Lemma D.2. *The set R_m^i is closed for each i and m.*

Proof. By induction on m.

$m = 1$: Let $E(s^i)$ be the set of $\mu \in \mathcal{M}(S^{-i} \times T^{-i})$ such that s^i is optimal under μ. It suffices to show the sets $E(s^i)$ are closed. If so, since λ^i is continuous, $(\lambda^i)^{-1}(E(s^i))$ is closed. The set R_1^i is the (finite) union over all sets $\{s^i\} \times (\lambda^i)^{-1}(E(s^i))$; so, R_1^i is closed.

First, notice that for each $s^{-i} \in S^{-i}$, the set $\{s^{-i}\} \times T^{-i}$ is clopen. It follows that

$$\mathrm{cl}(\{s^{-i}\} \times T^{-i}) \backslash \mathrm{int}(\{s^{-i}\} \times T^{-i}) = (\{s^{-i}\} \times T^{-i}) \backslash (\{s^{-i}\} \times T^{-i}) = \emptyset,$$

and so, for each $\mu \in \mathcal{M}(S^{-i} \times T^{-i})$, $\mathrm{cl}(\{s^{-i}\} \times T^{-i}) \backslash \mathrm{int}(\{s^{-i}\} \times T^{-i})$ is μ-null.

Now, take a sequence μ_n of measures in $E(s^i)$, with $\mu_n \to \mu$. The Portmanteau Theorem, together with the fact that each $\mathrm{cl}(\{s^{-i}\} \times T^{-i}) \backslash \mathrm{int}(\{s^{-i}\} \times T^{-i})$ is μ-null, implies that $\mu_n(\{s^{-i}\} \times T^{-i}) \to \mu(\{s^{-i}\} \times T^{-i})$.

For each $r^i \in S^i$ and integer n, define

$$x_n(r^i) = \sum_{s^{-i} \in S^{-i}} [\pi^i(s^i, s^{-i}) - \pi^i(r^i, s^{-i})] \mathrm{marg}_{S^{-i}} \mu_n(s^{-i}).$$

Note that $x_n(r^i) \geq 0$, and $x_n(r^i) \to x(r^i)$ where,

$$x(r^i) = \sum_{s^{-i} \in S^{-i}} [\pi^i(s^i, s^{-i}) - \pi^i(r^i, s^{-i})] \mathrm{marg}_{S^{-i}} \mu(s^{-i}).$$

Since, each $x_n(r^i) \geq 0$, $x(r^i) \geq 0$. With this, $\mu \in E(s^i)$ as desired.

$m \geq 2$: Assume the lemma holds for m. Then, using the induction hypothesis, it suffices to show that $S^i \times B^i(R_m^{-i})$ is closed, i.e., that $B^i(R_m^{-i})$ is closed. The induction hypothesis gives that R_m^{-i} is closed. So, by Lemma D.1, $\mathcal{M}(S^{-i} \times T^{-i}; R_m^{-i})$ is closed in $\mathcal{M}(S^{-i} \times T^{-i})$. Since λ^i is continuous, $B^i(R_m^{-i})$ is closed. $\square$

We note the following:

Corollary D.1. *If $t^i \in B^i(R_m^{-i})$ then Supp $\lambda^i(t^i) \subseteq R_m^{-i}$. Similarly, if $t^i \in B^i(\bigcap_m R_m^{-i})$ then Supp $\lambda^i(t^i) \subseteq \bigcap_m R_m^{-i}$.*

Proof of Proposition 10.1. Begin with part (i), and fix a type structure. We will show that the set $\text{proj}_S \bigcap_m R_m$ is a BRS. From this it follows that, for each $(s^1, t^1, \ldots, s^n, t^n) \in \bigcap_m R_m$, $(s^1, \ldots, s^n)$ is correlated rationalizable. To see that $\text{proj}_S \bigcap_m R_m$ is a BRS, fix $(s^i, t^i) \in \bigcap_m R_m^i$. Certainly s^i is optimal under $\text{marg}_{S^{-i}} \lambda^i(t^i)$, since $(s^i, t^i) \in R_1^i$. Also, for all $m, \lambda^i(t^i)(R_m^{-i}) = 1$, and so $\lambda^i(t^i)(\bigcap_m R_m^{-i}) = 1$. From this, $\lambda^i(t^i)(\text{proj}_{S^{-i}}(\bigcap_m R_m^{-i}) \times T^{-i}) = 1$, or

$$\text{marg}_{S^{-i}} \lambda^i(t^i) \left(\text{proj}_{S^{-i}} \bigcap_m R_m^{-i} \right) = 1,$$

as required.

Now part (ii). Construct a type structure as follows. For each i and $s^i \in S_M^i$, there is a measure $\mu(s^i) \in \mathcal{M}(S^{-i})$, with $\mu(s^i)(S_M^{-i}) = 1$, under which s^i is optimal. Fix such a measure $\mu(s^i)$ and define an equivalence relation $\sim^i$ on S_M^i, where $r^i \sim^i s^i$ if and only if $\mu(r^i) = \mu(s^i)$. Set $T^i = S_M^i / \sim^i$ (the quotient space). For $t^i \in T^i$, construct the measure $\lambda^i(t^i)$ on $S^{-i} \times T^{-i}$ as follows. Pick an $s^i \in t^i$ and set

$$\lambda^i(t^i)(s^{-i}, t^{-i}) = \begin{cases} \mu(s^i)(s^{-i}) & \text{if } s^j \in t^j \text{ for all } j \neq i, \\ 0 & \text{otehwise.} \end{cases}$$

(The definition is clearly independent of which $s^i \in t^i$ we choose.) Figure D1 depicts the construction of $\lambda^i(t^i)$.

We will show that $S_M \subseteq \text{proj}_S \bigcap_m R_m$. By construction, $\lambda^i(t^i)(\{(s^{-i}, t^{-i}) : s^{-i} \in S_M^{-i} \text{ and } s^{-i} \in t^{-i}\}) = 1$ for each $t^i \in T^i$. So, it suffices to show that, for all i and all m, if $s^i \in S_M^i$ and $s^i \in t^i$, then $(s^i, t^i) \in R_m^i$. For $m = 1$ this is immediate. Assume this is true for m. Then certainly $\lambda^i(t^i)(R_m^{-i}) = 1$ for all $t^i \in T^i$. Therefore, $(s^i, t^i) \in R_{m+1}^i$ when $s^i \in S_M^i$ and $s^i \in t^i$, as desired.   □

Appendix E. Proofs for Section 11

Fix a probability space $(\Omega, \mathcal{F}, \mu)$ and a measure space $(X, \mathcal{B}(X))$ where X is Polish. Note that each singleton is contained in $\mathcal{B}(X)$. Let $f : \Omega \to X$ be

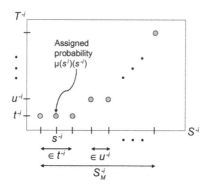

Figure D1.

a random variable. Also, fix $E \in \mathcal{F}$, and let $g : \Omega \to \mathbb{R}$ be a version of the conditional probability of E given $\sigma(f)$.

Lemma E.1. *If $\omega, \omega' \in f^{-1}(\{x\})$ then $g(\omega) = g(\omega')$.*

Proof. Fix $\omega \in f^{-1}(\{x\})$. Since $\{g(\omega)\}$ is closed and g is $\sigma(f)$-measurable, $g^{-1}(\{g(\omega)\}) \in \sigma(f)$. Also note that $\omega \in f^{-1}(\{x\}) \cap g^{-1}(\{g(\omega)\})$. From this, there is an event $G \in \mathcal{B}(X)$ with $x \in G$ and $f^{-1}(G) = g^{-1}(\{g(\omega)\})$ (Aliprantis and Border [1999, Lemma 4.22]). Using this, $\omega \in f^{-1}(\{x\}) \cap f^{-1}(G)$, from which it follows that $x \in G$. So, $f^{-1}(\{x\}) \subseteq f^{-1}(G) = g^{-1}(\{g(\omega)\})$, as required. $\square$

Let $\bar{g}$ be the constant value of g on $f^{-1}(\{x\})$.

Corollary E.1. $\mu(E \cap f^{-1}(\{x\})) = \bar{g} \times \mu(f^{-1}(\{x\}))$.

Proof. Using Lemma E.1 we have

$$\mu(E \cap f^{-1}(\{x\})) = \int_{f^{-1}(\{x\})} g(\omega) d\mu(\omega) = \bar{g} \times \mu(f^{-1}(\{x\})),$$

as required. $\square$

Corollary E.2. *If $\mu(f^{-1}(\{x\})) = 1$ then $g = \mu(E)$ a.s.*

Proof. By Corollary E.1 we have $\mu(E) = \bar{g}$, where $\bar{g}$ is the value of g on the probability-1 set $f^{-1}(\{x\})$. $\square$

Appendix F. A Finite-Levels Result

Suppose it is given that player i only reasons to m levels. In this case, the relevant variable associated with player i is his hierarchy of beliefs up to m levels.

To formalize this, begin by noticing that, if $\delta_m^i(t^i) = \delta_m^i(u^i)$ then $\delta_n^i(t^i) = \delta_n^i(u^i)$ for all $n \leq m$. Define composite maps $\eta_{i,m}^j = \delta_m^j \circ \vec{t}_i^j$ and $\eta_m^{-i} = \delta_m^{-i} \circ \vec{t}_i$.

Definition F.1. The random variables $\vec{s}_i^1, \ldots, \vec{s}_i^{i-1}, \vec{s}_i^{i+1}, \ldots, \vec{s}_i^n$ are $\lambda^i(t^i)$-*conditionally independent given the random variable* η_m^{-i} if, for all $j \neq i$ and $E^j \in \sigma(\vec{s}_i^j)$,

$$\lambda^i(t^i)\left(\bigcap_{j \neq i} E^j \| \sigma(\eta_m^{-i})\right) = \prod_{j \neq i} \lambda^i(t^i)(E^j \| \sigma(\eta_m^{-i})) \quad \text{a.s.}$$

Say the type t^i satisfies m-*conditional independence* (m-*CI*) if $\vec{s}_i^1, \ldots \vec{s}_i^{i-1}, \vec{s}_i^{i+1}, \ldots, \vec{s}_i^n$ are $\lambda^i(t^i)$-conditionally independent given η_m^{-i}.

Definition F.2. The random variable $\eta_{i,m}^j$ is $\lambda^i(t^i)$-*sufficient for the random variable* $\vec{s}_i^j$ if, for each $j \neq i$ and $E^j \in \sigma(\vec{s}_i^j)$,

$$\lambda^i(t^i)(E^j \| \sigma(\eta_m^{-i})) = \lambda^i(t^i)(E^j \| \sigma(\eta_{i,m}^j)) \quad \text{a.s.}$$

Say the type t^i satisfies m-*sufficiency* (m-*SUFF*) if, for each $j \neq i$, $\eta_{i,m}^j$ is $\lambda^i(t^i)$-sufficient for $\vec{s}_i^j$.

We then get the following corollary to Theorem 11.1:

Corollary F.1. *Fix* $m \geq 1$.

(i) *There is a game* G, *a player* i *in* G, *and a correlated rationalizable strategy* s^i *of* i, *such that the following holds: For any type structure* Φ, *there is no* $t^i \in T^i$ *such that* $(s^i, t^i) \in R_{m+1}^i$ *and* t^i *satisfies* m-*CI*.

(ii) *There is a game* G', *a player* i *in* G', *and a correlated rationalizable strategy* s^i *of* i, *such that the following holds: For any type structure* Φ, *there is no* $t^i \in T^i$ *such that* $(s^i, t^i) \in R_{m+1}^i$ *and* t^i *satisfies* m-*SUFF*.

This follows from taking player i to be Charlie in the game of Figure 5 or Figure 7, the strategy s^i to be her choice Y, and repeating the steps of the proof of Theorem 11.1.

Appendix G. Independent Rationalizability

Here we give a proof of the relationship stated in Section 2, and repeated below.

Proposition G.1. *Fix a game* $G = \langle S^1, \ldots, S^n; \pi^1, \ldots, \pi^n \rangle$. *There is an associated type structure* $\langle S^1, \ldots, S^n; T^1, \ldots, T^n; \lambda^1, \ldots, \lambda^n \rangle$ *such that each type satisfies CI and SUFF, and for each independent rationalizable strategy profile* $(s^1, \ldots, s^n)$, *there is a state* $(s^1, t^1, \ldots, s^n, t^n)$ *at which RCBR holds.*

First a definition: A set $\prod_{i=1}^{n} Q^i \subseteq S$ is an *independent best-response set (IBRS)* (cf. Pearce [1984]) if, for each i and every $s^i \in Q^i$, there is a $\mu \in \prod_{j \neq i} M(S^j)$ with $\mu(Q^{-i}) = 1$, under which s^i is optimal. It is well known that the set of independent rationalizable profiles is an IBRS, and every IBRS is contained in the independent rationalizable set.

To prove Proposition G.1, we follow exactly the proof of Proposition 10.1(ii) in Appendix D. Throughout, simply replace the set of player i's correlated rationalizable strategies with the set of i's independent rationalizable strategies. We have to show, in addition, that the type structure Φ constructed there satisfies CI and SUFF.

Using the IBRS property, for each independent rationalizable strategy s^i, there is a product measure $\mu(s^i)$ on S^{-i}, which assigns probability 1 to the independent rationalizable strategies of players $j \neq i$ and under which s^i is optimal. For $s^i \in t^i$, construct the measure $\lambda^i(t^i)$ as before.

We now give the intuition for why CI and SUFF hold, and then the formal proof. For each $j \neq i$, fix an independent rationalizable strategy s^j for player j. Consider the hierarchy of beliefs for j induced by the measure $\lambda^j(t^j)$ for $s^j \in t^j$. CI requires that the conditional of $\lambda^i(t^i)$, conditioned on the event that each player $j \neq i$ has the hierarchy induced by $\lambda^j(t^j)$, be a product measure. But this conditional comes from $\mu(s^i)$, conditioned on a certain rectangular subset of strategies for players $j \neq i$. (For each $j \neq i$, consider the other strategies r^j of player j with measures $\mu(r^j) = \mu(s^j)$. Take the product of these subsets.) Since $\mu(s^i)$ is a product measure, so is its conditional on any rectangular subset. The same argument establishes SUFF.

Recall from the text that $[t^j]$ is the subset $S^{-i} \times \{u^j \in T^j : \delta^j(u^j) = \delta^j(t^j)\} \times T^{-i-j}$ of $S^{-i} \times T^{-i}$.

Proof of Proposition G.1. Follow the proof of Proposition 10.1(ii) in Appendix D. Throughout, simply replace S^i_M with the set of i's independent rationalizable strategies. Then $S_M^i \subseteq \text{proj}_{S^i} \bigcap_m R_m^i$. We have to show, in addition, that each $t^i \in T^i$ satisfies CI and SUFF.

To do this, we will make use of a property the construction satisfies. Specifically, for each $t^i \in T^i$, $\delta^i(t^i) = \delta^i(u^i)$ only if $t^i = u^i$. To see this, fix $t^i \neq u^i$, $s^i \in t^i$, and $r^i \in u^i$. Note that $\text{marg}_{S^{-i}} \lambda^i(t^i) = \mu(s^i)$ and $\text{marg}_{S^{-i}} \lambda^i(u^i) = \mu(r^i)$ (the right-hand sides are independent of which $s^i \in t^i$ and $r^i \in u^i$ were chosen). If $\delta_1^i(t^i) = \delta_1^i(u^i)$ then $\mu(s^i) = \mu(r^i)$. It follows that $t^i = u^i$, as desired.

Fix $t^i \in T^i$ and also $(s^{-i}, t^{-i}) \in S^{-i} \times T^{-i}$. If, for some j, $s^j \notin t^j$, it is then immediate that t^i satisfies CI and SUFF, since

$$\lambda^i(t^i) \left(\bigcap_{k \neq i} [s^k] | \bigcap_{k \neq i} [t^k] \right) = 0 = \prod_{j \neq i} \lambda^i(t^i) \left([s^j] | \bigcap_{k \neq i} [t^k] \right),$$

$$\lambda^i(t^i)([s^j] | [t^j]) = 0 = \lambda^i(t^i) \left([s^j] | \bigcap_{k \neq i} [t^k] \right).$$

So, suppose $s^j \in t^j$ for all j. First note that

$$\lambda^i(t^i) \left(\bigcap_{k \neq i} [s^k] \cap \bigcap_{k \neq i} [t^k] \right)$$

$$= \mu(s^i)(s^{-i}) = \prod_{k \neq i} \mu(s^i)(\{s^k\} \times S^{-i-k}), \qquad \text{(G.1)}$$

where the second equality uses the fact that μ is a product measure. Write E^j for the set of all $s^j \in t^j$ and recall that $\delta^j(u^j) = \delta^j(t^j)$ only if $u^j = t^j$. Again using the fact that μ is a product measure, we have

$$\lambda^i(t^i) \left([s^j] \cap \bigcap_{k \neq i} [t^k] \right) = \sum_{r^k \in t^k} \mu(s^i) \left(\{s^j\} \times \prod_{k \neq i,j} E^k \right)$$

$$= \mu(s^i)(\{s^j\} \times S^{-i-j}) \times \prod_{k \neq i,j} \mu(s^i)(E^k \times S^{-i-k}),$$

$$\text{(G.2)}$$

where the first line is by construction and the second line follows from the fact that μ is a product measure. Similarly,

$$\lambda^i(s^i) \left(\bigcap_{k \neq i} [t^k] \right) = \mu(s^i) \left(\prod_{k \neq i} E^k \right) = \prod_{k \neq i} \mu(s^i)(E^k \times S^{-i-k}). \qquad \text{(G.3)}$$

Now note,

$$\prod_{j \neq i} \lambda^i(s^i) \left([s^j] | \bigcap_{k \neq i} [t^k] \right)$$

$$= \prod_{j \neq i} \frac{\mu(s^i)(\{s^j\} \times S^{-i-j}) \times \prod_{k \neq i,j} \mu(s^i)(E^k \times S^{-i-k})}{\mu(s^i)(E^j \times S^{-i-j}) \prod_{k \neq i,j} \mu(s^i)(E^k \times S^{-i-k})}$$

$$= \lambda^i(s^i) \left(\bigcap_{k \neq i} [s^k] | \bigcap_{k \neq i} [t^k] \right),$$

where the first line follows from Equations (G.2)–(G.3) and the second line follows from Equations (G.1)–(G.3). This establishes CI.

Finally, for each j,

$$\lambda^i(s^i)([t^j]) = \mu(s^i)(E^j \times S^{-i-j}). \tag{G.4}$$

So putting this together with Equations (G.2)–(G.3), we have

$$\lambda^i(s^i)([s^j]|[t^j]) = \frac{\mu(s^i)(\{s^j\} \times S^{-i-j})}{\mu(s^i)(E^j \times S^{-i-j})} = \lambda^i(s^i) \left([s^j] | \bigcap_{k \neq i} [t^k] \right),$$

establishing SUFF. □

Remark G.1. In the proof of Proposition G.1, for each i, the random variables $\eta_i^1, \ldots, \eta_i^{i-1}, \eta_i^{i+1}, \ldots, \eta_i^n$ are independent.

Proof. This is immediate from Equations (G.3)–(G.4) and the converse of Lemma C.2 (which follows immediately from the proof of the forward direction). □

Corollary G.1. *Consider a game* $G = \langle S^1, \ldots, S^n; \pi^1, \ldots, \pi^n \rangle$.

(i) *Fix a bimeasurable type structure* $\langle S^1, \ldots, S^n; T^1, \ldots, T^n; \lambda^1, \ldots, \lambda^n \rangle$ *where each type satisfies CI and SUFF, and has an independent assessment about the other players' hierarchies of beliefs. Suppose RCBR holds at the state* $(s^1, t^1, \ldots, s^n, t^n)$. *Then the strategy profile* $(s^1, \ldots, s^n)$ *is independent rationalizable in* G.

(ii) *There is a type structure* $\langle S^1, \ldots, S^n; T^1, \ldots, T^n; \lambda^1, \ldots, \lambda^n \rangle$ *such that each type satisfies CI and SUFF, and has an independent assessment about the other players' hierarchies of beliefs, and for each independent* rationalizable profile $(s^1, \ldots, s^n)$, there is a state $(s^1, t^1, \ldots, s^n, t^n)$ at which RCBR holds.

Proof. For part (i), repeat the proof of Proposition 10.1(i). Note that by Proposition 9.1, the set $\text{proj}_S \cap_m R_m$ is an IBRS. Part (ii) follows immediately from Proposition G.1 and Remark G.1.   □

Proposition G.1 should be distinguished from the following: Fix a game G and associated type structure Φ. Suppose that for each player i and type t^i, the marginal on S^{-i} of the measure $\lambda^i(t^i)$ is independent. Then: (i) if there is RCBR at the state $(s^1, t^1, \ldots, s^n, t^n)$, the strategy profile $(s^1, \ldots, s^n)$ is independent rationalizable in G; and (ii) the types $t^1, \ldots, t^n$ satisfy CI. Certainly (i) is true. (Just follow the proof of Proposition 10.1(i), noting that since each $\text{marg}_{S^{-i}} \lambda^i(t^i)$ is a product measure, the set $\text{proj}_S \cap_m R_m$ is an IBRS.) But (ii) may be false, as the next example shows. The reason is that within a given type structure, independence need not imply CI. (Of course, the fact that independence does not imply CI is well known in probability theory.)

Example G.1. Let $S^a = \{U, D\}, S^b = \{L, R\}$, and $S^c = \{Y\}$. The type spaces are $T^a = \{t^a, u^a\}, T^b = \{t^b, u^b\}$, and $T^c = \{t^c\}$, where:

$\lambda^a(t^a)$ assigns probability 1 to (L, t^b, Y, t^c);
$\lambda^a(u^a)$ assigns probability 1 to (R, u^b, Y, t^c);
$\lambda^b(t^b)$ assigns probability 1 to (U, t^a, Y, t^c);
$\lambda^b(u^b)$ assigns probability 1 to (D, u^a, Y, t^c);
$\lambda^c(t^c)$ assigns probability $\frac{1}{4}$ to each of $(U, t^a, L, t^b), (D, t^a, R, t^b), (D, u^a, L, u^b), (U, u^a, R, u^b)$.

Note that $\delta^a(t^a) \neq \delta^a(u^a)$ and $\delta^b(t^b) \neq \delta^b(u^b)$. Figure G1 depicts the measure $\lambda^c(t^c)$. Clearly, the marginal on the strategy sets of each type's

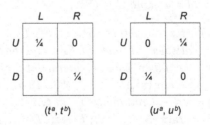

Figure G1.

measure is independent. But CI is violated. For example:

$$\frac{1}{2} = \lambda^c(t^c)([U] \cap [L]|[t^a] \cap [t^b])$$

$$\neq \lambda^c(t^c)([U]|[t^a] \cap [t^b]) \times \lambda^c(t^c)([L]|[t^a] \cap [t^b])$$

$$= \frac{1}{2} \times \frac{1}{2}.$$

(Note that SUFF is satisfied. As for RCBR, we can easily add payoffs for the players — just make them all 0's — so that RCBR holds at every state.)

Appendix H. Injectivity and Genericity

We start with a method of constructing measures that satisfy conditional independence and sufficiency. Fix finite sets $X^1, \ldots, X^m$, and a measure on $\prod_{i=1}^m X^i$. Suppose we can find associated finite sets $Y^1, \ldots, Y^m$ of additional variables, and for each i, an injection from X^i to Y^i. Then there is a natural way to construct a measure on $\prod_{i=1}^m (X^i \times Y^i)$ that agrees with the original measure on $\prod_{i=1}^m X^i$, and which satisfies conditional independence and sufficiency defined with respect to the additional variables.

Some notation: Let $[x^i] = \{x^i\} \times X^{-i} \times Y$, and define $[y^i]$ similarly.

Proposition H.1. *Let* $X^1, \ldots, X^m, Y^1, \ldots, Y^m$ *be finite sets and, for each* $i = 1, \ldots, m$, *let* $f^i : X^i \to Y^i$ *be an injection. Then, given a measure* $\mu \in \mathcal{M}(\prod_{i=1}^m X^i)$, *there is a measure* $\nu \in \mathcal{M}(\prod_{i=1}^m (X^i \times Y^i))$ *with:*

(i) $\mathrm{marg}_{\prod_{i=1}^m X^i} \nu = \mu$;

(ii) $\nu(\bigcap_{i=1}^m [x^i]| \bigcap_{i=1}^m [y^i]) = \prod_{i=1}^m \nu([x^i]| \bigcap_{i=1}^m [y^i])$ *whenever* $\nu(\bigcap_{i=1}^m [y^i]) > 0$;

(iii) *for each* $i = 1, \ldots, m$, $\nu[x^i]| \bigcap_{j=1}^m [y^j]) = \nu([x^i]|[y^i])$ *whenever* $\nu(\bigcap_{j=1}^m [y^i]) > 0$.

Figure H1 depicts the case $m = 2$. Since f^1 and f^2 are injective, the measure ν will assign positive probability to at most one point in each (y^1, y^2)-plane. (We will give it the probability $\mu((f^1)^{-1}(y^1), (f^2)^{-1}(y^2))$.) Conditions (i)–(iii) are then clear.

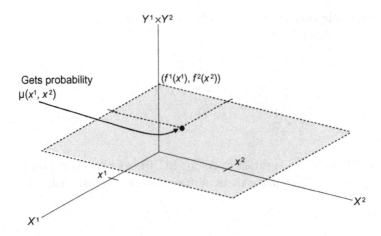

Figure H1.

Proof of Proposition H.1. Define $\nu \in \mathcal{M}(\prod_{i=1}^{m}(X^i \times Y^i))$ by

$$\nu(x^1, y^1, \ldots, x^m, y^m)$$
$$= \begin{cases} \mu(x^1, \ldots, x^m) & \text{if, for each } i = 1, \ldots, m, f^i(x^i) = y^i, \\ 0 & \text{otherwise.} \end{cases}$$

Clearly $\text{marg}_{\prod_{i=1}^{m} X^i} \nu = \mu$, establishing condition (i).

The proofs of (ii) and (iii) make repeated use of injectivity. For (ii), first assume $y^i = f^i(x^i)$ for all i. Then,

$$\nu\left(\bigcap_{i=1}^{m}[x^i] | \bigcap_{i=1}^{m}[y^i]\right) = \frac{\nu(x^1, f^1(x^1), \ldots, x^m, f^m(x^m))}{\nu\left(\bigcap_{i=1}^{m}[f^i(x^i)]\right)} = 1.$$

Also, for each i,

$$\nu\left([x^i] | \bigcap_{j=1}^{m}[y^j]\right) = \frac{\nu\left([x^i] \cap \bigcap_{j=1}^{m}[f^j(x^j)]\right)}{\nu\left(\bigcap_{j=1}^{m}[f^j(x^j)]\right)} = 1,$$

so (ii) holds. Next notice that if $y^i \neq f^i(x^i)$ for some i, then,

$$\nu\left(\bigcap_{j=1}^{m}[x^j] | \bigcap_{j=1}^{m}[y^j]\right) = \nu\left([x^i] | \bigcap_{j=1}^{m}[y^j]\right) = 0,$$

and (ii) again holds.

Turning to (iii), if $y^i = f^i(x^i)$, then,

$$\nu([x^i]|[y^i]) = \frac{\nu([x^i] \cap [f^i(x^i)])}{\nu([f^i(x^i)])} = 1,$$

and

$$\nu\left([x^i]| \bigcap_{j=1}^{m} [y^j]\right) = \frac{\nu\left([x^i] \cap \bigcap_{j=1}^{m} [y^j]\right)}{\nu\left(\bigcap_{j=1}^{m} [y^j]\right)} = 1,$$

so (iii) holds. If $y^i \neq f^i(x^i)$, then,

$$\nu\left([x^i]| \bigcap_{j=1}^{m} [y^j]\right) = \nu([x^i]|[y^i]) = 0.$$

and (iii) again holds. □

We now use Proposition H.1 to address the question in Section 6: Can we identify a class of games where the middle and outer sets in Figure 3 coincide? Here is one answer.

Fix a game G and a BRS $\prod_{i=1}^{n} Q^i$ of G. Then for every i and each $s^i \in Q^i$, there is a $\mu(s^i) \in \mathcal{M}(S^{-i})$ with $\mu(s^i)(Q^{-i}) = 1$, under which s^i is optimal. Say the BRS satisfies the *injectivity condition* if the measures $\mu(s^i)$ can be chosen so that $\mu(r^i) \neq \mu(s^i)$ if $r^i \neq s^i$, for $r^i, s^i \in Q^i$. That is, for every player i, each strategy in i's component of the BRS can be given a different support measure.

Proposition H.2. *Fix a game $G = \langle S^1, \ldots, S^n; \pi^1 \ldots, \pi^n \rangle$ and a BRS $\prod_{i=1}^{n} Q^i$ of G that satisfies the injectivity condition. Then there is a type structure $\langle S^1, \ldots, S^n; T^1, \ldots, T^n; \lambda^1, \ldots, \lambda^n \rangle$ such that each type satisfies CI and SUFF, and for each strategy profile $(s^1, \ldots, s^n) \in \prod_{i=1}^{n} Q^i$ there is a state $(s^1, t^1, \ldots, s^n, t^n)$ at which there is RCBR.*

Proof. For each i, let T^i be a copy of the set Q^i. We now apply Proposition H.1. Fix a player i. For each $j \neq i$, set $X^j = Q^j$ and $Y^j = T^j$. The identity map gives the injection f^j from X^j to Y^j.

For $t^i = s^i \in Q^i$, we construct $\lambda^i(t^i) \in \mathcal{M}(S^{-i} \times T^{-i})$ from $\mu(s^i)$, the same way that ν is constructed from μ in Proposition H.1. (For this, identify a measure on $\mathcal{M}(S^{-i} \times T^{-i})$ with support contained in $Q^{-i} \times T^{-i}$, with a measure on $\mathcal{M}(Q^{-i} \times T^{-i})$.)

Notice that if $t^j \neq u^j$ then $\text{marg}_{S^{-j}} \lambda^j(t^j) \neq \text{marg}_{S^{-j}} \lambda^j(u^j)$. From this it follows that, for any $t^j \neq u^j$, $\delta^j(t^j) \neq \delta^j(u^j)$. That is,

$$[t^j] = S^{-i} \times \{u^j \in T^j : \delta^j(u^j) = \delta^j(t^j)\} \times T^{-i-j} = S^{-i} \times \{t^j\} \times T^{-i-j}.$$

So, by Proposition H.1, t^i satisfies CI and SUFF.

It remains to show that $\prod_{i=1}^n Q^i \subseteq \text{pros}_S \bigcap_m R_m$. Fix $s^i \in T^i = Q^i$. By construction, $\text{marg}_{S^{-i}} \lambda(s^i) = \mu(s^i)$ and $\lambda(s^i)(\{(s^{-i}, s^{-i}) : s^{-i} \in Q^{-i}\}) = 1$. Certainly, if $s^i \in Q^i$ then $(s^i, s^i) \in R_1^i$. Assume inductively that, for all j, $(s^j, s^j) \in R_m^j$. Then certainly $\lambda^i(s^i)(R_m^{-i}) = 1$, so that $(s^i, s^i) \in R_{m+1}^i$. Thus, $(s^i, s^i) \in \bigcap_m R_m^i$, establishing the result. □

Recall that the correlated rationalizable set $\prod_{i=1}^n S_M^i$ is a BRS. So, Proposition H.2 tells us that if the correlated rationalizable set satisfies the injectivity condition, there is a type structure such that each type satisfies CI and SUFF, and for each correlated rationalizable profile $(s^1, \ldots, s^n)$ there is a state $(s^1, t^1, \ldots, s^n, t^n)$ at which RCBR holds. We conclude that if the correlated rationalizable set satisfies the injectivity condition, then the middle and outer sets in Figure 3 coincide.

Next, we show this condition holds generically (in the matrix). Fix an n-player strategic game form $\langle S^1, \ldots, S^n \rangle$. A particular game can then be identified with a point $(\pi^1, \ldots, \pi^n) \in \mathbb{R}^{n \times |S|}$. Following Battigalli–Siniscalchi (2003), say the game $(\pi^1, \ldots, \pi^n)$ satisfies the *strict best-response property* if for each $s^i \in S_M^i$, there exists $\mu \in \mathcal{M}(S^{-i})$ with $\mu(S_M^{-i}) = 1$ such that s^i is the unique strategy optimal under μ. Note, if a game $(\pi^1, \ldots, \pi^n)$ satisfies the strict best-response property, then the correlated rationalizable set satisfies the injectivity condition (but not vice versa).

Proposition H.3. *Let Γ be the set of games for which the strategies consistent with RCBR, CI, and SUFF are strictly contained in the correlated rationalizable strategies. The set Γ is nowhere dense in $\mathbb{R}^{n \times |S|}$.*

Proof. By Proposition H.2 and the above remarks, Γ is contained in the sets of games that fail the strict best-response property. Proposition 4.4 in Battigalli–Siniscalchi (2003) shows that for $n = 2$, the set of games that fail the strict best-response property is nowhere dense. Their argument readily extends to $n > 2$, giving our result. □

Of course, genericity in the matrix is usually viewed as too strong a condition: it is well understood that many games of applied interest are non-generic (even in the tree). (See the discussions in Mertens [1989, p. 582] and Marx–Swinkels [1997, pp. 224–225].) For this reason, we believe it is more illuminating to identify structural conditions on a game under which a particular statement — such as equality of the middle and outer sets in Figure 3 — holds. Injectivity is one such condition. No doubt, there are other conditions of interest.

Appendix I. Extrinsic correlation Contd.

Here we add a player (Nature) which does not have payoffs or types, and does not affect the payoffs of the other players. First, the definition of such an extended game: A finite n-player strategic-form game *with a (payoff-irrelevant) move by Nature* is a structure $G = \langle S^0, S^1, \ldots, S^n; \pi^1, \ldots, \pi^n \rangle$, where S^0 is the strategy set for Nature.

Definition I.1. Fix a game $G = \langle S^1, \ldots, S^n; \pi^1, \ldots, \pi^n \rangle$, and a game with a move by Nature, viz. $\bar{G} = \langle \bar{S}^0, \bar{S}^1, \ldots, \bar{S}^n; \bar{\pi}^1, \ldots, \bar{\pi}^n \rangle$. Say $\bar{G}$ is an extension of G if, for each $i = 1, \ldots, n$, $\bar{S}^i = S^i$ and

$$\bar{\pi}^i(s^0, s^1, \ldots, s^n) = \pi^i(s^1, \ldots, s^n)$$

for all $(s^0, s^1, \ldots, s^n) \in \prod_{j=0}^n S^j$.

For a game with a move by Nature, set $S = \prod_{i=1}^n S^i$ and, for $i = 1, \ldots, n$, $S^{-i} = \prod_{j \neq i: j=1, \ldots, n} S^j$. Define sets S_m^i inductively by $S_0^i = S^i$, and

$$S_{m+1}^i = \{ s^i \in S_m^i : \text{there exists } \mu \in \mathcal{M}(S^0 \times S^{-i}), \quad \text{with } \mu(S^0 \times S_m^{-i}) = 1,$$

$$\text{such that } \pi^i(s^i, \mu) \geq \pi^i(r^i, \mu) \quad \text{for every } r^i \in S_m^i \}.$$

Note the similarity to the definition in Section 10. As there, for each i, we write S_M^i for the set of correlated rationalizable strategies for player i. We also define analogs to the definitions of rationality and RCBR in Section 10. The next lemma is straightforward and we omit the proof.

Lemma I.1. *Fix a game G and an extension $\bar{G}$ of G. The correlated rationalizable strategies in G are the same as the correlated rationalizable strategies in $\bar{G}$.*

Now, we show that in a game with a nontrivial move by Nature, the correlated rationalizable strategies characterize CI, SUFF, and RCBR. For the definitions of CI and SUFF, redefine the random variables $\vec{s}_i^j, \vec{t}_i^j, \vec{s}_i$, and $\vec{t}_i$ from Section 9 so they are maps from the full state space (i.e., $S^0 \times S^{-i} \times T^{-i}$), and redefine δ^i, δ^{-i} so they are maps to hierarchies on

$S^0 \times S^{-i}$ (i.e., take $Y_1^i = S^0 \times S^{-i}$). With this modification, Definitions 9.1 and 9.2 are appropriate to analyze the extended game.

Proposition I.1. *Fix a game* $G = \langle S^0, S^1, \ldots, S^n ; \pi^1, \ldots, \pi^n \rangle$ *with* $|S^0| \geq$ 2. *There is a type structure for* G, *where:* (i) *each type satisfies CI and SUFF; and* (ii) *for each correlated rationalizable strategy profile* $(s^1, \ldots, s^n)$ *in* G *there is a state* $(s^0, s^1, t^1, \ldots, s^n, t^n)$ *at which there is RCBR.*

This is essentially a corollary to Proposition H.2: Fix the correlated rationalizable set of G. For each $s^i \in S_M^i$, there is a measure $\mu(s^i) \in \mathcal{M}(S^0 \times S^{-i})$ with $\mu(s^i)(S^0 \times S_M^{-i}) = 1$, under which s^i is optimal. Since $|S^0| \geq 2$, we can choose the measures so that $\mu(r^i) \neq (s^i)$ if $r^i \neq s^i$, for $r^i, s^i \in s_M^i$.

But, to use Proposition H.2, we must amend the proofs of Propositions H.1 and H.2 to apply to a game with a move by Nature.

Fix finite sets $X^0, X^1, \ldots, X^m, Y^1, \ldots, Y^m$. Let $[x^i] = X^0 \times \{x^i\} \times X^{-i} \times Y$, and define $[y^i]$ similarly. (We now set $X^{-i} = \prod_{j \neq i : j = 1, \ldots, n} X^j$.) Proposition H.1 can then be amended to say:

Proposition I.2. *Let* $X^0, X^1, \ldots, X^m, Y^1, \ldots, Y^m$ *be finite sets and, for each* $i = 1, \ldots, m$, *let* $f^i : X^i \to Y^i$ *be an injection. Then, given a measure* $\mu \in \mathcal{M}(\prod_{i=0}^m X^i)$, *there is a measure* $\nu \in \mathcal{M}(X^0 \times \prod_{i=1}^m (X^i \times Y^i))$ *with:*

(i) $\mathrm{marg}_{\prod_{i=0}^m X^i} \nu = \mu$;

(ii) $\nu(\bigcap_{i=1}^m [x^i] | \bigcap_{i=1}^m [y^i]) = \prod_{i=1}^m \nu([x^i] | \bigcap_{i=1}^m [y^i])$ *whenever* $\nu(\bigcap_{i=1}^m [y^i]) > 0$;

(iii) *for each* $i = 1, \ldots, m$, $\nu([x^i] | \bigcap_{j=1}^m [y^j]) = \nu([x^i] | [y^i])$ *whenever* $\nu(\bigcap_{j=1}^m [y^j]) > 0$.

Proof. Define $\nu \in \mathcal{M}(X^0 \times \prod_{i=1}^m (X^i \times Y^i))$ by

$$\nu(x^0, x^1, y^1, \ldots, x^m, y^m)$$
$$= \begin{cases} \mu(x^0, x^1, \ldots, x^m) & \text{if, for each } i = 1, \ldots, m, \ f^i(x^i) = y^i, \\ 0 & \text{otherwise,} \end{cases}$$

and the proof follows line-by-line from the proof of Proposition H.1. □

With this proposition, the proof of Proposition H.2 is readily amended. We have now shown:

Corollary I.1. *Fix a game* G *and an extension* $\bar{G} = \langle S^0, S^1, \ldots, S^n ; \bar{\pi}^1, \ldots, \bar{\pi}^n \rangle$ *of* G *with* $|S^0| \geq 2$. *There is a type structure for* $\bar{G}$ *where:* (i) *each type satisfies CI and SUFF; and* (ii) *for each correlated rationalizable strategy profile* $(s^1, \ldots, s^n)$ *in* G, *there is a state* $(s^0, s^1, t^1, \ldots, s^n, t^n)$ *in the type structure for* $\bar{G}$ *at which there is RCBR.*

Next, we comment briefly on two other sources of extrinsic correlation: (i) payoff uncertainty, and (ii) dummy players. We think both extensions are interesting. Note that both routes involve analyzing a given game G, by first changing G to a new game and then analyzing this new game. So these routes indeed involve extrinsic not intrinsic correlation.

(i) *Payoff uncertainty:* So far we have treated uncertainty over strategies. Uncertainty over payoffs is another potential source of correlation. In the text, the game was given — i.e., there was no uncertainty over the payoff functions $\pi^1, \ldots, \pi^n$. Now introduce a little uncertainty about payoff functions. Specifically, assume the payoff functions are common $(1 - \varepsilon)$-belief (Monderer and Samet [1989]), for some (small) $\varepsilon > 0$. (For short, say that a game G itself is common $(1 - \varepsilon)$-belief.)

Go back to the game of Figure 5 and the associated type structure in Figure 10 (but without the coin toss). The idea is that now Ann's types t^a and u^a will both give probability ε to Bob's having a different payoff function from the given one, and the types will differ in what Bob's alternative payoff function is. So t^a and u^a will induce different hierarchies of beliefs (now defined over strategies and payoff functions). We do a similar construction for Bob, and this way CI and SUFF hold in the new structure. We give a general construction in an online appendix.[13]

Yet, this is not a route to understanding the correlated rationalizable strategies in the original game. In introducing payoff uncertainty, we have changed the game from the original one, in which the payoff functions were given.

Even if we allow this change to the game, there is another difficulty. If we introduce payoff uncertainty, we lose a complete characterization of correlated rationalizability. In the online appendix, we show that given $\varepsilon > 0$, we can find a game G where the conditions of RCBR, CI, and SUFF (all redefined for the case of payoff uncertainty), and common $(1 - \varepsilon)$-belief of G, allow a strategy which is not correlated rationalizable to be played in G.[14] We conclude that while payoff uncertainty can — arguably — rescue the converse direction (part (ii)) of Proposition 10.1, we then lose the forward direction (part (i)).

---

[13] Available at http://www.stern.nyu.edu/~abranden.

[14] Note that we first fix ε, and common $(1 - \varepsilon)$-belief relative to this ε. Then we find a game where our conditions allow a strategy which is not correlated rationalizable. This order is important. Epistemic conditions should be stated independent of a particular game. If the conditions are allowed to depend on the game in question, then the condition could simply be that a strategy profile we are interested in is chosen. This would not be a useful epistemic analysis.

(ii) *Dummy players*: Here the idea is to add another player to the game. In the game of Figure 5, we add a fourth player ("Dummy") with a singleton strategy set. Ann's types t^a and u^a will differ in what they think Dummy thinks about the strategies chosen. Likewise with Bob, and we again get CI and SUFF. The online appendix again gives a general construction.

The same issue arises here. Adding a dummy player is changing the game. The basic question remains: What correlations can be understood in the original game?

References

Aliprantis, C and K Border (1999). *Infinite Dimensional Analysis: A Hitchhiker's Guide*. Berlin: Springer.

Aumann, R (1974). Subjectivity and correlation in randomized strategies. *Journal of mathematical Economics*, 1, 76–96.

Aumann, R (1987). Correlated equilibrium as an expression of Bayesian rationality. *Econometrica*, 55, 1–18.

Battigalli, P and M Siniscalchi (1999). Hierarchies of conditional beliefs and interactive epistemology in dynamic games. *Journal of Economic Theory*, 88, 188–230.

Battigalli, P and M Siniscalchi (2003). Rationalization and incomplete information. *Advances in Theoretical Economics*, 3, 1–46.

Bernheim, D (1984). Rationalizable strategic behavior. *Econometrica*, 52, 1007–1028.

Brandenburger, A and E Dekel (1987). Rationalizability and correlated equilibria. *Econometrica*, 55, 1391–1402.

Dekel, E, D Fudenberg, and S Morris (2007). Interim correlated rationalizability. *Theoretical Economics*, 2, 15–40.

Dellacherie, C and P-A Meyer (1978). *Probabilities and Potential*. Mathematics Studies 29. Amsterdam: North-Holland.

Ely, J and M Peski (2006). Hierarchies of belief and interim rationalizability. *Theoretical Economics*, 1, 19–65.

Kechris, A (1995). *Classical Descriptive Set Theory*. Berlin: Springer.

Kohlberg, E and J-F Mertens (1986). On the strategic stability of equilibria. *Econometrica*, 54, 1003–1038.

Liu, Q (2004). Representation of belief hierarchies in games with incomplete information.

Marx, L and J Swinkels (1997). Order independence for iterated weak dominance, *Games and Economic Behavior*, 18, 219–245.

Mauldin, R (1981). Bimeasurable functions. *Proceedings of the American Mathematical Society*, 83, 369–370.

Mertens, J-F (1989). Stable equilibria — A reformulation. Part 1. Definition and basic properties. *Mathematics of Operations Research*, 14, 575–624.

Mertens, J-F and S Zamir (1985). Formulation of Bayesian analysis for games with incomplete information. *International Journal of Game Theory*, 14, 1–29.

Monderer, D and D Samet (1989). Approximating common knowledge with common beliefs. *Games and Economic Behavior*, 1, 170–190.

Munkres, J (1975). *Topology: A First Course.* Englewood Cliffs, NJ: Prentice Hall.

Pearce, D (1984). Rationalizable strategic behavior and the problem of perfection. *Econometrica*, 52, 1029–1050.

Purves, R (1966). Bimeasurable functions. *Fundamenta Mathematica*, 58, 149–157.

Savage, L (1954). *The Foundations of Statistics.* New York, NY: Wiley.

Tan, T and S Werlang (1988). The Bayesian foundations of solution concepts of games. *Journal of Economic Theory*, 45, 370–391.

Chapter 5

Epistemic Conditions for Nash Equilibrium

Robert Aumann and Adam Brandenburger

Sufficient conditions for Nash equilibrium in an n-person game are given in terms of what the players know and believe — about the game, and about each other's rationality, actions, knowledge, and beliefs. Mixed strategies are treated not as conscious randomizations, but as conjectures, on the part of other players, as to what a player will do. Common knowledge plays a smaller role in characterizing Nash equilibrium than had been supposed. When $n = 2$, mutual knowledge of the payoff functions, of rationality, and of the conjectures implies that the conjectures form a Nash equilibrium. When $n \geq 3$ and there is a common prior, mutual knowledge of the payoff functions and of rationality, and common knowledge of the conjectures, imply that the conjectures form a Nash equilibrium. Examples show the results to be tight.

1. Introduction

In recent years, a literature has emerged that explores noncooperative game theory from a decision-theoretic viewpoint. This literature analyzes games in terms of the rationality[1] of the players and their *epistemic* state: what they know or believe about the game and about each other's rationality, actions, knowledge, and beliefs. As far as Nash's fundamental notion of

Originally published in *Econometrica*, 63, 1161–1180.

Keywords: Game theory; strategic games; equilibrium; Nash equilibrium; strategic equilibrium; knowledge; common knowledge; mutual knowledge; rationality; belief; belief systems; interactive belief systems; common prior; epistemic conditions; conjectures; mixed strategies.

Acknowledgments: We are grateful to Kenneth Arrow, John Geanakoplos, and Ben Polak for important discussions, and to a co-editor and the referees for very helpful editorial suggestions.

[1]We call a player *rational* if he maximizes his utility given his beliefs.

strategic equilibrium is concerned, the picture remains incomplete;[2] it is not clear just what epistemic conditions lead to Nash equilibrium. Here we aim to fill that gap. Specifically, we seek sufficient epistemic conditions for Nash equilibrium that are in a sense as "spare" as possible.

The stage is set by the following *Preliminary Observation: Suppose that each player is rational, knows his own payoff function, and knows the strategy choices of the others. Then the players' choices constitute a Nash equilibrium in the game being played.*[3]

Indeed, since each player knows the choices of the others, and is rational, his choice must be optimal given theirs; so by definition,[4] we are at a Nash equilibrium.

Though simple, the observation is not without interest. Note that it calls for *mutual* knowledge of the strategy choices — that each player know the choices of the others, with no need for the others to know that he knows (or for any higher order knowledge). It does *not* call for *common* knowledge, which requires that all know, all know that all know, and so on ad infinitum (Lewis [1969]). For rationality and for the payoff functions, not even mutual knowledge is needed; only that the players are in fact rational, and that each knows his own payoff function.[5]

The observation applies to pure strategies — henceforth called *actions*. It applies also to mixed actions, under the traditional view of mixtures as conscious randomizations; in that case it is the mixtures that must be mutually known, not their pure realizations.

In recent years, a different view of mixing has emerged.[6] According to this view, players do not randomize; each player chooses some definite action. But other players need not know which one, and the mixture represents their uncertainty, their conjecture about his choice. This is the context of our main results, which provide sufficient conditions for a profile of conjectures to constitute a Nash equilibrium.[7]

---

[2]See Section 7i.

[3]For a formal statement, see Section 4.

[4]Recall that a Nash equilibrium is a profile of strategies in which each player's strategy is optimal for him, given the strategies of the others.

[5]Knowledge of one's own payoff function may be considered tautologous. See Section 2.

[6]Harsanyi (1973), Armbruster and Böge (1979), Aumann (1987a), Tan and Werlang (1988), Brandenburger and Dekel (1989), among others.

[7]The preliminary observation, too, may be interpreted as referring to an equilibrium in conjectures rather than actions. When each player knows the actions of the others, then conjectures coincide with actions: what people do is the same as what others believe them to do. Therefore, the conjectures as well as the actions are in equilibrium.

Consider first the case of two players. Here the conjecture of each is a probability distribution on the other's actions — formally, a mixed action of the other. We then have the following (Theorem A): *Suppose that the game being played (i.e., both payoff functions), the rationality of the players, and their conjectures are all mutually known. Then the conjectures constitute a Nash equilibrium.*[8]

In Theorem A, as in the preliminary observation, common knowledge plays no role. This is worth noting, in view of suggestions that have been made that there is a close relation between Nash equilibrium and common knowledge — of the game, the players' rationality, their beliefs, and/or their choices.[9] On the face of it, such a relation sounds not implausible. One might have reasoned that each player plays his part of the equilibrium "because" the other does so; he, in turn, also does so "because" the first does so; and so on ad infinitum. This infinite regress does sound related to common knowledge; but the connection, if any, is murky.[10] Be that as it may, Theorem A shows that in two-person games, epistemic conditions not involving common knowledge in any way already imply Nash equilibrium.

When the number n of players exceeds 2, the conjecture of a player i is not a mixed action of another player, but a probability distribution on $(n-1)$-tuples of actions of all the other players. Though not itself a mixed

---

[8]The idea of the proof is not difficult. Call the players "Rowena" and "Colin"; let their conjectures and payoff functions be ϕ, g and ψ, h, respectively. Let a be an action of Rowena to which Colin's conjecture ψ assigns positive probability. Since Colin knows that Rowena is rational, he knows that a is optimal against her conjecture, which he knows to be ϕ, given her payoff function, which he knows to be g. Similarly any action b to which ϕ assigns positive probability is optimal against ψ given Colin's payoff function h. So (ψ, ϕ) is a Nash equilibrium in the game defined by (g, h).

[9]Thus, Kreps and Wilson (1982, p. 885): "An equilibrium in Nash's sense supposes that strategies are 'common knowledge' among the players." Or Geanakoplos, Pearce, and Stacchetti (1989, p. 62): "In traditional equilibrium analysis, the equilibrium strategy profile is taken to be common knowledge." Or Milgrom and Roberts (1991, p. 82): "Equilibrium analysis dominates the study of games of strategy, but... many... are troubled by its assumption that players... identify and play a particular vector of equilibrium strategies, that is,... that the equilibrium is common knowledge." See also, inter alia, Arrow (1986, p. S392), Binmore and Dasgupta (1986, pp. 2–5), Fudenberg and Kreps (1988, p. 2), Tan and Werlang (1988, pp. 381–385), Fudenberg and Tirole (1989, p. 267), Werlang (1989, p. 82), Binmore (1990, pp. 51, 61, 210), Rubinstein (1991, p. 915), Binmore (1992, p. 484), and Reny (1992, p. 628).

We ourselves have written in this vein (Aumann [1987b, p. 473] and Binmore and Brandenburger [1990, p. 119]); see Section 7f.

[10]Brandenburger and Dekel (1989) do state a relation between common knowledge and Nash equilibrium in the two-person case, but it is quite different from the simple sufficient conditions established here. See Section 7i.

action, i's conjecture does induce a mixed action[11] for each player j other than i; we call this i's *conjecture about j*. However, different players other than j may have different conjectures about j. Since j's component of the putative equilibrium is meant to represent the conjectures of other players i about j, and these may be different for different i, it is not clear how j's component should be defined.

To proceed, we need another definition. The players are said to have a *common prior*[12] if all differences between their probability assessments are due only to differences in their information; more precisely, if one can think of the situation as arising from one in which the players had the same information and probability assessments, and then got different information.

Theorem B, our n-person result, is now as follows: *In an n-player game, suppose that the players have a common prior, that their payoff functions and their rationality are mutually known, and that their conjectures are commonly known. Then for each player j, all the other players i agree on the same conjecture σ_j about j; and the resulting profile $(\sigma_1, \ldots, \sigma_n)$ of mixed actions is a Nash equilibrium.*

So common knowledge enters the picture after all, but in an unexpected way, and only when there are at least three players. Even then, what is needed is common knowledge of the players' conjectures, not of the game or of the players' rationality.

Theorems A and B are formally stated and proved in Section 4.

In the observation as well as the two results, the conditions are sufficient, not necessary. It is always possible for the players to blunder into a Nash equilibrium "by accident," so to speak, without anybody knowing much of anything. Nevertheless, the statements are "tight," in the sense that they cannot be improved upon; none of the conditions can be left out, or even significantly weakened. This is shown by a series of examples in Section 5, which, in addition, provide insight into the role played by the epistemic conditions.

One might suppose that one needs stronger hypotheses in Theorem B than in Theorem A only because when $n \geq 3$, the conjectures of two players about a third one may disagree. But that is not so. One of the examples in Section 5 shows that even when the necessary agreement is assumed

---

[11] The marginal on j's actions of i's overall conjecture.

[12] Aumann (1987a); for a formal definition, see Section 2. Harsanyi (1967–68) uses the term "consistency" to describe this situation.

outright, conditions similar to those of Theorem A do not suffice for Nash equilibrium when $n \geq 3$.

Summing up: With two players, mutual knowledge of the game, of the players' rationality, and of their conjectures implies that the conjectures constitute a Nash equilibrium. To reach the same conclusion when there are at least three players, one must also assume a common prior and common knowledge of the conjectures.

The above presentation, while correct, has been informal, and sometimes slightly ambiguous. For an unambiguous presentation, one needs a formal framework for discussing epistemic matters in game contexts; in which, for example, one can describe a situation where each player maximizes against the choices of the others, all know this, but not all know that all know this. In Section 2 we describe such a framework, called an *interactive belief system*; it is illustrated in Section 3. Section 6 defines infinite belief systems, and shows that our results apply to this case as well.

The paper concludes with Section 7, where we discuss conceptual matters and related work.

The reader wishing to understand just the main ideas should read Sections 1 and 5, and skim Sections 2 and 3.

2. Interactive Belief Systems

Let us be given a strategic *game form*; that is, a finite set $\{1, \ldots, n\}$ (the *players*), together with an *action set* A_i for each player i. Set $A := A_1 \times \cdots \times A_n$. An *interactive belief system* (or simply *belief system*) for this game form is defined to consist of:

(2.1)   for each player i, a set S_i, (i's *types*), and for each type s_i of i,

(2.2)   a probability distribution on the set S^{-i} of $(n-1)$-tuples of types of the other players (s_i's *theory*),

(2.3)   an action a_i of i (s_i's *action*), and

(2.4)   a function $g_i : A \to \mathbf{R}$ (s_i's *payoff function*).

The action sets A_i are assumed finite. One may also think of the type spaces S_i as finite throughout the paper; the ideas are then more transparent. For a general definition, where the S_i are measurable spaces and the theories are probability measures, see Section 6.

Set $S := S_1 \times \cdots \times S_n$. Call the members $s = (s_1, \ldots, s_n)$ of S *states of the world*, or simply *states*. An *event* is a subset E of S. By (2.2), s_i's theory has domain S^{-i}; define an extension $p(\cdot; s_i)$ of the theory to S, called i's *probability distribution on S at s_i*, as follows: If E is an event, define $p(E; s_i)$ as the probability that s_i's theory assigns to $\{s^{-i} \in S^{-i}: (s_i, s^{-i}) \in E\}$. Abusing our terminology a little, we will use "belief system" to refer also to the system consisting of S and the $p(\cdot; s_i)$; no confusion should result.[13]

A state is a formal description of the players' actions, payoff functions, and beliefs — about each other's actions and payoff functions, about these beliefs, and so on. Specifically, the theory of a type s_i represents the probabilities that s_i ascribes to the types of the other players, and so to their actions, their payoff functions, and their theories. It follows that a player's type determines his beliefs about the actions and payoff functions of the others, about their beliefs about these matters, about their beliefs about others' beliefs about these matters, and so on ad infinitum. The whole infinite hierarchy of beliefs about beliefs about beliefs... about the relevant variables is thus encoded in the belief system.[14]

A function $g : A \to \mathbf{R}^n$ (an n-tuple of payoff functions) is called a *game*.

Set $A^{-i} := A_1 \times \cdots \times A_{i-1} \times A_{i+1} \times \cdots \times A_n$; for a in A, set $a^{-i} := (a_1, \ldots, a_{i-1}, a_{i+1}, \ldots, a_n)$. When referring to a player i, the phrase "at s" means "at s_i." Thus, "i's action at s" means s_i's action (see (2.3)); we denote it $\mathbf{a}_i(s)$, and write $\mathbf{a}(s)$ for the n-tuple $(\mathbf{a}_1(s), \ldots, \mathbf{a}_n(s))$ of actions at s. Similarly, "i's payoff function at s" means s_i's payoff function (see (2.4)); we denote it $\mathbf{g}_i(s)$, and write $\mathbf{g}(s)$ for the n-tuple $(\mathbf{g}_1(s), \ldots, \mathbf{g}_n(s))$ of payoff functions[15] at s. Viewed as a function of a, we call $\mathbf{g}(s)$ "the game being played at s," or simply "the game at s."

Functions defined on S (like $\mathbf{a}_i$, $\mathbf{a}$, $\mathbf{g}_i$, and $\mathbf{g}$) may be viewed like random variables in probability theory. Thus, if $\mathbf{x}$ is such a function and x is one of its values, then $[\mathbf{x} = x]$, or simply $[x]$, denotes the event $\{s \in S : \mathbf{x}(s) = x\}$. For example, $[a_i]$ denotes the event that i chooses the action a_i, $[g]$ denotes

---

[13]The extension $p(\cdot; s_i)$ is uniquely determined by two conditions: first, that its marginal on S^{-i} be s_i's theory; second, that it assign probability 1 to i being of type s_i. We are thus implicitly assuming that a player of type s_i assigns probability 1 to being of type s_i. For a discussion, see Section 7c.

[14]Conversely, it may be shown that any such hierarchy satisfying certain minimal coherency requirements may be encoded in some belief system (Mertens and Zamir [1985]; also Armbruster and Böge [1979], Böge and Eisele [1979], and Brandenburger and Dekel [1993]).

[15]Thus, i's actual payoff at the state s is $\mathbf{g}_i(s)(\mathbf{a}(s))$.

the event that the game g is being played; and $[s_i]$ denotes the event that i's type is s_i.

A *conjecture* ϕ^i of i is a probability distribution on A^{-i}. For $j \neq i$, the marginal of ϕ^i on A_j is called the *conjecture of i about j induced by ϕ^i*. The theory of i at a state s yields a conjecture $\phi^i(s)$, called i's *conjecture at s*, given by $\phi^i(s)(a^{-i}) := p([a^{-i}]; s_i)$. We denote the n-tuple $(\phi^1(s), \ldots, \phi^n(s))$ of conjectures at s by $\phi(s)$.

Player i is called *rational at s* if his action at s maximizes his expected payoff given his information (i.e., his type s_i); formally, letting $g_i := \mathbf{g}_i(s)$ and $a_i := \mathbf{a}_i(s)$, this means that $\exp(g_i(a_i, \mathbf{a}^{-i}); s_i) \geq \exp(g_i(b_i, \mathbf{a}^{-i}); s_i)$ for all b_i in A_i. Another way of saying this is that i's actual choice a_i maximizes the expectation of his actual payoff g_i when the other players' actions are distributed according to his actual conjecture $\phi^i(s)$.

Player i is said to *know* an event E at s if at s, he ascribes probability 1 to E. Define $K_i E$ as the set of all those s at which i knows E. Set $K^1 E := K_1 E \cap \ldots \cap K_n E$; thus, $K^1 E$ is the event that all players know E. If $s \in K^1 E$, call E *mutually known at s*. Set $CKE := K^1 E \cap K^1 K^1 E \cap K^1 K^1 K^1 E \cap \ldots$; if $s \in CKE$, call E *commonly known at s*.

A probability distribution P on S is called a *common prior* if for all players i and all of their types s_i, the conditional distribution of P given s_i is $p(\cdot; s_i)$; this implies that for all i, all events E and F, and all numbers π,

(2.5)   if $p(E; s_i) = \pi p(F; s_i)$ for all $s_i \in S_i$, then, $P(E) = \pi P(F)$.

In words, (2.5) says that for each player i, if two events have proportional probabilities given any s_i, then they have proportional prior probabilities.[16]

Belief systems provide a formal language for stating epistemic conditions. When we say that a player knows some event E, or is rational, or has a certain conjecture ϕ^i or payoff function g_i, we mean that that is the case at some specific state s of the world. Some of these ideas are illustrated in Section 3.

We end this section with a lemma that is needed in the sequel.

Lemma 2.6. *Player i knows that he attributes probability π to an event E if and only if he indeed attributes probability π to E.*

Proof.   *If:* Let F be the event that i attributes probability π to E; that is, $F := \{t \in S : p(E; t_i) = \pi\}$. Thus, $s \in F$ if and only if $p(E; s_i) = \pi$.

---

[16]Note for specialists: We do not use "mutual absolute continuity."

Therefore, if $s \in F$, then all states u with $u_i = s_i$ are in F, and so $p(F; s_i) = 1$; that is, i knows F at s.

Only if: Suppose that i attributes probability $\rho \neq \pi$ to E. By the "if" part of the proof, he must know this, contrary to his knowing that he attributes probability π to E. $\qquad\square$

3. An Illustration

Consider a belief system in which all types of each player i have the same payoff function g_i, namely that depicted in Figure 1. Thus, the game being played is commonly known. Call the row and column players (Players 1 and 2) "Rowena" and "Colin" respectively. The theories are depicted in Figure 2; here C_1 denotes a type of Rowena whose action is C, whereas D_1 and D_2 denote two different types of Rowena whose actions are D. Similarly for Colin. Each square denotes a state, i.e., a pair of types. The two entries in each square denote the probabilities that the corresponding types of Rowena and Colin ascribe to that state. For example, Colin's type d_2 attributes $\frac{1}{2}$–$\frac{1}{2}$ probabilities to Rowena's type being D_1 or D_2. So at the state (D_2, d_2), he knows that Rowena will choose the action D. Similarly, Rowena knows at (D_2, d_2) that Colin will choose d. Since d and D are optimal against each other, both players are rational at (D_2, d_2) and (D, d) is a Nash equilibrium.

We have here a typical instance of the preliminary observation. At (D_2, d_2), there is mutual knowledge of the actions D and d, and both players are in fact rational. But the actions are not common knowledge. Though

	c	d
C	2, 2	0, 0
D	0, 0	1, 1

Figure 1.

	c_1	d_1	d_2
C_1	$\frac{1}{2}, \frac{1}{2}$	$\frac{1}{2}, \frac{1}{2}$	0, 0
D_1	$\frac{1}{2}, \frac{1}{2}$	0, 0	$\frac{1}{2}, \frac{1}{2}$
D_2	0, 0	$\frac{1}{2}, \frac{1}{2}$	$\frac{1}{2}, \frac{1}{2}$

Figure 2.

Colin knows that Rowena will play D, she does not know that he knows this; indeed, she attributes probability $\frac{1}{2}$ to his attributing probability $\frac{1}{2}$ to her playing C. Moreover, though both players are rational at (D_2, d_2), there is not even mutual knowledge of rationality there. For example, Colin's type d_1 chooses d, with an expected payoff of $\frac{1}{2}$, rather than c, with an expected payoff of 1; thus, this type is irrational. At (D_2, d_2), Rowena attributes probability $\frac{1}{2}$ to Colin being of this irrational type.

Note that the players have a common prior, which assigns probability $1/6$ to each of the six boxes not containing 0's. This, however, is not relevant to the above discussion.

4. Formal Statements and Proofs of the Results

We now state and prove Theorems A and B formally. For more transparent formulations, see Section 1. We also supply, for the record, a precise, unambiguous formulation of the preliminary observation.

Preliminary Observation: *Let a be an n-tuple of actions. Suppose that at some state s, all players are rational, and it is mutually known that $\mathbf{a} = a$. Then a is a Nash equilibrium.*

Theorem A: *With $n = 2$ (two players), let g be a game, ϕ a pair of conjectures. Suppose that at some state, it is mutually known that $\mathbf{g} = g$, that the players are rational, and that $\boldsymbol{\phi} = \phi$. Then (ϕ^2, ϕ^1) is a Nash equilibrium of g.*

The proof of Theorem A uses two lemmas.

Lemma 4.1. *Let ϕ be an n-tuple of conjectures. Suppose that at some state s, it is mutually known that $\boldsymbol{\phi} = \phi$. Then, $\boldsymbol{\phi}(s) = \phi$. (In words: if it is mutually known that the conjectures are ϕ, then they are indeed ϕ.)*

Proof.   Follows from Lemma 2.6.   □

Lemma 4.2. *Let g be a game, ϕ an n-tuple of conjectures. Suppose that at some state s, it is mutually known that $\mathbf{g} = g$, that the players are rational, and that $\boldsymbol{\phi} = \phi$. Let a_j be an action of a player j to which the conjecture ϕ^i of some other player i assigns positive probability. Then, a_j maximizes g_j against[17] ϕ^j.*

---

[17]That is, $\exp g_j(a_j, a^{-j}) \geq \exp g_j(b_j, a^{-j})$ for all b_j in A_j, when a^{-j} is distributed according to ϕ^j.

Proof. By Lemma 4.1, the conjecture of i at s is ϕ^i. So i attributes positive probability at s to $[a_j]$. Also, i attributes probability 1 at s to each of the three events $[j$ is rational$]$, $[\phi^j]$, and $[g_j]$. When one of four events has positive probability, and the other three each have probability 1, then their intersection is nonempty. So there is a state t at which all four events obtain: j is rational, he chooses a_j, his conjecture is ϕ^j, and his payoff function is g_j. So a_j maximizes g_j against ϕ^j. $\square$

Proof of Theorem A: By Lemma 4.2, every action a_1 with positive probability in ϕ^2 is optimal against ϕ^1 in g, and every action a_2 with positive probability in ϕ^1 is optimal against ϕ^2 in g. This implies that (ϕ^2, ϕ^1) is a Nash equilibrium of g. $\square$

Theorem B: *Let g be a game, ϕ an n-tuple of conjectures. Suppose that the players have a common prior, which assigns positive probability to it being mutually known that $\mathbf{g} = g$, mutually known that all players are rational, and commonly known that $\boldsymbol{\phi} = \phi$. Then for each j, all the conjectures ϕ^i of players i other than j induce the same conjecture σ_j about j, and $(\sigma_1, \ldots, \sigma_n)$ is a Nash equilibrium of g.*

The proof requires several more lemmas. Some of these are standard when "knowledge" means absolute certainty, but not quite as well known when it means probability 1 belief, as here.

Lemma 4.3. $K_i(E_1 \cap E_2 \cap \ldots) = K_i E_1 \cap K_i E_2 \cap \ldots$ *(a player knows each of several events if and only if he knows that they all obtain).*

Proof. At s, player i ascribes probability 1 to $E_1 \cap E_2 \cap \ldots$ if and only if he ascribes probability 1 to each of $E_1, E_2, \ldots$. $\square$

Lemma 4.4. $CKE \subset K_i CKE$ *(if something is commonly known, then each player knows that it is commonly known).*

Proof. Since $K_i K^1 F \supset K^1 K^1 F$ for all F, Lemma 4.3 yields $K_i CKE = K_i(K^1 E \cap K^1 K^1 E \cap \ldots) = K_i K^1 E \cap K_i K^1 K^1 E \cap \ldots \supset K^1 K^1 E \cap K^1 K^1 K^1 E \cap \ldots \supset CKE$. $\square$

Lemma 4.5. *Suppose P is a common prior, $K_i H \supset H$, and $p(E; s_i) = \pi$ for all $s \in$ H. Then, $P(E \cap H) = \pi P(H)$.*

Proof. Let H_i be the projection of H on S_i. From $K_i H \supset H$ it follows that $p(H; s_i) = 1$ or 0 according as to whether s_i is or is not[18] in H_i. So when

---

[18] In particular, i always knows whether or not H obtains.

$s_i \in H_i$, then $p(E \cap H; s_i) = p(E; s_i) = \pi = \pi p(H; s_i)$; and when $s_i \notin H_i$, then $p(E \cap H; s_i) = 0 = \pi p(H; s_i)$. The lemma now follows from (2.5). □

Lemma 4.6. *Let Q be a probability distribution on A with[19] $Q(a) = Q(a_i)Q(a^{-i})$ for all a in A and all i. Then, $Q(a) = Q(a_1) \ldots Q(a_n)$ for all a.*

Proof. By induction. For $n = 1$ and 2 the result is immediate. Suppose it is true for $n - 1$. From $Q(a) = Q(a_1)Q(a^{-1})$ we obtain, by summing over a_n, that $Q(a^{-n}) = Q(a_1)Q(a_2, \ldots, a_{n-1})$. Similarly, $Q(a^{-n}) = Q(a_i) Q(a_1, \ldots, a_{i-1}, a_{i+1}, \ldots, a_{n-1})$ whenever $i < n$. So the induction hypothesis yields $Q(a^{-n}) = Q(a_1)Q(a_2) \ldots Q(a_{n-1})$. Hence, $Q(a) = Q(a^{-n})Q(a_n) = Q(a_1)Q(a_2) \ldots Q(a_n)$. □

Proof of Theorem B: Set $F := CK[\phi]$, and let P be the common prior. By assumption, $P(F) > 0$. Set $Q(a) := P([a]|F)$. We show that for all a and i,

(4.7) $Q(a) = Q(a_i)Q(a^{-i})$.

Set $H := [a_i] \cap F$. By Lemmas 4.3 and 4.4, $K_i H \supset H$, since i knows his own action. If $s \in H$, it is commonly, and so mutually, known at s that $\phi = \phi$; so by Lemma 4.1, $\phi(s) = \phi$; that is, $p([a^{-i}]; s_i) = \phi^i(a^{-i})$. So Lemma 4.5 (with $E = [a^{-i}]$) yields $P([a] \cap F) = P([a^{-i}] \cap H) = \phi^i(a^{-i})P(H) = \phi^i(a^{-i})P([a_i] \cap F)$. Dividing by $P(F)$ yields $Q(a) = \phi^i(a^{-i})Q(a_i)$; then summing over a_i, we get

(4.8) $Q(a^{-i}) = \phi^i(a^{-i})$.

Thus, $Q(a) = Q(a^{-i})Q(a_i)$, which is (4.7).

For each j, define a probability distribution σ_j on A_j by $\sigma_j(a_j) := Q(a_j)$. Then (4.8) yields $\phi^i(a_j) = Q(a_j) = \sigma_j(a_j)$ for $j \neq i$. Thus, for all i, the conjecture about j induced by ϕ^i is σ_j, which does not depend on i. Lemma 4.6, (4.7), and (4.8) then yield

(4.9) $\phi^i(a^{-i}) = \sigma_1(a_1) \ldots \sigma_{i-1}(a_{i-1})\sigma_{i+1}(a_{i+1}) \ldots \sigma_n(a_n)$;

that is, the distribution ϕ^i is the product of the distributions σ_j with $j \neq i$.

Since common knowledge implies mutual knowledge, the hypothesis of the theorem implies that there is a state at which it is mutually known

---

[19] We denote $Q(a^{-i}) := Q(A_i \times \{a^{-i}\})$, $Q(a_i) := Q(A^{-i} \times \{a_i\})$, and so on.

that $\mathbf{g} = g$, that the players are rational, and that $\phi = \phi$. So by Lemma 4.2, each action a_j with $\phi^i(a_j) > 0$ for some $i \neq j$ maximizes g_j against ϕ^j. By (4.9), these a_j are precisely the ones that appear with positive probability in σ_j. Again using (4.9), we conclude that each action appearing with positive probability in σ_j maxmizes g_j against the product of the distributions σ_k with $k \neq j$. This implies that $(\sigma_1, \ldots, \sigma_n)$ is a Nash equilibrium of g.

5. Tightness of the Results

This section explores possible variations on Theorem B. For simplicity, let $n = 3$ (three players). Each player's "overall" conjecture is then a distribution on pairs of actions of the other two players; so the three conjectures form a triple of probability mixtures of action pairs. On the other hand, an equilibrium is a triple of mixed actions. Our discussion hinges on the relation between these two kinds of objects.

First, since our real concern is with mixtures of *actions* rather than of action *pairs*, could we not formulate conditions that deal directly with each player's "individual" conjectures — his conjectures about each of the other players — rather than with his overall conjecture? For example, one might hope that it would be sufficient to assume common knowledge of each player's individual conjectures.

Example 5.1 shows that this hope is vain, even when there is a common prior, and rationality and payoff functions are commonly known. Overall conjectures do play an essential role.

Nevertheless, *common* knowledge of the overall conjectures seems a rather strong assumption. Could not we get away with less — say, with mutual knowledge of the overall conjectures, or with mutual knowledge to a high order?[20]

Again, the answer is no. Recall that people with a common prior but with different information may disagree on their posterior probabilities for some event E, even though these posteriors are mutually known to an arbitrarily high order (Geanakoplos and Polemarchakis [1982]). Using this, one may construct an example with arbitrarily high order mutual knowledge of the overall conjectures, common knowledge of rationality and payoff functions, and a common prior, where different players have different

---

[20]Set $K^2E := K^1K^1E$, $K^3E := K^1K^2E$, and so on. If $s \in K^mE$, call E *mutually known to order m at s*.

individual conjectures about some particular player j. Thus, there is not even a clear *candidate* for a Nash equilibrium.[21]

The question remains whether (sufficiently high order) mutual knowledge of the overall conjectures implies Nash equilibrium of the individual conjectures when the players do happen to agree on them. Do we then get Nash equilibrium? Again, the answer is no; this is shown in Example 5.2.

Finally, Example 5.3 shows that the common prior assumption is really needed: Rationality, payoff functions, and the overall conjectures are commonly known, and the individual conjectures agree; but there is no common prior, and the agreed-upon individual conjectures do *not* form a Nash equilibrium.

Summing up, one *must* consider the overall conjectures; and nothing less than common knowledge of these conjectures, together with a common prior, will do.

Also, one may construct examples showing that in Theorems A and B, mutual knowledge of rationality cannot be replaced by the simple fact of rationality, and that knowing one's own payoff function does not suffice — all payoff functions must be mutually known.

Except in Example 5.3, the belief systems in this section have common priors, and these are used to describe them. In all the examples, the game being played is (as in Section 3) fixed throughout the belief system, and so is commonly known. Each example has three players, Rowena, Colin, and Matt, who choose the row, column, and matrix (west or east), respectively. As in Section 3, each type is denoted by the same letter as its action, and a subscript is added.

Example 5.1. Here the individual conjectures are commonly known and agreed upon, rationality is commonly known, and there is a common prior, and yet we do not get Nash equilibrium.[22] Consider the game of Figure 3, with theories induced by the common prior in Figure 4. At each state, Colin and Matt agree on the conjecture $\frac{1}{2}U + \frac{1}{2}D$ about

---

[21]This is *not* what drives Example 5.1; since the individual conjectures are commonly known there, they must agree (Aumann [1976]).

[22]Examples 2.5, 2.6, and 2.7 of Aumann (1974) display correlated equilibria that are not Nash, but they are quite different from Example 5.1. First, the context there is *global*, as opposed to the *local* context considered here (Section 7i). Second, even if we do adapt those examples to the local context, we find that the individual conjectures are not even mutually known, to say nothing of being commonly known; and when there are more than two players (Examples 2.5 and 2.6), the individual conjectures are not agreed upon either.

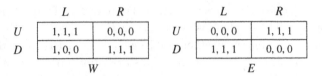

Figure 3.

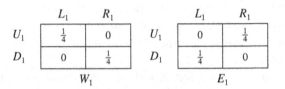

Figure 4.

Rowena, and this is commonly known. Similarly, it is commonly known that Rowena and Matt agree on the conjecture $\frac{1}{2}L + \frac{1}{2}R$ about Colin, and that Rowena and Colin agree on $\frac{1}{2}W + \frac{1}{2}E$ about Matt. All players are rational at all states, so rationality is common knowledge at all states. But $(\frac{1}{2}U + \frac{1}{2}D, \frac{1}{2}L + \frac{1}{2}R, \frac{1}{2}W + \frac{1}{2}E)$ is not a Nash equilibrium, because if these were independent mixed strategies, Rowena could gain by moving to D.

Note that the overall conjectures are not commonly (nor even mutually) known at any state. For example, at (U_1, L_1, W_1), Rowena's conjecture is $(\frac{1}{2}LW + \frac{1}{2}RE)$, but nobody else knows that that is her conjecture.

Example 5.2. Here we have mutual knowledge of the overall conjectures, agreement of individual conjectures, common knowledge of rationality, and a common prior, and yet the individual conjectures do not form a Nash equilibrium. Consider the game of Figure 5. For Rowena and Colin, this is simply "matching pennies"; their payoffs are not affected by Matt's choice. So at a Nash equilibrium, they must play $\frac{1}{2}H + \frac{1}{2}T$ and $\frac{1}{2}h + \frac{1}{2}t$ respectively. Thus, Matt's expected payoff is $\frac{3}{2}$ for W, and 2 for E; so he must play E. Hence, $(\frac{1}{2}H + \frac{1}{2}T, \frac{1}{2}h + \frac{1}{2}t, E)$ is the unique Nash equilibrium of this game.

Consider now the theories induced by the common prior in Figure 6. Rowena and Colin know which of the three large boxes contains the true state, and in fact this is commonly known between the two of them. In each box, Rowena and Colin "play matching pennies optimally"; their conjectures about each other are $\frac{1}{2}H + \frac{1}{2}T$ and $\frac{1}{2}h + \frac{1}{2}t$. Since these conjectures obtain at each state, they are commonly known (among all

	h	t		h	t
H	1, 0, 3	0, 1, 0	H	1, 0, 2	0, 1, 2
T	0, 1, 0	1, 0, 3	T	0, 1, 2	1, 0, 2
	W			E	

Figure 5.

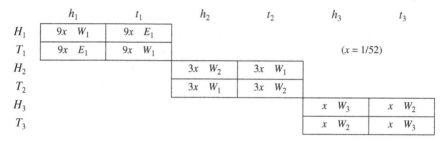

	h_1	t_1	h_2	t_2	h_3	t_3
H_1	$9x$ W_1	$9x$ E_1				
T_1	$9x$ E_1	$9x$ W_1		$(x = 1/52)$		
H_2			$3x$ W_2	$3x$ W_1		
T_2			$3x$ W_1	$3x$ W_2		
H_3					x W_3	x W_2
T_3					x W_2	x W_3

Figure 6.

three players); so it is also commonly known that Rowena and Colin are rational.

As for Matt, suppose first that he is of type W_1 or W_2. Each of these types intersects two adjacent boxes in Figure 6; it consists of the diagonal states in the left box and the off-diagonal ones in the right box. The diagonal states on the left have equal probability, as do the off-diagonal ones on the right; but on the left it is three times on the right. So Matt assigns the diagonal states three times the probability of the off-diagonal states; i.e., his conjecture is $\frac{3}{8}Hh + \frac{3}{8}Tt + \frac{1}{8}Th + \frac{1}{8}Ht$. Therefore, his expected payoff from choosing W is $\frac{3}{8} \cdot 3 + \frac{3}{8} \cdot 3 + \frac{1}{8} \cdot 0 + \frac{1}{8} \cdot 0 = 2\frac{1}{4}$, whereas from E it is only 2 (as all his payoffs in the eastern matrix are 2). So W is indeed the optimal action of these types; so they are rational. It may be checked that also E_1 and W_3 are rational. Thus, the rationality of all players is commonly known at all states.

Consider now the state $s := (H_2, h_2, W_2)$ (the top left state in the middle box). Rowena and Colin know at s that they are in the middle box, so they know that Matt's type is W_1 or W_2. We have just seen that these two types have the same conjecture, so it follows that Matt's conjecture is mutually known at s. Also Rowena's and Colin's conjectures are mutually known at s (Rowena's is $\frac{1}{2}hW + \frac{1}{2}tW$, Colin's is $\frac{1}{2}HW + \frac{1}{2}TW$).

Finally, the individual conjectures derived from Matt's overall conjecture $\frac{3}{8}Hh + \frac{3}{8}Tt + \frac{1}{8}Th + \frac{1}{8}Ht$ are $\frac{1}{2}H + \frac{1}{2}T$ for Rowena and $\frac{1}{2}h + \frac{1}{2}t$ for Colin. These are the same as Rowena's and Colin's conjectures about each other. Since Matt plays W throughout the middle box, both Rowena and Colin conjecture W for Colin there. Thus, throughout the middle box, individual conjectures are agreed upon.

To sum up: There is a common prior; at all states, the game is commonly known and all players are commonly known to be rational. At the top left state in the middle box, the overall conjectures of all players are mutually known, and the individual conjectures are agreed: $\sigma_R = \frac{1}{2}H + \frac{1}{2}T$, $\sigma_c = \frac{1}{2}h + \frac{1}{2}t$, $\sigma_M = W$. But $(\sigma_R, \sigma_C, \sigma_M)$ is not a Nash equilibrium.

One can construct similar examples in which the mutual knowledge of the conjectures is of arbitrarily high order, simply by using more boxes; the result follows as before.

Example 5.3. Here we show that one cannot dispense with common priors in Theorem B. Consider again the game of Figure 5, with the theories depicted in Figure 7 (presented in the style of Figure 2). At each state there is common knowledge of rationality, of the overall conjectures (which are the same as in the previous example), and of the game. As before, individual conjectures are in agreement. And as before, the individual conjectures $(\frac{1}{2}H + \frac{1}{2}T, \frac{1}{2}h + \frac{1}{2}t, W)$ do not constitute a Nash equilibrium.

6. General (Infinite) Belief Systems

For a general definition of a belief system, which allows it to be infinite,[23] we specify that the type spaces S_i be measurable spaces. As before, a *theory* is a probability measure on $S^{-i} = \times_{j \neq i} S_j$, which is now endowed with the

	h_1	t_1
H_1	$\frac{1}{2}, \frac{1}{2}, \frac{3}{8}$	$\frac{1}{2}, \frac{1}{2}, \frac{1}{8}$
T_1	$\frac{1}{2}, \frac{1}{2}, \frac{1}{8}$	$\frac{1}{2}, \frac{1}{2}, \frac{3}{8}$

$$W_1$$

Figure 7.

---

[23]Infinite belief systems are essential for a complete treatment: The belief system required to encode a given coherent hierarchy of beliefs (see footnote 14) is often uncountably infinite.

standard product structure.[24] The state space $S = \times_j S_j$, too, is endowed with the product structure. An *event* is now a measurable subset of S. The "action functions" $\mathbf{a}_i$ ((2.3)) are assumed measurable; so are the payoff functions $\mathbf{g}_i$ ((2.4)), as functions of s_i, for each action n-tuple a separately. Also the "theory functions" ((2.2)) are assumed measurable, in the sense that for each event E and player i, the probability $p(E; s_i)$ is measurable as a function of the type s_i. It follows that the conjectures ϕ^i are also measurable functions of s_i.

With these definitions, the statements of the results make sense, and the proofs remain correct, without any change.

7. Discussion

a. *Belief Systems.* An interactive belief system is not a prescriptive model; it does not suggest actions to the players. Rather, it is a formal framework — a language — for *talking* about actions, payoffs, and beliefs. For example, it enables us to say whether a given player is behaving rationally at a given state, whether this is known to another player, and so on. But it does not prescribe or even suggest rationality; the players do whatever they do. Like the disk operating system of a personal computer, the belief system simply organizes things, so that we can coherently discuss what they do.

Though entirely apt, use of the term "state of the world" to include the actions of the players has perhaps caused confusion. In Savage (1954), the decision maker cannot affect the state; he can only react to it. While convenient in Savage's one-person context, this is not appropriate in the interactive, many-person world under study here. To describe the state of such a world, it is fitting to consider all the players simultaneously; and then since each player must take into account the actions of the others, the actions should be included in the description of the state. Also the plain, everyday meaning of the term "state of the world" includes one's actions: Our world is shaped by what we do.

It has been objected that since the players' actions are determined by the state, they have no freedom of choice. But this is a misunderstanding. Each player may do whatever he wants. It is simply that whatever he does is part of the description of the state. If he wishes to do something else, he is heartily welcome to it; but then the state is different.

---

[24]The σ-field of measurable sets is the smallest σ-field containing all the "rectangles" $\times_{j \neq i} T_j$, where T_j is measurable in S_j.

Though including one's own action in the state is not a new idea,[25] it may still leave some readers uncomfortable. Perhaps this discomfort stems from a notion that actions should be part of the solution, whereas, including them in the state might suggest that they are part of the problem.

The "problem-solution" viewpoint is the older, classical approach of game theory. The viewpoint adopted here is different — it is *descriptive*. Not *why* the players do what they do, not what *should* they do; just what *do* they do, what *do* they believe. Are they rational, are they irrational, are their actions — or beliefs — in equilibrium? Not "why," not "should," just *what*. Not that *i* does *a because* he believes *b*; simply that he does *a*, and believes *b*.

The idea of belief system is due to John Harsanyi (1967–68), who introduced the concept of *I-game* to enable a coherent formulation of games in which the players need not know each other's payoff functions. I-games are just like belief systems, except that in I-games a player's type does not determine his action (only his payoff function).

As indicated, belief systems are primarily a convenient framework to enable *us* — the analysts — to discuss the things we want to discuss: actions, payoffs, beliefs, rationality, equilibrium, and so on. As for the players themselves, it is not clear that they need concern themselves with the structure of the model. But if they, too, want to talk about the things that we want to talk about, that is OK; it is just as convenient a framework for them as it is for us. In this connection, we note that the belief system itself may always be considered common knowledge among the players. Formally, this follows from the work of Mertens and Zamir (1985); for an informal discussion, see Aumann (1987a, p. 9 ff.).

b. *Knowledge and Belief:* In this chapter, "know" means "ascribe probability 1 to." This is sometimes called "believe," while "know" is reserved for absolute certainty with no possibility at all for error. Since our conditions are sufficient, the results are stronger with probability 1 than with absolute certainty. If probability 1 knowledge of certain events implies that σ is a Nash equilibrium, then a fortiori, so does absolute certainty of those events.

c. *Knowledge of One's Own Type:* It is implicit in our set-up that each player i knows his own type s_i — that is, he knows his theory, his payoff function g_i, and his action a_i.

---

[25] See, e.g., Aumann (1987a).

Knowledge of one's theory is not a substantive restriction; the theory consists of beliefs, and it is tautologous that one knows what one's beliefs are (a formal expression of this is Lemma 2.6).

Knowledge of one's payoff function is a more subtle matter. On the face of it, it would seem quite possible for a player's payoff to depend on circumstances known to others but not to himself. In our set-up, this could be expressed by saying that a player's payoff might depend on the types of other players as well as his own. To avoid this, one may interpret $g_i(a)$ as expressing the payoff that i *expects* when a is played, rather than what he actually gets. And since one always knows one's own expectation, one may as well construct the system so that knowledge of one's own payoff is tautological.

We come finally to knowledge of one's own action. If one thinks of actions as conscious choices, as we do here, this is very natural — one might almost say tautologous. That players are aware of — "know" — their own conscious choices is implicit in the word "conscious."

Of course, if explicit randomization is allowed, then the players need *not* be aware of their own pure actions. But even then, they *are* aware of the mixtures they choose; so *mutatis mutandis*, our analysis applies to the mixtures. See Section 1 for a brief discussion of this case; it is not our main concern here, where we think of i's mixed actions as representing the beliefs of other players about what i will do.

d. *Knowledge of Conjectures*: Both our theorems assume some form of knowledge (mutual or common) of the players' *conjectures*. Though knowledge of what others will *do* is undoubtedly a strong assumption, one can imagine circumstances in which it would obtain. But can one know what others *think?* And if so, can this happen in contexts of economic interest?

In fact, it might happen in several ways. One has to do with players who are members of well-defined economic populations, like insurance companies and customers, or sellers and buyers in general. For example, someone is buying a car. She knows that the salesman has statistical information about customers' bargaining behavior, and she even knows what that statistical information is. So she knows the salesman's conjecture about her. The conjecture may even be commonly known by the two players. But it is more likely that though the customer knows the salesman's conjecture about her, she does not know that he knows that she knows, and indeed perhaps he does not; then the knowledge of the salesman's conjecture is only mutual.

No doubt, this story has its pros and cons; we do not want to make too much of it. It is meant only to show that a player may well know another's conjecture in situations of economic interest.

e. *Knowledge of Equilibrium:* Our results state that a specified (mixed) strategy n-tuple σ is an equilibrium; they do not state that the players know it to be an equilibrium, or that this is commonly known. In Theorems A and B, though, it is in fact mutual knowledge to order 1 — but not necessarily to any higher order — that σ is a Nash equilibrium. In the preliminary observation, it need not even be mutual knowledge to order 1 that σ is a Nash equilibrium; but this does follow if, in addition to the stated assumptions, one assumes mutual knowledge of the payoff functions.

f. *Common Knowledge of the Model:* Binmore and Brandenburger (1990, p. 119) have written that "in game theory, it is typically understood that the structure of the game... is common knowledge." In the same vein, Aumann (1987b, p. 473) has written that "the common knowledge assumption underlies all of game theory and much of economic theory. Whatever be the model under discussion,... the model itself must be common knowledge; otherwise the model is insufficiently specified, and the analysis incoherent." This seemed sound when written, but in the light of recent developments — including the present work[26] — it no longer does. Admittedly, we do use a belief system, which is, in fact, commonly known. But the belief system is not a "model" in the sense of being exogenously given; it is merely a language for discussing the situation (see Section 7a). There is nothing about the real world that must be commonly known among the players. When writing the above, we thought that some real exogenous framework must be commonly known; this no longer seems appropriate.

g. *Independent Conjectures:* The proof of Theorem B implies that the individual conjectures of each player i about the other players j are independent. Alternatively, one could *assume* independence, as in the following:

Remark 7.1. Let σ be an n-tuple of mixed strategies. Suppose that at some state, it is mutually known that the players are rational, that the game g is being played, that the conjecture of each player i about each other player j is σ_j, and that it is independent of i's conjecture about all other players. Then, σ is a Nash equilibrium in g.

---

[26] Specifically, that common knowledge of the payoff functions plays no role in our theorems.

Here, we assume mutual rather than common knowledge of conjectures and do not assume a common prior. On the other hand, we assume outright that the individual conjectures are agreed upon, and that each player's conjectures about the others are independent. We consider this result of limited interest in the context of this chapter; neither assumption has the epistemic flavor that we are seeking. Moreover, in the current subjectivist context, we find independence dubious as an assumption (though not necessarily as a conclusion). See Aumann (1987a, p. 16).

h. *Converses*: We have already mentioned (at the end of Section 1) that our conditions are not necessary, in the sense that it is quite possible to have a Nash equilibrium even when they are not fulfilled. Nevertheless, there is a sense in which the converses hold: Given a Nash equilibrium in a game g, one can construct a belief system in which the conditions are fulfilled. For the preliminary observation, this is immediate: Choose a belief system where each player i has just one type, whose action is i's component of the equilibrium and whose payoff function is g_i. For Theorems A and B, we may suppose that as in the traditional interpretation of mixed strategies, each player chooses an action by an independent conscious randomization according to his component σ_i of the given equilibrium σ. The types of each player correspond to the different possible outcomes of the randomization; each type chooses a different action. All types of player i have the same theory, namely, the product of the mixed strategies of the other $n - 1$ players appearing in σ, and the same payoff function, namely g_i. It may then be verified that the conditions of Theorems A and B are met.

These "converses" show that the sufficient conditions for Nash equilibrium in our theorems are not too strong, in the sense that they do not imply more than Nash equilibrium; every Nash equilibrium is attainable with these conditions. Another sense in which they are not too strong — that the conditions cannot be dispensed with or even appreciably weakened — was discussed in Section 5.

i. *Related Work:* This chapter joins a growing epistemic literature in noncooperative game theory. For two-person games, Tan and Werlang (1988) show that if the players' payoff functions, conjectures, and rationality are all commonly known, then the conjectures constitute a Nash equilibrium.[27] Brandenburger and Dekel (1989) take this further. They ask,

---

[27]This is a restatement of their result in our formalism and terminology, which differ substantially from theirs. In particular, two players' "knowing each other" in the sense of Tan and Werlang implies that in our sense, their conjectures are common knowledge.

"when is Nash equilibrium equivalent to common knowledge of rationality? When do these two basic ideas, one from game theory and the other from decision theory, coincide?" The answer they provide is that in two-person games, a sufficient condition for this is that the payoff functions and conjectures be commonly known.[28] That is, *if* the payoff functions and the conjectures are commonly known, *then* rationality is commonly known if and only if the conjectures constitute a Nash equilibrium. The "only if" part of this is precisely the above result of Tan–Werlang.[29,30]

Our Theorem A improves on the Tan–Werlang result in that it assumes only mutual knowledge where they assume common knowledge.

Aumann (1987a) is also part of the epistemic literature, but the question it addresses is quite different from that of this chapter. It asks about the distribution of action profiles over the entire state space when all players are assumed rational at all states of the world and there is a common prior. The answer is that it represents a correlated (not Nash!) equilibrium. Conceptually, that paper takes a *global* point of view; its result concerns the distribution of action profiles as a whole. Correlated equilibrium itself is an essentially global concept; it has no natural local formulation. In contrast, the viewpoint of the current chapter is *local*. It concerns the information of the players, at some specific state of the world; and it asks whether the players' actions or conjectures *at that state* constitute a Nash equilibrium. Matters like knowledge that is mutual but not common, or players who are rational but may ascribe irrationality to one another, do not come under the purview of Aumann (1987a).

---

[28] For example, if there is an accepted convention as to how to act in a certain situation (like driving on the right), then that convention constitutes a Nash equilibrium if and only if it is commonly known that everyone is acting rationally.

[29] We may add that Armbruster and Böge (1979) also treat two-person games in this spirit.

[30] One may ask whether our theorems can be extended to equivalence results in the style of Brandenburger–Dekel. The answer is yes. For two-person games, it can be shown that if the payoff functions and the conjectures are mutually known, then rationality is mutually known if and only if the conjectures constitute a Nash equilibrium; this extends Theorem A. Theorem B may be extended in a similar fashion, as may the preliminary observation. The "only if" parts of these extensions coincide with the theorems established in the body of this chapter.

The "if" parts of these extensions are interesting, but perhaps not as much as the "if" part of the Brandenburger–Dekel result: Mutual knowledge of rationality is weaker than common knowledge, and so it is less appealing as a conclusion. That is one reason that we have not made more of these extensions.

Brandenburger and Dekel (1987, Proposition 4.1) derive Nash equilibrium in n-person games from independence assumptions, as we do[31] in Proposition 7.1. As we have already noted (Section 7g), such assumptions lack the epistemic flavor that interests us here.

In brief: The aim of this chapter has been to identify sufficient epistemic conditions for Nash equilibrium that are as spare as possible; to isolate just what the players themselves must know in order for their conjectures about each other to constitute a Nash equilibrium. For two-person games, our Theorem A goes significantly beyond that done previously in the literature on this issue. For n-person games, little had been done before.

References

Armbruster, W and W Böge (1979). Bayesian game theory. In Moeschlin, O and D Pallaschke (Eds.), *Game Theory and Related Topics*. Amsterdam: North-Holland.

Arrow, K (1986). Rationality of self and others in an economic system. *Journal of Business*, 59, S385–S399.

Aumann, R (1974). Subjectivity and correlation in randomized strategies. *Journal of Mathematical Economics*, 1, 67–96.

Aumann, R (1976). Agreeing to disagree. *Annals of Statistics*, 4, 1236–1239.

Aumann, R (1987a). Correlated equilibrium as an expression of Bayesian rationality. *Econometrica*, 55, 1–18.

Aumann, R (1987b). Game theory. In Eatwell, J, M Milgate and P Newman (Eds.), *The New Palgrave: A Dictionary of Economics*. London, UK: MacMillan.

Binmore, K (1990). *Essays on the Foundations of Game Theory*. Oxford, UK: Basil Blackwell.

Binmore, K (1992). *Fun and Games*. Lexington: D.C. Heath and Company.

Binmore, K and A Brandenburger (1990). Common knowledge and game theory. In Binmore, K (Ed.), *Essays on the Foundation of Game Theory*, pp. 105–150. Oxford, UK: Basil Blackwell.

Binmore, K and P Dasgupta (1986). Game theory: A survey. In Binmore, K and P Dasgupta (Eds.), *Economic Organizations as Games*, pp. 1–45. Oxford, UK: Basil Blackwell.

Böge, W and Th Eisele (1979). On solutions of Bayesian games. *International Journal of Game Theory*, 8, 193–215.

---

[31] Though their result (henceforth B&D4) differs from our Proposition 7.1 (hereforth P7) in several ways. First, B&D4 makes an assumption tantamount to common knowledge of conjectures, while P7 asks only for mutual knowledge. Second, P7 directly assumes agreement among the individual conjectures, which B&D4 does not. Finally, B&D4 requires "concordant" priors (a weaker form of common priors), while P7 does not.

Brandenburger, A and E Dekel (1987). Rationalizability and correlated equilibria. *Econometrica*, 55, 1391–1402.

Brandenburger, A and E Dekel (1989). The role of common knowledge assumptions in game theory. In Hahn, F (Ed.), *The Economics of Missing Markets, Information, and Games*, pp. 46–61. Oxford, UK: Oxford University Press.

Brandenburger, A and E Dekel (1993). Hierarchies of beliefs and common knowledge. *Journal of Economic Theory*, 59, 189–198.

Fudenberg, D and D Kreps (1988). A theory of learning, experimentation, and equilibrium in games. Unpublished, Graduate School of Business, Stanford University.

Fudenberg, D and J Tirole (1989). Noncooperative game theory for industrial organization: an introduction and overview. In Schmalensee, R and R Willig (Eds.), *Handbook of Industrial Organization*, Vol. 1, pp. 259–327. Amsterdam: North-Holland.

Geanakoplos, J, D Pearce, and E Stacchetti (1989). Psychological games and sequential rationality. *Games and Economic Behavior*, 1, 60–79.

Geanakoplos, J and H Polemarchakis (1982). We can't disagree forever. *Journal of Economic Theory*, 28, 192–200.

Harsanyi, J (1967–68). Games with incomplete information played by 'Bayesian' players. Parts I–III, *Management Science*, 8, 159–182, 320–334, 486–502.

Harsanyi, J (1973). Games with randomly disturbed payoffs: A new rationale for mixed strategy equilibrium points. *International Journal of Game Theory*, 2, 1–23.

Kreps, D and R Wilson (1982). Sequential equilibria. *Econometrica*, 50, 863–894.

Lewis, D (1969). *Convention: A Philosophical Study*. Cambridge, MA: Harvard University Press.

Mertens, J-F and S Zamir (1985). Formulation of Bayesian analysis for games with incomplete information. *International Journal of Game Theory*, 14, 1–29.

Milgrom, P and J Roberts (1991). Adaptive and sophisticated learning in normal form games. *Games and Economic Behavior*, 3, 82–100.

Nash, J (1951). Non-cooperative games. *Annals of Mathematics*, 54, 286–295.

Reny, P (1992). Backward induction, normal form perfection and explicable equilibria. *Econometrica*, 60, 627–649.

Rubinstein, A (1991). Comments on the interpretation of game theory. *Econometrica*, 59, 909–924.

Savage, L (1954). *The foundations of statistics*. New York, NY: Wiley.

Tan, T and S Werlang (1988). The Bayesian foundations of solution concepts of games. *Journal of Economic Theory*, 45, 370–391.

Werlang, S (1989). Common knowledge. In Eatwell, J, M Milgate, and P Newman (Eds.), *The New Palgrave: Game Theory*, pp. 74–85. New York, NY: W. W. Norton.

Chapter 6

Lexicographic Probabilities and Choice Under Uncertainty

Lawrence Blume, Adam Brandenburger, and Eddie Dekel

Two properties of preferences and representations for choice under uncertainty which play an important role in decision theory are: (i) admissibility, the requirement that weakly dominated actions should not be chosen; and (ii) the existence of well defined conditional probabilities, that is, given any event a conditional probability which is concentrated on that event and which corresponds to the individual's preferences. The conventional Bayesian theory of choice under uncertainty, subjective expected utility (SEU) theory, fails to satisfy these properties — weakly dominated acts may be chosen, and the usual definition of conditional probabilities applies only to nonnull events. This chapter develops a non-Archimedean variant of SEU where decision makers have lexicographic beliefs; that is, there are (first-order) likely events as well as (higher-order) events which are infinitely less likely but not necessarily impossible. This generalization of preferences, from those having an SEU representation to those having a representation with lexicographic beliefs, can be made to satisfy admissibility and yield well defined conditional probabilities and at the same time to allow for "null" events. The need for a synthesis of expected utility with admissibility, and to provide a ranking of null events, has often been stressed in the decision theory literature. Furthermore,

Originally published in *Econometrica*, 59, 61–79.

Keywords: admissibility; weak dominance; conditional probabilities; lexicographic probabilities; non-Archimedean preferences; subjective expected utility.

Financial support: Harvard Business School Division of Research, Miller Institute for Basic Research in Science, NSF Grants IRI-8608964 and SES-8808133.

Acknowledgments: We wish to thank Bob Anderson, Ken Arrow, Mark Bagnoli, John Geanakoplos, Ehud Kalai, Andreu Mas-Colell, Klaus Nehring, Martin Osborne, Ket Richter, and Bill Zame for helpful comments. We are especially indebted to David Kreps and three referees.

lexicographic beliefs are appropriate for characterizing refinements of Nash equilibrium. In this chapter we discuss: axioms on, and behavioral properties of, individual preferences which characterize lexicographic beliefs; probability-theoretic properties of the representations; and the relationships with other recent extensions of Bayesian SEU theory.

1. Introduction

There are two important properties of preferences and representations for choice under uncertainty. The first is the criterion of admissibility, namely, that a decision maker should not select a weakly dominated action (Luce and Raiffa [1957, Chapter 13]). The second property is that for *any* event there is a conditional probability that is concentrated on that event and that represents the decision maker's conditional preferences given that event. We call such conditional probabilities "well defined." The importance for a complete and intuitive theory to provide conditional probabilities given any event has long been discussed in the context of probability theory[1] and philosophy.[2] Moreover, the criterion of backward induction, which specifies that at every choice node in a decision tree choices maximize expected utility with respect to "beliefs" at that node, requires the use of well defined conditional probabilities in representing conditional preferences at every node in the tree. Nonetheless, conventional subjective expected utility (SEU) theory does not satisfy these properties.[3] For both properties, the source of the problem is the same: conditional on events which are not expected to occur, SEU theory leads to trivial choice problems — all acts are indifferent.

Obviously, conventional SEU theory can be refined to satisfy admissibility and to determine well defined conditional probabilities by ruling out null events. But this method is too restrictive. For example, such preferences could not characterize pure strategy equilibria in games. In this chapter, we

---

[1]For example, de Finetti (1972, p. 82) says "there seems to be no justification... for introducing the restriction $P(H) \neq 0$," where H is a conditioning event, and Blackwell and Dubins (1975, p. 741) are concerned with when "conditional distributions... satisfy the intuitive desideratum... of being *proper*," that is, concentrated on the conditioning event.

[2]See Harper, Stalnacker, and Pearce (1981) as well as the references cited there for a discussion of the relationship between linguistic intuition, counterfactuals, and conditional probabilities.

[3]Some early work in statistical decision theory was concerned with the problematic relationship between Bayes procedures and admissible procedures. See Blackwell and Girshick (1954, Section 5.2) and Arrow, Barankin, and Blackwell (1953).

axiomatize preferences under uncertainty which both satisfy admissibility and yield well defined conditional probabilities, yet allow for events which are "null," although not in the sense of Savage (1954). We develop a non-Archimedean SEU theory starting from Fishburn's (1982) version of the SEU framework due to Anscombe and Aumann (1963). A key feature of our representation is the introduction of a lexicographic hierarchy of beliefs. Such beliefs can capture the idea that a die landing on its side is infinitely more likely than its landing on an edge, which in turn is infinitely more likely than its landing on a corner. These considerations of "unlikely" events might seem rather arcane, but are nevertheless crucial in a game-theoretic context, as the discussion suggests and as is shown in the sequel to this chapter (Blume, Brandenburger, and Dekel [1991]). In particular, a player in a game may be unwilling to exclude *entirely* from consideration any action of an opponent, and, moreover, which actions are unlikely is in a sense "endogenous" (i.e., depends on the equilibrium under consideration).

These objectives motivate our first and main departure from the axioms of SEU — weakening the Archimedean axiom. After a review of SEU in Section 2, we present a general representation theorem for preferences with the weakened Archimedean axiom in Section 3. This representation allows for events which are "null," yet are taken into consideration by the decision maker, as well as events which are null in the sense of Savage. We then strengthen the state independence axiom to rule out Savage-null events, and in Section 4 we show that these preferences satisfy admissibility and determine well defined conditional probabilities. Section 5 discusses an Archimedean axiom intermediate in strength between the standard Archimedean axiom and the Archimedean axiom of Section 3. This intermediate axiom leads to a representation of choice which is closely related to conditional probability systems (Myerson [1986a, b]). In Section 6 an alternative representation of preferences, which is equivalent to that of Section 3, is provided using infinitesimal numbers instead of a lexicographic order of vectors. Section 7 discusses the surprisingly delicate issue of modelling stochastic independence with lexicographic probabilities.

It is worth emphasizing that the notion of lexicographic beliefs arises in a very natural fashion — as a consequence of satisfying the decision-theoretic properties of admissibility and the existence of well defined conditional probabilities on all events, together with allowing for some kind of null events. Lexicographic models have been used to explore other issues in decision theory. An early example is Chipman (1960, 1971a,b), who developed and applied them to provide an alternative to the Friedman and Savage (1948) explanation of gambling and insurance purchases, and to discuss portfolio choice and other economic applications. Fishburn (1974) provides a comprehensive survey. Kreps and Wilson (1982) introduced a lexicographic method of updating beliefs in game trees in the context of sequential equilibrium.[4] Hausner (1954) and Richter (1971) provide the technical foundations for the work we present here.

2. Subjective Expected Utility on Finite State Spaces

There are two distinct approaches to the theory of subjective expected utility. In Savage's (1954) framework, individuals have preferences over acts which map a state space into consequences. Anscombe and Aumann (1963) (as well as Chernoff [1954], Suppes [1956], and Pratt, Raiffa, and Schlaifer [1964]) use axioms which refer to objective probabilities. Although the Savage (1954) framework is perhaps more appealing, for reasons of tractability and because we will want to apply our results to finite games, we employ the Anscombe and Aumann framework in Section 3 to develop our non-Archimedean SEU theory. To facilitate subsequent comparisons, a brief review of Anscombe and Aumann's SEU theory follows.

The decision maker faces a finite set of states Ω and a set of (pure) consequences C. Let $\mathscr{P}$ denote the set of simple (i.e., finite support) probability distributions on consequences. The objective lotteries in $\mathscr{P}$ provide a scale for measuring the utilities of consequences and the subjective probabilities of states. The decision maker has preferences over acts, which are maps from the state space Ω into $\mathscr{P}$. Thus, the set of acts is the product space $\mathscr{P}^{\Omega}$. The ωth coordinate of act x is denoted x_{ω}. The interpretation of an act x is that when it is chosen, the consequence for the decision maker if state ω occurs is determined by the lottery x_{ω}. The set $\mathscr{P}^{\Omega}$ is a mixture space; in particular, for $0 \le \alpha \le 1$ and $x, y \in \mathscr{P}^{\Omega}$, $\alpha x + (1 - \alpha)y$ is the act that in state ω assigns probability $\alpha x_{\omega}(c) + (1 - \alpha)y_{\omega}(c)$ to each $c \in C$. Nonempty subsets of Ω are termed events. For any event $S \subset \Omega$, x_S denotes the tuple $(x_{\omega})_{\omega \in S}$. We will denote $x_{\Omega - S}$ by x_{-S}. A constant act maps each state into the same lottery on consequences: $x_{\omega} = x_{\omega'}$ for all $\omega, \omega' \in \Omega$. For notational simplicity we often write ω to represent the event $\{\omega\} \subset \Omega$.

---

[4]The relationship between this chapter and the decision-theoretic underpinnings of sequential equilibrium can be seen in our discussion in Section 5 of conditional probability systems. McLennan (1989b) uses conditional probability systems to characterize sequential equilibrium.

The decision maker's weak preference relation over pairs of acts is denoted by $\succcurlyeq$. The relations of strict preference, denoted $\succ$, and indifference, denoted $\sim$, are defined by: $x \succ y$ if $x \succcurlyeq y$ and not $y \succcurlyeq x$; and $x \sim y$ if $x \succcurlyeq y$ and $y \succcurlyeq x$. The following axioms characterize those preference orders with (Archimedean) SEU representations.

Axiom 1. (Order): $\succcurlyeq$ *is a complete and transitive binary relation on* $\mathscr{P}^{\Omega}$.

Axiom 2. (Objective Independence): *For all* $x, y, z \in \mathscr{P}^{\Omega}$ *and* $0 < \alpha \leq 1$, *if* $x \succ$ *(respectively* $\sim$*)* y, *then* $\alpha x + (1 - \alpha)z \succ$ *(respectively* $\sim$*)* $\alpha y + (1 - \alpha)z$.

Axiom 3. (Nontriviality): *There are* $x, y \in \mathscr{P}^{\Omega}$ *such that* $x \succ y$.

Axiom 4. (Archimedean Property): *If* $x \succ y \succ z$, *then there exists* $0 < \alpha < \beta < 1$ *such that* $\beta x + (1 - \beta)z \succ y \succ \alpha x + (1 - \alpha)z$.

A definition of null events requires the notion of conditional preferences $\succcurlyeq_S$ for each $S \subset \Omega$, as in Savage (1954).

Definition 2.1. $x \succcurlyeq_S y$ if, for some $z \in \mathscr{P}^{\Omega}$, $(x_S, z_{-S}) \succcurlyeq (y_S, z_{-S})$.

By Axioms 1 and 2, this definition is independent of the choice of z. (This can be seen by assuming to the contrary that: (i) $(x_S, z_{-S}) \succcurlyeq (y_S, z_{-S})$; while (ii) $(y_S, w_{-S}) \succ (x_S, w_{-S})$. Then taking $\frac{1}{2} : \frac{1}{2}$ mixtures of (x_S, w_{-S}) with (i) and of (x_S, z_{-S}) with (ii) and applying Axioms 1 and 2 yields a contradiction.) An event S is Savage-null if its conditional preference relation is "trivial."

Definition 2.2. The event $S \subset \Omega$ is *Savage-null* if $x \sim_S y$ for all $x, y \in \mathscr{P}^{\Omega}$.

Axiom 5. (Nonnull State Independence): *For all states* $\omega, \omega' \in \Omega$ *which are not Savage-null and for any two constant acts* $x, y \in \mathscr{P}^{\Omega}$, $x \succcurlyeq_{\omega} y$ *if and only if* $x \succcurlyeq_{\omega'} y$.

The following representation theorem can be found in Anscombe and Aumann (1963) and Fishburn (1982, p. 111, Theorem 9.2).

Theorem 2.1. *Axioms 1–5 hold if and only if there is an affine function* $u: \mathscr{P} \to \mathbb{R}$ *and a probability measure* p *on* Ω *such that, for all* $x, y \in \mathscr{P}^{\Omega}$,

$$x \succcurlyeq y \Leftrightarrow \sum_{\omega \in \Omega} p(\omega)u(x_{\omega}) \geq \sum_{\omega \in \Omega} p(\omega)u(y_{\omega}).$$

Furthermore, u is unique up to positive affine transformations, p is unique, and $p(S) = 0$ if and only if the event S is Savage-null.

Since, u is an affine function, and x_ω has finite support, $u(x_\omega) = \Sigma_{c \in C} u(\delta_c) x_\omega(c)$ where, δ_c denotes the measure assigning probability one to c. In order to focus on the subjective probabilities, which are our main concern, and for clarity of the equations, we write u as an affine function on $\mathscr{P}$ as above, rather than including this latter summation explicitly.

Corollary 2.1. *If the event S is not Savage-null, then for all $x, y \in \mathscr{P}^\Omega$,*

$$x \succsim_S y \Leftrightarrow \sum_{\omega \in S} p(\omega|S) u(x_\omega) \geq \sum_{\omega \in S} p(\omega|S) u(y_\omega).$$

In this corollary, which is immediate from Definition 2.1 and Theorem 2.1, $p(\omega|S)$ is given by the usual definition of conditional probability: $p(\omega|S) = p(\omega \cap S)/p(S)$. Corollary 2.1 applies only to events which are not Savage-null since conditional preferences on Savage-null events are trivial and conditional expected utility given any Savage-null event is not denned. To guarantee admissibility and well defined conditional probabilities, Savage-null events must be ruled out. This can be done by strengthening the nonnull state independence axiom.

Axiom 5′. (State Independence): *For all states $\omega, \omega' \in \Omega$ and for any two constant acts $x, y \in \mathscr{P}^\Omega$, $x \succsim_\omega y$ if and only if $x \succsim_{\omega'} y$.*

Under Axioms 1–4 and 5′ the same representation as in Theorem 2.1 obtains, with the additional feature that $p(\omega) > 0$ for all $\omega \in \Omega$. The consequence is that all odds ratios are finite. The decision maker *must* trade off utility gains in any one state against utility gains in *any* other state. Our formulation of non-Archimedean SEU theory avoids this. We will have states which are not Savage-null, and yet which are infinitely less likely than other states.

3. Lexicographic Probability Systems and Non-Archimedean SEU Theory

In this section, we undertake the promised weakening of the Archimedean property (Axiom 4). The consequence of weakening this axiom is the introduction of a new class of null events distinct from the class of Savage-null events. The weakened Archimedean axiom does not eliminate the Savage-null events; that is the consequence of strengthening state independence. Thus, the decision theory we introduce in this section, non-Archimedean SEU theory, is strictly weaker than the conventional

Archimedean SEU theory in that it can rationalize a strictly larger set of choices. Our new axiom is a restriction of the Archimedean property to those triples of acts x, y, z such that $x_{-\omega} = y_{-\omega} = z_{-\omega}$ for some state $\omega \in \Omega$.

Axiom 4'. (Conditional Archimedean Property): *For each $\omega \in \Omega$, if $x \succ_\omega y \succ_\omega z$, then there exists $0 < \alpha < \beta < 1$ such that $\beta x + (1 - \beta)z \succ_\omega y \succ_\omega \alpha x + (1 - \alpha)z$.*

As a consequence of this weakening of Axiom 4, a numerical representation of preferences is not always possible. (However, in Section 6 we show that a numerical representation *is* possible if one is willing to interpret "numerical" as including infinitesimals.) Here, we assign to each act a vector of expected utilities in a Euclidean space, and order these vectors using the lexicographic ordering, which we denote $\geq_L$.[5] The expected utility vectors are calculated by taking expectations of a single utility function with respect to a lexicographic hierarchy of probability distributions.

Definition 3.1. A *lexicographic probability system* (LPS) is a K-tuple $\rho = (p_1, \ldots, p_K)$, for some integer K, of probability distributions on Ω.

Theorem 3.1. *Axioms 1–3, 4', and 5 hold if and only if there is an affine function $u: \mathscr{P} \to \mathbb{R}$ and an LPS $(p_1, \ldots, p_K)$ on Ω such that, for all $x, y \in \mathscr{P}^\Omega$,*

$$x \succcurlyeq y \Leftrightarrow \left(\sum_{\omega \in \Omega} p_k(\omega)u(x_\omega) \right)_{k=1}^{K} \geq_L \left(\sum_{\omega \in \Omega} p_k(\omega)u(y_\omega) \right)_{k=1}^{K}.$$

Furthermore, u is unique up to positive affine transformations. There is a minimal K less than or equal to the cardinality of Ω. Among LPS's of minimal length K, each p_k is unique up to linear combinations of $p_1, \ldots, p_k$ which assign positive weight to p_k. Finally, $p_k(S) = 0$ for all k if and only if the event S is Savage-null.

The proof of Theorem 3.1, together with proofs of all subsequent results in the chapter, can be found in the Appendix. The restriction in the uniqueness part of the theorem to LPS's of minimal length is made in order to avoid redundancies such as the duplication of levels in the hierarchy. (For example, the LPS's $(p_1, p_2, \ldots, p_K)$ and $(p_1, p_1, p_2, \ldots, p_K)$

---

[5]For $a, b \in \mathbb{R}^k$, $a \geq_L b$ if and only if whenever $b_k > a_k$, there exists a $j < k$ such that $a_j > b_j$.

obviously represent the same preferences.) Among LPS's of minimal length K, an LPS $(q_1, \ldots, q_K)$ will generate the same preferences as $(p_1, \ldots, p_K)$ if and only if each $q_k = \sum_{i=1}^{k} \alpha_i p_i$ where, the α_i's are numbers such that $\sum_{i=1}^{k} \alpha_i p_i$ is a probability distribution on Ω and $\alpha_k > 0$. In particular, p_1 is unique.

These preferences include $K = 1$, Archimedean theory, as a special case. The following is an example of non-Archimedean behavior allowed by Axiom 4' but not by Axiom 4. The decision maker will bet on the throw of a die. She has two levels of beliefs, represented by the probability distributions p_1 and p_2. The state space Ω contains the 6 faces of the die, the 12 edges, and the 8 corners. Let

$$p_1(\omega) = \begin{cases} 1/6 & \text{if } \omega \text{ is a face,} \\ 0 & \text{otherwise,} \end{cases} \qquad p_2(\omega) = \begin{cases} 1/12 & \text{if } \omega \text{ is an edge,} \\ 0 & \text{otherwise.} \end{cases}$$

Consider now two bets. Bet x pays off \$$v$ if the die lands on the face labelled 1, and nothing otherwise. Bet y pays off \$1 if the die lands on the face labelled 2 or on any edge, and 0 otherwise. The decision maker's utility function is $u(w) = w$. With these preferences and lexicographic beliefs, $x \succ y$ whenever $v > 1$, and $y \succ x$ whenever $v \leq 1$. Notice that there is no v such that $x \sim y$. This type of behavior cannot be rationalized by Archimedean SEU theory since it explicitly violates Axiom 4. Each face occurs with positive first order probability, and each edge occurs with positive second order probability. However, the die landing on a corner is a Savage-null event, hence is assigned probability 0 by *both* p_1 and p_2. The point of the example is to demonstrate how, even though the die landing on an edge is not a Savage-null event, it is "infinitely less likely" than its landing on a face. This terminology is made precise in Sections 5 and 6.

As we mentioned in Section 2, strengthening state independence from Axiom 5 to Axiom 5' rules out Savage-null events. However, Section 5 shows that in the more general lexicographic framework not all notions of null events are ruled out.

Corollary 3.1. *Axioms 1–3, 4', and 5' hold if and only if there is an affine function $u \colon \mathscr{P} \to \mathbb{R}$ and an LPS $\rho = (p_1, \ldots, p_k)$ on Ω such that, for all $x, y \in \mathscr{P}^{\Omega}$,*

$$x \succeq y \Leftrightarrow \left(\sum_{\omega \in \Omega} p_k(\omega) u(x_\omega) \right)_{k=1}^{K} \geq_L \left(\sum_{\omega \in \Omega} p_k(\omega) u(y_\omega) \right)_{k=1}^{K}.$$

Furthermore, u is unique up to positive affine transformations, the LPS ρ has the same uniqueness properties as in Theorem 3.1, and for each ω there is a k such that $p_k(\omega) > 0$.

The preferences of the decision maker described above for betting on the roll of a die do not satisfy Axiom 5′ — landing on a corner is Savage-null. But now suppose the decision maker has third order beliefs

$$p_3(\omega) = \begin{cases} 1/8 & \text{if } s \text{ is a corner,} \\ 0 & \text{otherwise.} \end{cases}$$

As before, face landings are infinitely more likely than edge landings, which in turn are infinitely more likely than corner landings. But now there are no Savage-null events.

4. Admissibility and Conditional Probabilities

The notion of admissibility and the issues underlying the existence of well defined conditional probabilities are both related to the representation of conditional preferences. In this section, we investigate admissibility and prove a result, analogous to Corollary 2.1, on the representation of conditional preferences for non-Archimedean SEU theory.

Definition 4.1. Let u be a utility function. The preference relation $\succcurlyeq$ is *admissible with respect to u* if whenever $u(x_\omega) \geq u(y_\omega)$ for all $\omega \in \Omega$, with strict inequality for at least one ω, then $x \succ y$.

This definition is a statement about the behavior of a utility function representing $\succcurlyeq$, rather than a statement about conditional preferences. It is helpful to contrast this definition with Theorem 4.1 below which is stated solely in terms of the conditional preferences. Consider therefore the following class of decision problems. Let $\mathscr{S}$ be a partition of Ω, and for each S in $\mathscr{S}$ let $X(S) \subset \mathscr{P}^S$ be the subset of acts in $\mathscr{P}^S$ that the decision maker can choose among if S occurs. A strategy is then an element of $X \equiv \Pi_{S \in \mathscr{S}} X(S) \subset \mathscr{P}^\Omega$, and preferences $\succcurlyeq$ are defined on the set of strategies.

Theorem 4.1. *Suppose $\succcurlyeq$ satisfies Axioms 1 and 2.* (i) *For $x, y \in X$, if $x \succcurlyeq_S y$ for all $S \in \mathscr{S}$ and $x \succ_S y$ for some $S \in \mathscr{S}$, then $x \succ y$.* (ii) *For $x \in X$, if $x \succcurlyeq y$ for all $y \in X$, then $x \succcurlyeq_S z$ for all z such that $z_S \in X(S)$.*

Theorem 4.1 (i) is often referred to as the "sure thing principle." It states that if an act x is conditionally (weakly) preferred to y given any information cell in $\mathscr{S}$, and is strictly preferred given some cell, then x is unconditionally strictly preferred to y. Taking $\mathscr{S}$ to be the finest possible partition leads to a result that clearly resembles admissibility, but whose hypothesis is a claim about conditional preferences rather than utility functions. The second part of Theorem 4.1 states that an optimal strategy must be conditionally optimal on all cells in $\mathscr{S}$, and this bears a resemblance to the logic of backward induction. The point of this theorem is that both properties are satisfied by Archimedean SEU theory. Admissibility and backwards induction are best understood *not* as conditions on the preferences, but in terms of the *representation* of conditional preferences.[6] Admissibility was defined in terms of the representation. Similarly, we interpret backwards induction rationality to be the restriction that an optimal strategy maximize conditional expected utility on each cell $S \in \mathscr{S}$; hence, well defined conditional probabilities are required.

In this section, we suppose henceforth that Axiom 5' is satisfied. The main purpose is to show that in the non-Archimedean framework, this implies admissibility and the existence of LPS's which represent conditional preferences and which are concentrated on the conditioning event.

Theorem 4.2. (Admissibility):.*Suppose $\succcurlyeq$ satisfies Axioms 1–3, 4', and 5', and let u and ρ denote a utility function and an LPS which represent $\succcurlyeq$. Then $\succcurlyeq$ is admissible with respect to u.*

Theorem 4.2 is an immediate consequence of Corollary 3.1. We now turn to the definition of conditional probabilities for lexicographic hierarchies of beliefs.

Definition 4.2. Let $\rho = (p_1, \ldots, p_K)$ be an LPS on the state space Ω. For any nonempty event S, the *conditional LPS given S* is $\rho_S \equiv (p_{k_1}(\cdot|S), \ldots, p_{k_L}(\cdot|S))$, where the indices k_l are given by $k_0 = 0$, $k_l = \min\{k : p_k(S) > 0 \text{ and } k > k_{l-1}\}$ for $l > 0$, and $p_{k_l}(\cdot|S)$ is given by the usual definition of conditional probabilities.

This notion of conditional probability is intuitively appealing — the conditional LPS is obtained by taking conditional probabilities of all p_k's in the LPS p for which conditionals are defined ($p_k(S) > 0$) and discarding the other p_k's. Axiom 5' implies that at least one p_k will not be discarded so

---

[6]Since the preferences determine the representation, these conditions can be stated in terms of the preferences alone — however as Theorems 4.1–4.3 show, it may be more insightful to think of these properties using the representation.

that $L \geq 1$. In Section 6, where probabilities are allowed to be infinitesimals, it is seen that this definition is equivalent to an exact analog of the usual definition of conditional probabilities. Clearly Definition 4.2 satisfies two of our objectives: it is an LPS (hence a "subjective probability" in the lexicographic framework); and it is concentrated on the conditioning event S. The main issue which remains to be verified is that these conditional probabilities represent conditional preferences, as is shown in Theorem 4.3, which is a non-Archimedean version of Corollary 2.1.

Theorem 4.3. *Suppose* $\succcurlyeq$ *satisfies Axioms* 1–3, 4′, *and* 5′, *and let* u *and* ρ *denote a utility function and an LPS which represent* $\succcurlyeq$. *Then for any nonempty event* S, *the utility function* u *and the conditional LPS* $\rho_S \equiv (p_{k_1}(\cdot|S), \ldots, p_{k_L}(\cdot|S))$ *represent the conditional preferences* $\succcurlyeq_S$:

$$x \succcurlyeq_S y \Leftrightarrow \left(\sum_{\omega \in S} p_{k_l}(\omega|S) u(x_\omega) \right)_{l=1}^{L} \geq_L \left(\sum_{\omega \in S} p_{k_l}(\omega|S) u(y_\omega) \right)_{l=1}^{L}.$$

5. Lexicographic Conditional Probability Systems

In this section, we discuss in more detail the ways in which events can be null in lexicographic probability systems, and examine the relationship between the characterization of Section 3 and other recent developments (Myerson [1986a,b], McLennan [1989a,b], Hammond [1987]). This relationship will be clarified by axiomatizing lexicographic *conditional* probability systems (not to be confused with the conditional LPS's of Definition 4.2), using an Archimedean property intermediate in strength between Axioms 4 and 4′. This intermediate Archimedean axiom will arise naturally from an understanding of null events in the lexicographic framework.

In the Archimedean framework, an event S is infinitely more likely than another event T (in terms of probability ratios) if and only if T is Savage-null. Non-Archimedean theory admits a richer likelihood order on events. We will investigate a partial order on events, $S \gg T$, to be read as "S is infinitely more likely than T." One could proceed to define such a notion in terms of the representation or the preferences. We adopt the latter approach, which provides a more primitive characterization, in order to better understand the relationship with Savage-null events. The following characterization of Savage-null events (Definition 2.2) will be useful.

Theorem 5.1. *Assume that $\succeq$ satisfies Axioms 1 and 2. An event T is Savage-null if and only if there exists a nonempty disjoint event S such that*

$$x \succ_S \ (\text{respectively} \ \sim_S) \ y$$

implies

$$(x_{-T}, w_T) \succ_{S \cup T} \ (\text{respectively} \ \sim_{S \cup T}) \ (y_{-T}, z_T)$$

for all w_T, z_T.

That is, an event T is Savage-null if for some disjoint event S, when comparing (x_S, w_T) and (y_S, z_T), the consequences in the event S are determining for both $\succ_{S \cup T}$ and $\sim_{S \cup T}$. An intuitively weaker order on events arises from supposing that consequences in the event S are determining for $\succ_{S \cup T}$ alone. In the remainder of this section, we assume that there are no Savage-null events, and examine the properties of an alternative likelihood ordering on events.

Definition 5.1. *For disjoint events $S, T \subset \Omega$ with $S \neq \varnothing$, $S \gg T$ if*

$$x \succ_S y \text{ implies } (x_{-T}, w_T) \succ_{S \cup T} (y_{-T}, z_T)$$

for all w_T, z_T.

Theorem 5.2. *Assume that $\succeq$ is represented by a utility function u and an LPS $\rho = (p_1, \ldots, p_K)$. For a pair of states ω^1 and ω^2, $\omega^1 \gg \omega^2$ if and only if u and the LPS $((1,0), (0,1))$ on $\{\omega^1, \omega^2\}$ represent $\succeq \{\omega^1, \omega^2\}$.*

Theorem 5.2 says that for a pair of *states* the order $\gg$ corresponds to the conditional probabilities (which represent the conditional preferences given that pair of states).[7] More generally, for events, if $S \gg T$ then $p_{k_1}(T | S \cup T) = 0$ (where $k_1 = \min\{k : p_k(S \cup T) > 0\}$). However, the converse to this is in general false, so zero probabilities in the representation do not correspond to the ranking $\gg$. A related difficulty with Definition 5.1 is that $S \gg T$ and $S' \gg T$ need not imply $S \cup S' \gg T$. Both these difficulties can be seen in the following example. Consider a state space $\Omega = \{Heads, Tails, Edge, Heads^*\}$ and the LPS $p_1 = (1/2, 1/2, 0)$, $p_2 = (1/2, 0, 1/2)$. Even though $\{Tails\} \gg \{Edge\}$, $\{Heads\} \gg \{Edge\}$,

---

[7] If the strict order $\gg$ is restricted to states, then the induced weak order can be characterized as follows: ω' is not infinitely more likely than $\omega \Leftrightarrow \min\{k : p_k(\omega) > 0\} \geq \min\{k : p_k(\omega') > 0\}$. This weak order turns out to be useful in characterizing proper equilibrium (Myerson [1978]) — see the sequel to this chapter (Blume, Brandenburger, and Dekel [1991]).

and $p_1(Edge) = 0$, it is not the case that $\{Heads,\ Tails\} \gg \{Edge\}$ since $x \equiv (2, 0, 0) \succ (1, 1, 0) \equiv y$, but $(2, 0, 0) \prec (1, 1, 2)$.[8] These problems result from the fact that the supports of p_1 and p_2 overlap. Hence, we will now distinguish the subset of LPS's whose component probability measures have disjoint supports, and introduce the Archimedean axiom which is used in Theorem 5.3 below to characterize this subset.

Definition 5.2. An LPS $\rho = (p_1, \ldots, p_K)$, where the supports of the p_k's are disjoint, is a *lexicographic conditional probability system* (LCPS).

Axiom 4″. *There is a partition $\{\Pi_1, \ldots, \Pi_K\}$ of Ω such that: (a) for each k, if $x \succ_{\Pi_k} y \succ_{\Pi_k} z$, then there exists $0 < \alpha < \beta < 1$ such that $\beta x + (1 - \beta)z \succ_{\Pi_k} y \succ_{\Pi_k} \alpha x + (1 - \alpha)z$; (b) $\Pi_1 \gg \cdots \gg \Pi_K$.*

Theorem 5.3. *Axioms 1–3, 4″, and 5′ hold if and only if there is an affine function $u : \mathscr{P} \to \mathbb{R}$ and an LCPS $\rho = (p_1, \ldots, p_K)$ on Ω such that, for all $x, y \in \mathscr{P}^\Omega$,*

$$x \succeq y \Leftrightarrow \left(\sum_{\omega \in \Omega} p_k(\omega)u(x_\omega) \right)_{k=1}^{K} \geq_L \left(\sum_{\omega \in \Omega} p_k(\omega)u(y_\omega) \right)_{k=1}^{K}.$$

Furthermore, u is unique up to positive affine transformations, ρ is unique, and for each $k = 1, \ldots, K$, the support of p_k is Π_k.

Corollary 5.1. *Assume that $\succeq$ is represented by a utility function u and an LCPS $\rho = (p_1, \ldots, p_K)$ For a disjoint nonempty pair of events S, T,*

$$S \gg T \Leftrightarrow k' < l' \quad \text{for all } k' \in \{k : p_k(S|S \cup T) > 0\}$$
$$\text{and } l' \in \{l : p_l(T|S \cup T) > 0\}.$$

Theorem 5.3 and Corollary 5.1 show that as a result of strengthening the Archimedean property from Axiom 4′ to 4″, the relation $\gg$ corresponds to zero probabilities in the representation, and Theorem 5.2 can be strengthened to hold for events as well as states.

There is an interesting interpretation of an LCPS. The first-order belief p_1 can be thought of as a prior distribution. If the expected utilities under p_1 of two acts are the same, then the decision maker considers the event $\Omega - \Pi_1$, where by Theorem 5.3, Π_1 is the support of p_1. The second-order belief p_2 is then the "posterior" conditional on the event $\Omega - \Pi_1$. More generally, higher order beliefs can also be thought of as conditional probability distributions.

---

[8]The numbers in these triplets are in utility payoffs.

Return to the example of the coin toss discussed earlier. We now describe how the LPS (p_1, p_2) can be reinterpreted as an LCPS. Suppose there is a possibility that the coin is being tossed in a "dishonest" fashion which guarantees heads. If we denote this event by $\{Heads^*\}$ then the expanded state space is $\Omega^* = \{Heads, Tails, Edge, Heads^*\}$. The beliefs $p_1^* = (1/2, 1/2, 0, 0), p_2^* = (0, 0, 1/2, 1/2)$ on Ω^* have nonoverlapping supports. Moreover, if the events $\{Heads\}$ and $\{Heads^*\}$ are indistinguishable, then (p_1^*, p_2^*) induces the same lexicographic probabilities over the payoff-relevant outcomes — namely, the coin lands on heads, tails, or the edge — as does (p_1, p_2). This example shows how to map an LPS into an LCPS on an expanded state space. A more general treatment is developed in Hammond (1987). However, this reinterpretation does not mean that it suffices to work with LCPS's alone. If in fact $\{Heads\}$ and $\{Heads^*\}$ are indistinguishable, then bets on $\{Heads\}$ versus $\{Heads^*\}$ cannot be made and so the subjective probabilities (p_1^*, p_2^*) cannot be derived; including such payoff-irrelevant states contradicts our basic model which admits *all* possible acts in $\mathscr{P}^\Omega$.

LCPS's provide a bridge between the work in this chapter and some ideas in Myerson (1986a,b). Myerson also starts from the existing SEU theory, but his modification leads in a different direction to that pursued here. Myerson augments the basic preference relation $\succcurlyeq$ by postulating the existence of a distinct preference relation corresponding to each nonempty subset S of the state space Ω. Although interpreted as conditional preferences, these preferences differ from $\succcurlyeq_S$ as defined by Savage (1954) (and Definition 2.1 in this chapter). Using this preference structure, Myerson derives the notion of a *conditional probability system* (CPS). The reader is referred to Myerson (1986a,b) for the definition of a CPS, which can be shown to be isomorphic to an LCPS. An important distinction between Myerson's preference structure and ours is that the latter satisfies admissibility (Theorem 4.2) whereas, the former does not. McLennan (1989b) uses CPS's to provide an existence proof, and a characterization, of sequential equilibrium.

6. A "Numerical" Representation for Non-Archimedean SEU

This section provides a "numerical" representation for the preferences described by Axioms 1–3, 4', and 5', where the "numbers" are elements in a non-Archimedean ordered field $\mathbb{F}$ which is a strict extension of the real number field $\mathbb{R}$. The field $\mathbb{F}$ is non-Archimedean: it contains both infinite numbers (larger than any real number) and infinitesimal numbers (smaller than any real number) in addition to the reals.[9]

The basic result on the existence of utility functions states that a complete, transitive, and reflexive preference relation on a set X has a real-valued representation if and only if X contains a countable order-dense subset. Without this Archimedean restriction on X, a representation is still possible. A complete, transitive, and reflexive preference relation on *any* set X has a numerical representation taking values in a non-Archimedean ordered field (Richter [1971]). In this chapter we have weakened the Archimedean property (Axiom 4) of SEU to Axiom 4′. This weakening still permits a real-valued utility function on consequences, but requires the subjective probability measure to be non-Archimedean. By analogy with the real-valued case, a non-Archimedean probability measure on Ω is a function $p : \Omega \to \mathbb{F}$ such that $p(\omega) \geq 0$ for each $\omega \in \Omega$, and $\Sigma_{\omega \in \Omega} \, p(\omega) = 1$.

Theorem 6.1. *Axioms 1–3, 4′, and 5′ hold if and only if there is an affine function $u : \mathscr{P}^{\Omega} \to \mathbb{R}$ and an $\mathbb{F}$-valued probability measure p on Ω, where $\mathbb{F}$ is a non-Archimedean ordered-field extension of $\mathbb{R}$, such that, for all $x, y \in \mathscr{P}^{\Omega}$,*

$$x \succcurlyeq y \Leftrightarrow \sum_{\omega \in \Omega} p(\omega)u(x_{\omega}) \geq \sum_{\omega \in \Omega} p(\omega)u(y_{\omega}).$$

Furthermore, u is unique up to positive affine transformations. If p' is another $\mathbb{F}$-valued probability measure such that u and p' represent $\succcurlyeq$, then for all $\omega \in \Omega, p(\omega) - p'(\omega)$ is infinitesimal. Finally, $p(\omega) > 0$ for all $\omega \in \Omega$.

The disadvantage of the representation of Theorem 6.1 is that its proof requires less familiar techniques. Its advantage is that, for most purposes, it really does provide a numerical representation. The probabilities of states can be added, multiplied, divided, and compared just like real numbers. For example, the usual definition of conditional probabilities applies for an $\mathbb{F}$-valued probability measure as well: $p(T|S) \equiv p(T \cap S)/p(S)$ for events $S, T \subset \Omega$ (Theorem 6.1 implies that $p(S) \neq 0$). By analogy with the real-valued case, it is easy to show that

$$x \succcurlyeq_S y \Leftrightarrow \sum_{\omega \in S} p(\omega)u(x_{\omega}) \geq \sum_{\omega \in S} p(\omega)u(y_{\omega}).$$

---

[9]We do not distinguish between the subfield of $\mathbb{F}$ which is order-isomorphic to $\mathbb{R}$ and $\mathbb{R}$ itself. Likewise, $\geq$ will be used to denote the order on both $\mathbb{F}$ and $\mathbb{R}$.

Dividing both sides of the inequality by $p(S)$ shows that u and $p(\cdot|S)$ do indeed represent the conditional preference relation $\succeq_S$.

We have found that some results are more easily understood by employing the LPS's of Section 3, while others are best seen using the representation of this section. An example of the former is the statement of the uniqueness of the subjective beliefs in the representation theorems. An example of the latter is the issue of stochastic independence discussed in Section 7.

The sufficiency part of Theorem 6.1 can be proved using arguments analogous to those when preferences are Archimedean. The necessity part can be proved in a number of ways, including compactness arguments from logic or by using ultrafilters. A sketch of the ultrafilter argument is provided in the Appendix.[10]

7. Stochastic Independence and Product Measures

This section considers three possible definitions of stochastic independence. The first requires that the (Archimedean or non-Archimedean) probability measure p be a product measure; the second requires that a decision-theoretic stochastic independence axiom be satisfied; and the third that p be an "approximate product measure." In the Archimedean setting, all three definitions are equivalent, but in the non-Archimedean case the definitions are successively weaker. The notion of independence implicit in refinements of Nash equilibrium such as perfect equilibrium (Selten [1975]) is the first one above. The fact that it is *not* equivalent, in the non-Archimedean setting, to the stochastic independence axiom suggests that alternative (weaker) refinements may also be worthy of consideration.[11]

Suppose in the following that the set of states Ω is a product space $\Omega^1 \times \cdots \times \Omega^N$, and for $n = 1, \ldots, N$ let $\Omega^{-n} = \Pi_{m \neq n}\Omega^m$. An act x is said to be constant across Ω^n if for every $\omega^n, \tilde{\omega}^n$ in Ω^n and all ω^{-n} in Ω^{-n}, $x_{(\omega^n,\omega^{-n})} = x_{(\tilde{\omega}^n,\omega^{-n})}$. Assume that the preference relation $\succeq$ satisfies Axioms 1–3 and 5'. We are interested in focusing on the distinction between

---

[10]Hammond (1987) has discussed LPS's that can be represented by a probability measure taking values in the non-Archimedean ordered field of rational functions over $\mathbb{R}$.

[11]Such refinements may involve a sequence of *correlated* trembles converging to a Nash equilibrium — see the sequel to this chapter (Blume, Brandenburger, and Dekel [1991]). Related issues are discussed in Binmore (1987, 1988), Dekel and Fudenberg (1990), Fudenberg, Kreps, and Levine (1988), and Kreps and Ramey (1987).

the Archimedean case in which Axiom 4 holds and the non-Archimedean case in which Axiom 4′ holds.

Definition 7.1. An (Archimedean or non-Archimedean) probability measure p on Ω is a *product measure* if there are probability measures p^n on Ω^n, for $n = 1, \ldots, N$, such that for all $\omega = (\omega^1, \ldots, \omega^N) \in \Omega$, $p(\omega) = p^1(\omega^1) \times \cdots \times p^N(\omega^N)$.

Axiom 6. (Stochastic Independence): *For any $n = 1, \ldots, N$ and every pair of acts x, y that are constant across Ω^n,*

$$x \succcurlyeq_{\{\omega^n\} \times \Omega^{-n}} y \Leftrightarrow x \succcurlyeq_{\{\tilde{\omega}^n\} \times \Omega^{-n}} y.$$

Roughly speaking, Axiom 6 requires that the conditional preferences $\succcurlyeq_{\{\omega^n\} \times \Omega^{-n}}$ and $\succcurlyeq_{\{\tilde{\omega}^n\} \times \Omega^{-n}}$ (viewed as preferences on $\mathscr{P}^{\Omega^{-n}}$) be identical for all ω^n and $\tilde{\omega}^n$ in Ω^n. Suppose the preference relation $\succcurlyeq$ satisfies the Archimedean property (Axiom 4), and let u and p denote a utility function and (Archimedean) probability measure which represent the preferences. It is routine to show that p is a product measure if and only if stochastic independence (Axiom 6) holds. This equivalence breaks down in the non-Archimedean setting. Any preference relation represented by some utility function u and non-Archimedean probability measure p satisfies Axiom 6 if p is a product measure. But the converse is false, as the following example due to Roger Myerson (private communication) shows. Let $\Omega = \{\omega_1^1, \omega_2^1, \omega_3^1\} \times \{\omega_1^2, \omega_2^2, \omega_3^2\}$ where the non-Archimedean probability measure is as depicted in Figure 7.1 (in which $\varepsilon > 0$ is an infinitesimal). Fix a utility function u. By Theorem 6.1, the probability measure p and utility function u determine a preference relation which satisfies Axioms 1–3, 4′, and 5′. One can verify that the conditional preference relation given any row is the same and that, likewise, the conditional preference relation given any column is the same. Hence, Axiom 6 is satisfied. However, p is not a product measure and, moreover, there is no measure which *is* a product measure and which represents the same preference relation.

In the non-Archimedean case (Axiom 4′ rather than Axiom 4), Axiom 6 is sufficient for the existence of a weaker kind of product measure — a concept known in nonstandard probability theory as S-independence.

Definition 7.2. An (Archimedean or non-Archimedean) probability measure p is an *approximate product measure* if there are probability measures p^n on Ω^n, for $n = 1, \ldots, N$, such that for all $\omega = (\omega^1, \ldots, \omega^N) \in \Omega$, $p(\omega) - (p^1(\omega^1) \times \cdots \times p^N(\omega^N))$ is infinitesimal.

	ω_1^2	ω_2^2	ω_3^2
ω_1^1	$1\text{-}2\epsilon - 4\epsilon^2 - 2\epsilon^3 - \epsilon^4$	ϵ	$2\epsilon^2$
ω_2^1	ϵ	ϵ^2	ϵ^3
ω_3^1	ϵ^2	ϵ^3	ϵ^4

Figure 7.1.

Suppose the preference relation $\succcurlyeq$ satisfies Axiom 4′, and let u and p denote a utility function and non-Archimedean probability measure which represent the preferences. It is straightforward to see that if Axiom 6 holds, then p is an approximate probability measure.[12] However, requiring a non-Archimedean probability measure p to be an approximate product measure is strictly weaker than demanding Axiom 6 to hold, as the following example demonstrates. Let $\Omega = \{\omega_1^1, \omega_2^1\} \times \{\omega_1^2, \omega_2^2\}$ where the non-Archimedean probability measure p is as depicted in Figure 7.2 (in which ε and δ are positive infinitesimals with ε/δ infinite). The measure p is an approximate product measure but it is clear that, for any utility function u, the conditional preference relation given the top row differs from that given the bottom row.

Appendix

This Appendix contains proofs of the results in Sections 3–6. The first two lemmas are used in the proofs of Theorems 3.1 and 6.1.

---

[12]Equivalently, in term of the lexicographic representation of preferences derived in Section 3, Axiom 6 implies that the first-level probability measure p_1 of an LPS $(p_1, \ldots, p_K)$ representing $\succcurlyeq$ is a product measure.

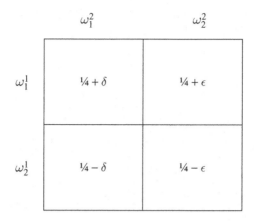

Figure 7.2.

Lemma A.1. *Given a preference relation $\succeq$ on $\mathscr{P}^{\Omega}$ satisfying Axioms 1, 2, 4′, and 5′, there is an affine function $u\colon \mathscr{P} \to \mathbb{R}$ such that for every $\omega \in \Omega, x \succeq_{\omega} y$ if and only if $u(x_{\omega}) \geq u(y_{\omega})$.*

Proof. For each ω, the conditional preference relation $\succeq_{\omega}$ satisfies the usual order, independence, and Archimedean axioms of von Neumann–Morgenstern expected utility theory. This follows from the fact that $\succeq$ satisfies Axioms 1, 2, and 4′, respectively. Hence, each $\succeq$ can be represented by an affine utility function $u_{\omega}\colon \mathscr{P} \to [0,1]$. Under Axiom 5′, the conditional preference relation $\succeq_{\omega}$ is independent of ω, hence every $\succeq_{\omega}$ can be represented by a *common* utility function $u : \mathscr{P} \to [0,1]$. □

The next step is to use the utility function u to scale acts in utiles: an act $x \in \mathscr{P}^{\Omega}$ is represented by the tuple $(u(x_{\omega}))_{\omega \in \Omega} \in [0,1]^{\Omega}$. The preference relation $\succeq$ on $\mathscr{P}^{\Omega}$ induces a preference relation $\geq^{*}$ on $[0,1]^{\Omega}$: given $a, b \in [0,1]^{\Omega}$, define a $\geq^{*} b$ if and only if $x \succeq y$ for some $x, y \in \mathscr{P}^{\Omega}$ with $u(x_{\omega}) = a_{\omega}, u(y_{\omega}) = b_{\omega}$ for each $\omega \in \Omega$. This definition is meaningful since by Axioms 1 and 2, it is independent of the particular choice of x and y.

Lemma A.2. *The preference relation $\geq^{*}$ on $[0,1]^{\Omega}$ satisfies the order and independence axioms.*

Proof. This follows immediately from the fact that $\succeq$ satisfies Axioms 1 and 2. □

Proof of Theorem 3.1. Given the preference relation $\succeq$ on $\mathscr{P}^{\Omega}$, construct the induced preference relation $\geq^{*}$ on $[0,1]^{\Omega}$ of Lemma A.2. By a result of

Hausner (1954, Theorem 5.6), there are K (where K is equal to or less than the cardinality of Ω) affine functions $U_k \colon [0,1]^\Omega \to \mathbb{R}$, $k = 1, \ldots, K$, such that for $a, b \in [0,1]^\Omega$,

$$a \geq^* b \Leftrightarrow (U_k(a))_{k=1}^K \geq_L (U_k(b))_{k=1}^K.$$

The next step is to derive subjective probabilities. By linearity, $U_k(a) = \sum_{\omega \in \Omega} U_k(e^\omega) a_\omega$ where e^ω is the vector with 1 in the ωth position and 0's elsewhere. By nontriviality (Axiom 3), each U_k can be chosen to satisfy $\sum_{\omega \in \Omega} U_k(e^\omega) > 0$. So,

$$V_k(a) = \sum_{\omega \in \Omega} r_k(\omega) a_\omega, \quad \text{where } r_k(\omega) = \frac{U_k(e^\omega)}{\sum_{\omega' \in \Omega} U_k(e^{\omega'})},$$

is a positive affine transformation of $U_k(a)$. For each $\omega, r_1(\omega) \geq 0$ since otherwise $e^\omega <^* (0, \ldots, 0)$, contradicting Axiom 3. So, we can define a probability measure p_1 on Ω by $p_1 = r_1$. For $k > 1$, find numbers α_i, $i = 1, \ldots, k$, with $\alpha_i \geq 0$, $\alpha_k > 0$, and $\sum_{i=1}^k \alpha_i = 1$, such that for each ω, $p_k(\omega) = \sum_{i=1}^k \alpha_i r_i(\omega) \geq 0$. (Again, such α_i's exist since otherwise $e^\omega <^* (0, \ldots, 0)$.) The p_k's defined in this way are probability measures on Ω.

To sum up, we have derived probability measures $p_1, \ldots, p_K$ on Ω such that for $a, b \in [0,1]^\Omega$,

$$a \geq^* b \Leftrightarrow \left(\sum_{\omega \in \Omega} p_k(\omega) a_\omega\right)_{k=1}^K \geq_L \left(\sum_{\omega \in \Omega} p_k(\omega) b_\omega\right)_{k=1}^K.$$

On recalling that $x \succcurlyeq y$ if and only if $a \geq^* b$, where $u(x_\omega) = a_\omega$ and $u(y_\omega) = b_\omega$ for each ω, the representation of Theorem 3.1 is established. The uniqueness properties and the "if" direction of Theorem 3.1 follow easily from routine arguments. $\quad\square$

Proof of Corollary 3.1. For each $\omega \in \Omega$, there must be a k such that $p_k(\omega) > 0$ since otherwise $\succcurlyeq_\omega$ would be trivial, contradicting the fact that under Axiom 5' there are no Savage-null events. $\quad\square$

Proof of Theorem 4.1. (1) It is shown that $x \succcurlyeq_S y$, and $x \succ_T$ (respectively $\succcurlyeq_T$) y for disjoint S and T implies that $x \succ_{S \cup T}$ (respectively $\succcurlyeq_{S \cup T}$) y. A simple induction argument, which is omitted, would complete the proof. By Definition 2.1: $x \succcurlyeq_S y$ implies that $(x_S, x_T, z_{-(S \cup T)}) \succcurlyeq (y_S, x_T, z_{-(S \cup T)})$; and $x \succ_T y$ implies that $(y_S, x_T, z_{-(S \cup T)}) \succ (y_S, y_T, z_{-(S \cup T)})$. By transitivity this implies $(x_S, x_T, z_{-(S \cup T)}) \succ (y_S, y_T, z_{-(S \cup T)})$, which in turn implies $x \succ_{S \cup T} y$. The same argument holds for weak preference, proving the assertion.

(2) If this result is false then there is an event $S \in \mathscr{S}$ and an act z such that $z \succ_S x$ and $z_S \in X(S)$. But then $(z_S, x_{-S}) \in X$ and $(z_S, x_{-S}) \succ x$ by (1) above, so that x is not optimal. $\qquad\square$

Proof of Theorem 4.2. If $u(x_\omega) \geq u(y_\omega)$ for all ω, then for every k, $\Sigma_{\omega \in \Omega} p_k(\omega)u(x_\omega) \geq \Sigma_{\omega \in \Omega} p_k(\omega)u(y_\omega)$. Suppose $u(x_{\omega'}) > u(y_{\omega'})$, and let k' be the first k such that $p_k(\omega') > 0$. Then, $\Sigma_{\omega \in \Omega} p_{k'}(\omega)u(x_\omega) > \Sigma_{\omega \in \Omega} p_{k'}(\omega)u(y_\omega)$, so $x \succ y$. $\qquad\square$

Proof of Theorem 4.3. The proof is analogous to that for the SEU case. Clearly $x \succsim_S y$ if and only if $(x_S, x_{-S}) \succsim (y_S, x_{-S})$ which in turn is true if and only if

$$\left(\sum_{\omega \in S} p_k(\omega)u(x_\omega) + \sum_{\omega \in \Omega - S} p_k(\omega)u(x_\omega) \right)^K_{k=1}$$

$$\geq_L \left(\sum_{\omega \in S} p_k(\omega)u(y_\omega) + \sum_{\omega \in \Omega - S} p_k(\omega)u(x_\omega) \right)^K_{k=1}.$$

Subtracting,

$$\left(\sum_{\omega \in S} p_k(\omega)u(x_\omega) \right)^K_{k=1} \geq_L \left(\sum_{\omega \in S} p_k(\omega)u(y_\omega) \right)^K_{k=1}.$$

Finally, for each k, if $p_k(S) > 0$ divide the kth component of both sides by $p_k(S)$. $\qquad\square$

Proof of Theorem 5.1. *Only if*: Should the displayed equation in Theorem 5.1 fail to hold for any S, then either: $x \succ_S y$ and (y_{-T}, z_T) $\succsim_{S \cup T} (x_{-T}, w_T)$; or $x \sim_S y$ and $(y_{-T}, z_T) \succ_{S \cup T} (x_{-T}, w_T)$; or $x \sim_S y$ and $(x_{-T}, w_T) \succ_{S \cup T} (y_{-T}, z_T)$. By Theorem 4.1 this implies either $z \succ_T w$, or $w \succ_T z$, so that T is not Savage-null.

If: By Definition 2.2, if T is not Savage-null, then there exist w, z such that $w \succ_T z$, but then $x \sim_S x$ while $(x_{-T}, w_T) \succ_{S \cup T} (x_{-T}, z_T)$. $\qquad\square$

Proof of Theorem 5.2. In an LPS $\rho = (p_1, \ldots, p_K)$, K can be taken to be less than or equal to the cardinality of Ω without loss of generality, so attention can be restricted to LPS's (p_1, p_2). Furthermore, the uniqueness results of Theorem 3.1 imply that for any u, the LPS's $((1,0),(0,1))$ and

$((1,0),(\alpha, 1 - \alpha))$, for $\alpha < 1$, represent the same preferences. Finally, if $\succcurlyeq \{\omega^1, \omega^2\}$ is represented by $((\beta, 1 - \beta), (\gamma, 1 - \gamma))$ for some $\beta < 1$ then there exist x, y such that $x \succ_{\omega^1} y$ but $\beta u(x_{\omega^1}) + (1 - \beta) u(x_{\omega^2}) < \beta u(y_{\omega^1}) + (1 - \beta) u(y_{\omega^2})$, so that $\omega^1 \not\gg \omega^2$ The "if" direction is immediate. $\square$

Proof of Theorem 5.3. Since Axiom 4″(a) implies Axiom 4′, it follows from Corollary 3.1 that there is an affine function $u : \mathscr{P} \to \mathbb{R}$ and an LPS $\rho = (p_1, \ldots, p_K)$ such that

$$x \succcurlyeq y \Leftrightarrow \left(\sum_{\omega \in \Omega} p_k(\omega) u(x_\omega) \right)_{k=1}^K \geq_L \left(\sum_{\omega \in \Omega} p_k(\omega) u(y_\omega) \right)_{k=1}^K.$$

It remains to show that ρ can be chosen so that the p_k's have disjoint supports. Let $K_1 = \min\{k : p_k(\Pi_1) > 0\}$. By Axioms 4″(a) and 5′, $p_{K_1}(\omega) > 0$ for all $\omega \in \Pi_1$. By Axiom 4″(a), the p_k's can be chosen so that $p_k(\Pi_1) = 0$ for all $k > K_1$. By Axiom 4″(b), $p_k(\Pi_2) = 0$ for all $k \leq K_1$. Next, let $K_2 > K_1$ be defined by $K_2 = \min\{k : p_k(\Pi_2) > 0\}$. Continuing in this fashion shows that ρ can be chosen so that the supports of its component measures are disjoint. The uniqueness properties and the "if" direction follow easily from routine arguments. $\square$

Proof of Corollary 5.1. The proof follows immediately from Theorem 5.3
$\square$

Proof of Theorem 6.1. Given the preference relation $\succcurlyeq$ on $\mathscr{P}^\Omega$, construct the induced relation $\geq^*$ on $[0, 1]^\Omega$ of Lemma A.2. Since $\geq^*$ satisfies the order axiom, it follows from general arguments using ultrafilters (see, e.g., Richter [1971, Theorem 9]) that there is a representation for $\geq^*$ taking values in a non-Archimedean ordered field $\mathbb{F}$. That is, there exists such a field $\mathbb{F}$ and a utility function $U : [0, 1]^\Omega \to \mathbb{F}$ such that for $a, b \in [0, 1]^\Omega$,

$$a \geq^* b \Leftrightarrow U(a) \geq U(b).$$

Furthermore, since $\geq^*$ satisfies the von Neumann–Morgenstern independence axiom, it follows from routine separating hyperplane arguments that for every finite set $A \subset [0, 1]^\Omega$, there is an affine function $U^A : [0, 1] \to \mathbb{R}$ representing $\geq^*$ on A. Consequently, the ultrafilter argument can be extended to conclude that the utility function U may be taken to be affine. To summarize, we have shown that there is an affine function $U : [0, 1]^\Omega \to \mathbb{F}$ representing $\geq^*$ on $[0,1]^\Omega$.

The next step is to derive subjective probabilities. By linearity, $U(a) = \Sigma_{\omega \in \Omega} U(e^\omega) a_\omega$ (where e^ω is the ωth unit vector). By nontriviality (Axiom 3),

$\Sigma_{\omega \in \Omega} U(e^{\omega}) > 0$. So define

$$V(a) = \sum_{\omega \in \Omega} p(\omega) a_{\omega}, \quad \text{where } p(\omega) = \frac{U(e^{\omega})}{\sum_{\omega' \in \Omega} U(e^{\omega'})}.$$

Since, V is a positive affine transformation of U, it also represents $\geq^*$. The $p(\omega)$'s defined this way constitute an $\mathbb{F}$-valued probability measure on Ω. On recalling that $x \succcurlyeq y$ if and only if $a \geq^* b$, where $u(x_{\omega}) = a_{\omega}$ and $u(y_{\omega}) = b_{\omega}$ for each ω, the representation of Theorem 6.1 is established. $\qquad \square$

References

Anscombe, F and R Aumann (1963). A definition of subjective probability. *Annals of Mathematical Statistics*, 34, 199–205.

Arrow, K, E Barankin, and D Blackwell (1953). Admissible points of convex sets. In Kuhn, H and A Tucker (Eds.), *Contributions to the Theory of Games*, Vol. 2. Princeton, NJ: Princeton University Press.

Binmore, K (1987). Modeling rational players I. *Journal of Economics and Philosophy*, 3, 179–214.

Binmore, K (1988). Modeling rational players II. *Journal of Economics and Philosophy*, 4, 9–55.

Blackwell, D and L Dubins (1975). On existence and non-existence of proper, regular, conditional distributions. *The Annals of Probability*, 3, 741–752.

Blackwell, D and M Girshick (1954). *Theory of Games and Statistical Decisions*. New York, NY: Wiley.

Blume, L, A Brandenburger, and E Dekel (1991). Lexicographic probabilities and equilibrium refinements. *Econometrica*, 59, 81–98.

Chernoff, H (1954). Rational selection of decision functions. *Econometrica*, 22, 422–443.

Chipman, J (1960). The foundations of utility. *Econometrica*, 28, 193–224.

Chipman, J (1971a). On the lexicographic representation of preference orderings. In Chipman, J, L Hurwicz, M Richter, and H Sonnenschein (Eds.), *Preference Utility and Demand*. New York, NY: Harcourt Brace Jovanovich.

Chipman, J (1971b). Non-Archimedean behavior under risk: An elementary analysis — with application to the theory of assets. In Chipman, J, L Hurwicz, M Richter, and H Sonnenschein (Eds.), *Preferences, Utility and Demand*. New York, NY: Harcourt Brace Jovanovich.

de Finetti, B (1972). *Probability, Induction and Statistics*. New York, NY: Wiley.

Dekel, E and D Fudenberg (1990). Rational behavior with payoff uncertainty. *Journal of Eonomic Theory*, 52, 243–267.

Fishburn, P. (1974). Lexicographic orders, utilities, and decision rules: A survey. *Management Science*, 20, 1442–1471.

Fishburn, P (1982). *The Foundations of Expected Utility*. Dordrecht: Reidel.

Friedman, M and L Savage (1948). The utility analysis of choices involving risk. *Journal of Political Economy*, 56, 279–304.

Fudenberg, D, D Kreps, and D Levine (1988). On the robustness of equilibrium refinements. *Journal of Economic Theory*, 44, 354–380.

Hammond, P (1987). Extended probabilities for decision theory and games. Department of Economics, Stanford University.

Harper, W, R Stalnacker, and G Pearce (Eds.) (1981). *IFs: Conditionals, Belief, Decisions, Chance and Time*. Boston, MA: D. Reidel.

Hausner, M (1954). Multidimensional utilities. In Thrall, R, C Coombs, and R Davis (Eds.), *Decision Processes*. New York, NY: Wiley.

Kreps, D and G Ramey (1987). Structural consistency, consistency and sequential rationality. *Econometrica*, 55, 1331–1348.

Kreps, D and R Wilson (1982). Sequential equilibria. *Econometrica*, 50, 863–894.

Luce, R and H Raiffa (1957). *Games and Decisions*. New York, NY: Wiley.

McLennan, A (1989a). The space of conditional systems is a ball. *International Journal of Game Thoery*, 18, 125–139.

McLennan, A (1989b). Consistent conditional systems in noncooperative game theory. *International Journal of Game Theory*, 18, 140–174.

Myerson, R (1978): Refinements of the Nash equilibrium concept. *International Journal of Game Theory*, 7, 73–80.

Myerson, R (1986a). Multistage games with communication. *Econometrica*, 54, 323–358.

Myerson, R (1986b). Axiomatic foundations of Bayesian decision theory. Discussion Paper No. 671, J.L. Kellogg Graduate School of Management, Morthwestern University.

Pratt, J, H Raiffa, and R Schlaifer (1964). The foundations of decision under uncertainty: an elementary exposition. *Journal of American Statistical Association*, 59, 353–375.

Richter, M (1971). Rational choice. In Chipman, J, L Hurwicz, M Richter, and H Sonnenschein (Eds.), *Preferences, Utility and Demand*. New York, NY: Harcourt Brace Jovanovich.

Savage, L (1954). *The Foundations of Statistics*. New York, NY: Wiley.

Selten, R (1975). Reexamination of the perfectness concept of equilibrium points in extensive games. *International Journal of Game Theory*, 4, 25–55.

Suppes, P (1956). A set of axioms for paired comparisons. Unpublished, Center for Behavioral Sciences.

Chapter 7

Admissibility in Games

Adam Brandenburger, Amanda Friedenberg,
and H. Jerome Keisler

Suppose that each player in a game is rational, each player thinks the other players are rational, and so on. Also, suppose that rationality is taken to incorporate an admissibility requirement — that is, the avoidance of weakly dominated strategies. Which strategies can be played? We provide an epistemic framework in which to address this question. Specifically, we formulate conditions of rationality and mth-order assumption of rationality (RmAR) and rationality and common assumption of rationality (RCAR). We show that (i) RCAR is characterized by a solution concept we call a "self-admissible set"; (ii) in a "complete" type structure, RmAR is characterized by the set of strategies that survive $m + 1$ rounds of elimination of inadmissible strategies; (iii) under certain conditions, RCAR is impossible in a complete structure.

Originally published in *Econometrica*, 76, 307–352.

Keywords: Epistemic game theory; rationality; admissibility; iterated weak dominance; self-admissible sets; assumption; completeness.

Note: This chapter combines two papers, "Epistemic Conditions for Iterated Admissibility" (by Brandenburger and Keisler, June 2000) and "Common Assumption of Rationality in Games" (by Brandenburger and Friedenberg, January 2002).

Financial support: Harvard Business School, Stern School of Business CMS-EMS at Northwestern University, Department of Economics at Yale University, Olin School of Business, National Science Foundation and the Vilas Trust Fund.

Acknowledgments: We are indebted to Bob Aumann, Pierpaolo Battigalli, Martin Cripps, Joe Halpern, Johannes Hörner, Martin Osborne, Marciano Siniscalchi, and Gus Stuart for important input. Geir Asheim, Chris Avery, Oliver Board, Giacomo Bonanno, Ken Corts, Lisa DeLucia, Christian Ewerhart, Konrad Grabiszewski, Rena Henderson, Elon Kohlberg, Stephen Morris, Ben Polak, Phil Reny, Dov Samet, Michael Schwarz, Jeroen Swinkels, and participants in various seminars gave valuable comments. Eddie Dekel and the referees made very helpful observations and suggestions.

1. Introduction

What is the implication of supposing that each player in a game is rational, each player thinks the other players are rational, and so on? The natural first answer to this question is that the players will choose *iteratively undominated* (IU) strategies — that is, strategies that survive iterated deletion of strongly dominated strategies. Bernheim (1984) and Pearce (1984) gave essentially this answer, via their concept of rationalizability.[1] Pearce (1984) also defined the concept of a *best-response set* (BRS) and gave this as a more complete answer.

In this chapter we ask: What is the answer to the above question when rationality of a player is taken to incorporate an admissibility requirement — that is, the avoidance of weakly dominated strategies?

Our analysis will identify a weak-dominance analog to Pearce's concept of a BRS, which we call a self-admissible set (SAS). We will also identify conditions under which players will choose iteratively admissible (IA) strategies — that is, strategies that survive iterated deletion of weakly dominated strategies.

The case of weak dominance is important. Weak-dominance concepts give sharp predictions in many games of applied interest. Separate from its power in applications, admissibility is a prima facie reasonable criterion: It captures the idea that a player takes all strategies for the other players into consideration; none is entirely ruled out. It also has a long heritage in decision and game theory. (See the discussion in Kohlberg and Mertens [1986, Section 2.7].)

But there are significant conceptual hurdles to overcome in order understand admissibility in games. Below, we review some issues that have already been identified in the literature, add new ones, and offer a resolution.

The chapter is organized as follows. The next section is an informal discussion of the issues and results to follow. The formal treatment is in Sections 3–10. Section 11 discusses some open questions. The heuristic treatment of the next section can be read either before or in parallel with the formal treatment.[2]

---

[1] Under the original definition, which makes an independence assumption, the rationalizable strategies can be a strict subset of the iteratively undominated strategies. Recent definitions (e.g., Osborne and Rubinstein [1994]) allow for correlation; in this case, the two sets are equal.

[2] Online Supplemental material can be found in Brandenburger, Friedenberg, and Keisler (2008).

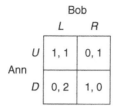

Figure 2.1.

2. Heuristic Treatment

We begin with the standard equivalence: Strategy s is admissible if and only if there is a strictly positive probability measure on the strategy profiles for the other players, under which s is optimal. In an influential paper, Samuelson (1992) pointed out that this poses a basic challenge for an analysis of admissibility in games. Consider the game in Figure 2.1, which is essentially Example 8 in Samuelson (1992).

Suppose rationality incorporates admissibility. Then, if Bob is rational, he should assign positive probability to both U and D, and so will play L. Likewise, if Ann is rational, presumably she should assign positive probability to both L and R. But if Ann thinks Bob is rational, should she not assign probability 1 to L? (We deliberately use the loose term "thinks." We will be more precise below.) The condition that Ann is rational appears to conflict with the condition that she thinks Bob is rational.

2.1. *Lexicographic probabilities*

Our method for overcoming this hurdle will be to allow Ann at the same time both to include and to exclude a strategy of Bob's. Ann will consider some of Bob's strategies infinitely less likely than others, but still possible. The strategies that get infinitesimal weight can be viewed as both included (because they do not get zero weight) and excluded (because they get only infinitesimal weight).

In Figure 2.2, Ann has a *lexicographic probability system* (LPS) on Bob's strategies. (LPS's were introduced in Blume, Brandenburger, and Dekel [1991a].) Ann's primary measure ("hypothesis") assigns probability 1 to L. Her secondary measure (depicted in square brackets) assigns probability 1 to R. Ann considers it infinitely more likely that Bob plays L than that Bob plays R, but does not entirely exclude R from consideration.

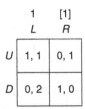

Figure 2.2.

In our lexicographic decision theory, Ann will choose strategy s over strategy s' if s yields a sequence of expected payoffs lexicographically greater than the sequence s' yields. So, with the LPS shown, she will choose U (not D).

Can we say that Ann believes Bob is rational? Customarily, we would say yes if Ann assigns probability 1 to the event that Bob is rational, but now Ann has an LPS that is not a single measure, so we need to look at the question at a deeper level. Recall, at the level of preferences, Ann believes an event E if her preference over acts, conditional on not-E, is trivial. (In short, not-E is Savage-null.) But, clearly, Ann's preference conditional on the event that Bob is irrational (plays R) is not trivial: under her secondary hypothesis, she chooses D over U.

We will settle for the weaker condition that Ann considers the event E infinitely more likely than not-E and, in this case, we will say Ann *assumes* E. (Later, we give assumption a preference basis.) In Figure 2.2, Ann considers it infinitely more likely that Bob is rational than irrational. This is our resolution of the tension between requiring Ann to be rational — in the sense of admissibility — and requiring her to think Bob is rational.

LPS's are a basic tool for dealing with the idea of unexpected events in the context of a strategic-form analysis. There is an analogous tool for analyzing the extensive form, namely conditional probability systems (CPS's). (The concept goes back to Rényi [1955].) CPS's are a key element of the Battigalli and Siniscalchi (2002) extensive-form epistemic analysis. Our (LPS-based) assumption concept is closely related to their (CPS-based) concept of "strong belief." In fact, there will be a close parallel between many of the ingredients in this chapter and in Battigalli and Siniscalchi (2002). Section 2.8 returns to discuss these connections and to highlight the big debt we owe to Battigalli and Siniscalchi.

2.2. *Rationality and common assumption of rationality*

In the game in Figure 2.2, the conditions that Bob is rational, and that Ann is rational and assumes Bob is rational, imply a unique strategy for each player.

In general, we can formulate an infinite sequence of conditions:

(a1) Ann is rational;

(a2) Ann is rational and assumes (b1);

(a3) Ann is rational, assumes (b1), assumes (b2);

...

(b1) Bob is rational;

(b2) Bob is rational and assumes (a1);

(b3) Bob is rational, assumes (a1), assumes (a2);

...

There is *rationality and common assumption of rationality* (RCAR) if this sequence holds. RCAR is a natural "baseline" epistemic condition on a game when rationality incorporates admissibility. We want to know what strategies can be played under RCAR.

To answer, we need some more epistemic apparatus. Let T^a and T^b be spaces of *types* for Ann and Bob, respectively. Each type t^a for Ann is associated with an LPS on the product of Bob's strategy and type spaces (i.e., on $S^b \times T^b$). Likewise for Bob. A state of the world is a 4-tuple (s^a, t^a, s^b, t^b), where s^a and t^a are Ann's actual strategy and type, and likewise for Bob. This is a standard *type structure* in the epistemic literature, with the difference that types are associated with LPS's, not single probability measures.

In these structures, rationality is a property of a strategy–type pair. A pair (s^a, t^a) is *rational* if it satisfies the following *admissibility* requirement: The LPS σ associated with t^a has full support (rules nothing out), and s^a lexicographically maximizes Ann's expected payoff under σ. (In particular, s^a is not weakly dominated.) Otherwise the pair is irrational. Likewise for Bob.

Start with a game and an associated type structure. We get a picture like Figure 2.3, where the outer rectangle is $S^b \times T^b$ and the shaded area is the strategy–type pairs satisfying "RCAR for Bob."

Now fix a strategy–type pair (s^a, t^a) that satisfies RCAR for Ann. Then Ann assumes (b1), assumes (b2), ... By a conjunction property of assumption, it follows that Ann assumes the joint event (b1) and (b2) and ..., that is, Ann assumes "RCAR for Bob." This gives a picture

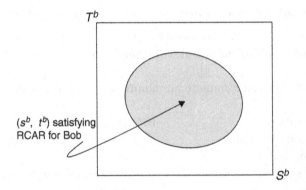

Figure 2.3.

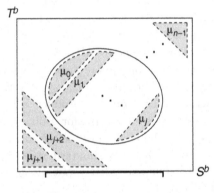

Figure 2.4.

like Figure 2.4, where the sequence of measures $(\mu_0, \ldots, \mu_{n-1})$ is the LPS associated with t^a. There is an initial segment $(\mu_0, \ldots, \mu_j)$ of this sequence which concentrates exactly on the event "RCAR for Bob." This is because Ann considers pairs (s^b, t^b) inside this event infinitely more likely than pairs outside the event.

Since (s^a, t^a) is rational, strategy s^a lexicographically maximizes Ann's expected payoff, under $(\mu_0, \ldots, \mu_{n-1})$. This establishes (by taking a convex combination of the marginals on S^b) that there is a strictly positive measure on S^b under which s^a is optimal. That is, s^a must be admissible. Strategy s^a must also lexicographically maximize Ann's expected payoff under the initial segment $(\mu_0, \ldots, \mu_j)$. It follows (again taking a convex combination of the marginals) that there is a strictly positive measure on the projection of the event "RCAR for Bob" under which s^a is optimal. That is, s^a must be admissible with respect to the projection.

Take the set of all states (s^a, t^a, s^b, t^b) satisfying RCAR and let $Q^a \times Q^b$ be its projection into $S^a \times S^b$. By the discussion, the product $Q^a \times Q^b$ has the following two properties:

(i) *each $s^a \in Q^a$ is admissible (i.e., is admissible with respect to S^b),*
(ii) *each $s^a \in Q^a$ is admissible with respect to Q^b;*
and likewise with a and b interchanged.

(Note the similarity of these properties to the definition of a best-response set (Pearce [1984]) — a concept based, of course, on strong dominance.) But these two properties are not yet enough to characterize RCAR, as the next example shows.

2.3. Convex combinations

Consider the game in Figure 2.5. The set $\{U\} \times \{L, R\}$ has properties (i) and (ii), but U cannot be played under RCAR. Indeed, fix a type structure and suppose (U, t^a) is rational. In terms of Ann's payoffs, $U = \frac{1}{2}N + \frac{1}{2}D$. It follows that (N, t^a) and (D, t^a) will also be rational. Next consider a strategy–type pair (s^b, t^b) for Bob, which is rational and assumes Ann is rational (i.e., Bob assumes the event (a1) defined in Section 2.2). So Bob considers rational pairs (s^a, t^a) for Ann infinitely more likely than irrational pairs. But then $s^b = R$, since the LPS associated with t^b must give each of U, N, and D positive probability before giving M positive probability. Now consider a strategy–type pair (s^a, t^a) for Ann, which is rational and such

Bob

		L	R
	U	1, 4	1, 4
	M	−1, 3	−1, 0
Ann	N	2, 0	0, 3
	D	0, 0	2, 3

Figure 2.5.

that Ann assumes Bob is rational and assumes Bob is rational and assumes Ann is rational (that is, Ann assumes the events (b1) and (b2)). We get $s^a = D$ (not U), since the LPS associated with t^a must give R positive probability before giving L positive probability.

The key to the example is that U is a convex combination for Ann of N and D, so that (N, t^a) and (D, t^a) are rational whenever (U, t^a) is. This suggests that the projection of the RCAR set should have the following property:

(iii) *if $s^a \in Q^a$ and r^a is part of a convex combination of strategies for Ann that is equivalent for her to s^a, then $r^a \in Q^a$;*
and likewise for Bob.

We define a *self-admissible set* (SAS) to be a set $Q^a \times Q^b \subseteq S^a \times S^b$ of strategy pairs which has properties (i), (ii), and (iii).[3] The strategies played under RCAR always constitute an SAS (Theorem 8.1(i)).

2.4. *Irrationality*

Does the converse hold? That is, given an SAS, is there an associated type structure so that the strategies played under RCAR correspond to this SAS?

To address the converse direction, we need to consider a further aspect of admissibility in games. Under admissibility, Ann considers everything possible. But this is only a decision-theoretic statement. Ann is in a game, so we imagine she asks herself: "What about Bob? What does he consider possible?" If Ann truly considers everything possible, then it seems she should, in particular, allow for the possibility that Bob does not! Alternatively put, it seems that a full analysis of the admissibility requirement should include the idea that other players do not conform to the requirement.

More precisely, we know that if a strategy–type pair (s^a, t^a) for Ann is rational, then the LPS associated with t^a has full support. But we are going to allow Ann to consider the possibility that there are types t^b for Bob associated with LPS's that do not have full support. (Ann allows that Bob does not consider everything.) Of course, by definition, if (s^b, t^b) is a rational pair for Bob, then the LPS associated with t^b will have full support.

---

[3]Brandenburger and Friedenberg (2004) investigated properties of SAS's.

Bob

		L	C	R
	U	4, 0	4, 1	0, 1
Ann	M	0, 0	0, 1	4, 1
	D	3, 0	2, 1	2, 1

Figure 2.6.

But there may be other strategy–type pairs present too. Our argument is that the presence of such pairs is conceptually appropriate if the topic is admissibility in games.

To see the significance of this, consider the game in Figure 2.6 (kindly provided by Pierpaolo Battigalli). The set $\{U, M, D\} \times \{C, R\}$ is an SAS. (It is also the IA set.) With the converse direction in mind, let us understand why D is consistent with RCAR.

Fix a type structure. Notice that L is (strongly) dominated, so all pairs (L, t^b) for Bob are irrational. A pair (C, t^b) or (R, t^b) will be rational if the LPS associated with t^b has full support, and irrational otherwise. We use this in the following.

Turn to Ann. Notice that if D is optimal under a measure, then the measure either assigns probability $\frac{1}{2}$ to C and $\frac{1}{2}$ to R or assigns positive probability to both L and R. Moreover, in the first case, U and M will necessarily be optimal too.

Fix a rational pair (D, t^a), where t^a assumes Bob is rational. Let $(\mu_0, \ldots, \mu_{n-1})$ be the full-support LPS associated with t^a. By the full-support condition, there is some measure that gives $\{L\} \times T^b$ positive probability. Let μ_i be the first such measure. Also, since t^a assumes "Bob is rational," the rational strategy–type pairs for Bob must be infinitely more likely than the irrational pairs. Therefore, $i \neq 0$. Using the rationality of (D, t^a), we now have that for each measure μ_k with $k < i$: (i) μ_k assigns probability $\frac{1}{2}$ to $\{C\} \times T^b$ and probability $\frac{1}{2}$ to $\{R\} \times T^b$, and (ii) U, M, and D are each optimal under μ_k. It follows that D must also be optimal under μ_i, and so μ_i must assign positive probability to both $\{L\} \times T^b$ and $\{R\} \times T^b$. Now use again the fact that rational strategy–type pairs for Bob must be infinitely more likely than the irrational pairs. Since each point in $\{L\} \times T^b$ is irrational, μ_i must assign strictly positive probability to the irrational pairs in $\{R\} \times T^b$. This is possible if there are non-full-support types for Bob.

It is important to understand that we have two forms of irrationality in this chapter. One is more or less standard: A strategy–type pair is irrational if s^a is not optimal under the LPS associated with t^a. This is just the usual notion of irrationality, but now optimality is defined lexicographically. In a type structure, some strategy–type pairs "do their sums right" and optimize, and others do not. Both kinds of pairs are present, but the latter kind do not play a special role in our analysis.

There is a second form of irrationality which is new. For us, a player is rational if he optimizes and also rules nothing out. So irrationality might mean not optimizing. But it can also mean optimizing while not considering everything possible (the LPS associated with t^a does not have full support). This form of irrationality is present in the example above, and it plays a central role in our analysis. (See also Section 11C.)

To keep things simple, we use the one term "irrationality" to cover both situations, but we repeat that there are these two cases.

2.5. Characterization of RCAR

We can now state our characterization of RCAR in games (Theorem 8.1):

> Start with a game and an associated type structure. Let $Q^a \times Q^b$ be the projection into $S^a \times S^b$ of the states (s^a, t^a, s^b, t^b) satisfying RCAR. Then $Q^a \times Q^b$ is an SAS of the game.

We also have:

> Start with a game and an SAS $Q^a \times Q^b$. There is a type structure (with non-full-support types) such that $Q^a \times Q^b$ is the projection into $S^a \times S^b$ of the states (s^a, t^a, s^b, t^b) satisfying RCAR.

It is easy to check that the IA strategies constitute an SAS of a game. So, in particular, every game possesses an SAS, and RCAR is possible in every game. But a game may possess other SAS's too. In the game in Figure 2.7, there are three SAS's: $\{(U, L)\}, \{U\} \times \{L, R\}$, and $\{(D, R)\}$. (The third is the IA set. Note that the other two SAS's are not contained in the IA set. This is different from the case of strong dominance: It is well known that any Pearce best-response set is contained in the set of strategies that survives iterated strong dominance).

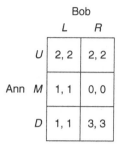

Figure 2.7.

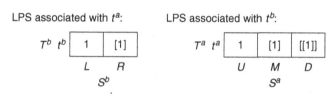

Figure 2.8.

2.6. *Iterated admissibility*

So the question remains: What epistemic conditions select the IA set in a game from among the family of SAS's? To investigate this, consider Figure 2.8, which gives a type structure for the game in Figure 2.7. Ann and Bob each have a single type. Ann's LPS assigns primary probability 1 to (L, t^b), and secondary probability 1 (in square brackets) to (R, t^b). Bob's LPS assigns primary probability 1 to (U, t^a), secondary probability 1 (in square brackets) to (M, t^a), and tertiary probability 1 (in double square brackets) to (D, t^a). Ann (resp. Bob) has just one rational strategy–type pair, namely (U, t^a) (resp. (L, t^b)). Ann's unique type t^a assumes Bob is rational (the rational pair (L, t^b) is considered infinitely more likely than the irrational pair (R, t^b)). Likewise, Bob's unique type t^b assumes Ann is rational (the rational pair (U, t^a) is considered infinitely more likely than the irrational pairs (M, t^a) and (D, t^a)). By induction, the RCAR set is then the singleton $\{(U, t^a, L, t^b)\}$. This is an instance of Theorem 8.1: The projection into $S^a \times S^b$ of $\{(U, t^a, L, t^b)\}$ is an SAS, namely $\{(U, L)\}$.

In this structure, Ann assumes Bob plays L, making U her unique rational choice. Both M and D are irrational for her. In fact, Bob considers

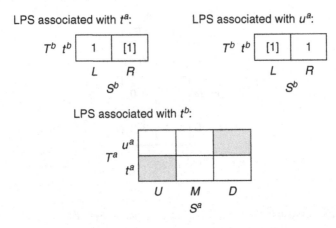

Figure 2.9.

it infinitely more likely that Ann plays M than D, which is why he plays L. Bob is free to assign the probabilities this way. To assume Ann is rational, it is enough that Bob considers U infinitely more likely than both M and D, as he does.

What if Bob considered D infinitely more likely than M? Then he would rationally play R not L. Presumably Ann would then play D and the IA set would result. Figure 2.9 gives a scenario under which Bob will, in fact, consider D infinitely more likely than M. We have added a type u^a for Ann that assumes Bob plays R. Now there is a second rational pair for Ann, namely (D, u^a). (Note there is no type v^a for Ann which we could add to the structure to make (M, v^a) rational for Ann, since M is inadmissible.) If Bob assumes Ann is rational, then he must consider the shaded pairs in Figure 2.9 infinitely more likely than the unshaded pairs. If rational, he must play R, as desired.

Call a type structure *complete* if the range of the map from T^a (Ann's type space) to the space of LPS's on $S^b \times T^b$ (Bob's strategy space cross Bob's type space) properly contains the set of full-support LPS's on $S^b \times T^b$, and similarly with Ann and Bob interchanged. More loosely, a type structure is complete if it contains all possible full-support types and at least one nonfull-support type (as per Section 2.4). Complete type structures exist for every finite game (Proposition 7.2). Figure 2.9 suggests that with this setup, we should be able to identify the IA strategies.

For $m \geq 0$, say there is *rationality and mth-order assumption of rationality* (RmAR) if conditions (a($m + 1$)) and (b($m + 1$)) of Section 2.2

hold. We have (Theorem 9.1):

> *Start with a game and an associated complete type structure. Let $Q^a \times Q^b$ be the projection into $S^a \times S^b$ of the states (s^a, t^a, s^b, t^b) satisfying RmAR. Then $Q^a \times Q^b$ is the set of strategies that survive $(m+1)$ rounds of IA.*

2.7. A negative result

Note that our Theorem 9.1 actually identifies, for any m, the $(m+1)$-iteratively admissible strategies, not the IA strategies. Of course, for a given (finite) game, there is a number M such that for all $m \geq M$, the m-iteratively admissible strategies coincide with the IA strategies. Nevertheless, our result is not quite an epistemic condition for IA in all finite games. That would be one common condition — across all games — that yields IA. For example, one might hope to characterize the IA set as the projection of a set of states which is constructed in a uniform way in all complete type structures.

One would expect the RCAR set to be a natural candidate for this set of states. But the following negative result (Theorem 10.1) shows that RCAR will not work and is the reason for our limited statement of Theorem 9.1:

> *Start with a game in which Ann has more than one "strategically distinct" strategy, and an associated continuous complete type structure. Then no state satisfies RCAR.*

For the meaning of a continuous type structure, see Definition 7.8. The complete type structure we get from our existence result (Proposition 7.2) is continuous.

In a certain sense, the result says that players cannot "reason all the way." Here is an intuition for the result. Suppose the RCAR set is nonempty. Then there must be a type t^a for Ann that assumes each of the decreasing sequence of events (b1), (b2), ... (these events were defined in Section 2.2). That is, strategy–type pairs not in (b1) must be considered infinitely less likely than pairs in (b1); pairs not in (b2) must be considered infinitely less likely than pairs in (b2), and so on. Let $(\mu_0, \dots, \mu_{n-1})$ be the LPS associated with t^a. Figure 2.10 shows the most parsimonious way to arrange the measures μ_i, so that Ann indeed assumes each of (b1), (b2), ... But even in this case, we will run out of measures and Ann will not be able to assume any of the events (bn), (b(n+1)), ... More loosely, at some point

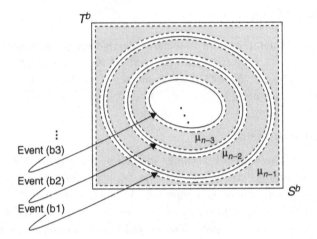

Figure 2.10.

Ann will hit her primary hypothesis μ_0, at which point there is no next (more likely) order of likelihood.

In the complete type structure we get from Proposition 7.2, each event $(b(m+1))$ is "significantly" smaller than event (bm). This is because Bob has many types that assume the event $(a(m-1))$ but not the event (am). So the measures μ_i do indeed have to be arranged as shown. This was not true in the incomplete structure of Figure 2.8, where these events do not shrink at all. That is why we had a state there satisfying RCAR, while no such state exists for the complete structure of Proposition 7.2.

2.8. The ingredients

To recap: We begin with the fundamental inclusion–exclusion challenge identified by Samuelson (1992). Our resolution is to allow some states to be infinitely more likely than others. (We do this by using LPS's and the concept of assumption.) Then we characterize the strategies consistent with RCAR as the SAS's of a game. In a complete type structure, the strategies consistent with $RmAR$ are those that survive $(m+1)$ rounds of iterated admissibility. However, under certain conditions, RCAR in a complete structure is impossible.

Our examination of admissibility builds on fundamental work on the tree by Battigalli and Siniscalchi (2002). They study the solution concept of extensive-form rationalizability (EFR), an extensive-form analog to the iteratively undominated (IU) strategies. (The concept was defined by Pearce [1984] and later simplified by Battigalli [1997].)

Battigalli and Siniscalchi used conditional probability systems (CPS's) to describe what players believe given what they observe in the tree. They next introduced the concept of strong belief. (This is the requirement that a player assign probability 1 to an event at each information set that is consistent with the event.) With rationality defined for the tree, they showed:

Start with a game and an associated (CPS-based) complete type structure. Let $Q^a \times Q^b$ be the projection into $S^a \times S^b$ of the states (s^a, t^a, s^b, t^b) satisfying rationality and mth-order strong belief of rationality. Then $Q^a \times Q^b$ is the set of strategies that survive $(m+1)$ rounds of elimination of EFR.

Clearly, our Theorem 9.1 (previewed in Section 2.6) is closely related. In terms of ingredients, LPS's and CPS's can be formally related. See Halpern (2007) for a general treatment. Assumption can be viewed as a strategic-form analog to strong belief. Asheim and Søvik (2005) explored this connection; see also our companion piece (Brandenburger, Friedenberg, and Keisler [2006]). The role of completeness in our analysis is similar to its role in Battigalli and Siniscalchi.

There are also similarities in terms of output. IA and EFR are outcome equivalent in generic trees.[4] Of course, many games of interest are non-generic.[5] In simultaneous-move games, EFR reduces to IU. IA and EFR will then differ whenever IA and IU do.

There are other differences between the two analyses. In Battigalli and Siniscalchi, there is no analog to our negative result (Theorem 10.1). The reason is that full-support LPS's are, in a particular sense, more informative than CPS's on the tree. The online Supplemental material gives an exact treatment of this difference.

Also, we cover the case of incomplete type structures via our Theorem 8.1 (previewed in Section 2.5). We think of a particular incomplete structure as giving the "context" in which the game is played. In line with Savage's Small-Worlds idea in decision theory (Savage [1954, pp. 82–91]),

---

[4]See Brandenburger and Friedenberg (2003). Shimoji (2004) has a result relating IA and EFR, where EFR is defined relative to "normal-form information sets" (Mailath, Samuelson, and Swinkels [1993]).

[5]Examples include auction games, voting games, Bertrand, and zero-sum games. See Mertens (1989) and Marx and Swinkels (1997) for the same observation on nongenericity, and lists of examples.

who the players are in the given game can be seen as a shorthand for their experiences before the game. The players' possible characteristics — including their possible types — then reflect the prior history or context. (Seen in this light, complete structures represent a special "context-free" case, in which there has been no narrowing down of types.) SAS's are our characterization of the epistemic condition of RCAR in the contextual case.[6]

3. SAS's and the IA Set

We now begin the formal treatment. Fix a two-player finite strategic-form game $\langle S^a, S^b, \pi^a, \pi^b \rangle$, where S^a, S^b are the (finite) strategy sets, and π^a, π^b are payoff functions for Ann and Bob, respectively.[7] Given a finite set X, let $\mathcal{M}(X)$ denote the set of all probability measures on X. The definitions to come all have counterparts with a and b reversed. We extend π^a to $\mathcal{M}(S^a) \times \mathcal{M}(S^b)$ in the usual way: $\pi^a(\sigma^a, \sigma^b) = \sum_{(s^a, s^b) \in S^a \times S^b} \sigma^a(s^a)\sigma^b(s^b)\pi^a(s^a, s^b)$. Throughout, we adopt the convention that in a product $X \times Y$, if $X = \emptyset$, then $Y = \emptyset$ (and *vice versa*).

Definition 3.1. Fix $X \times Y \subseteq S^a \times S^b$. A strategy $s^a \in X$ is *weakly dominated with respect to* $X \times Y$ if there exists $\sigma^a \in \mathcal{M}(S^a)$, with $\sigma^a(X) = 1$, such that $\pi^a(\sigma^a, s^b) \geq \pi^a(s^a, s^b)$ for every $s^b \in Y$ and $\pi^a(\sigma^a, s^b) > \pi^a(s^a, s^b)$ for some $s^b \in Y$. Otherwise, say s^a is *admissible with respect to* $X \times Y$. If s^a is admissible with respect to $S^a \times S^b$, simply say that s^a is *admissible*.

Write Supp σ for the support of σ. We have the usual equivalence:

Lemma 3.1. *A strategy $s^a \in X$ is admissible with respect to $X \times Y$ if and only if there exists $\sigma^b \in \mathcal{M}(S^b)$, with Supp $\sigma^b = Y$, such that $\pi^a(s^a, \sigma^b) \geq \pi^a(r^a, \sigma^b)$ for every $r^a \in X$.*

Definition 3.2. Say r^a *supports* s^a if there exists some $\sigma^a \in \mathcal{M}(S^a)$ with $r^a \in$ Supp σ^a and $\pi^a(\sigma^a, s^b) = \pi^a(s^a, s^b)$ for all $s^b \in S^b$. Write su(s^a) for the set of $r^a \in S^a$ that support s^a.

In words, the strategy r^a is contained in su(s^a) if it is part of a convex combination of Ann's strategies that is equivalent for her to s^a.

---

[6]The online Supplemental material (Brandenburger, Friedenberg, and Keisler [2008]) contains discussion of other related work.

[7]For notational simplicity, we restrict attention throughout to two-player games. But the analysis can be extended without change to games with three or more players.

We can now define SAS's and the IA set:

Definition 3.3. Fix $Q^a \times Q^b \subseteq S^a \times S^b$. The set $Q^a \times Q^b$ is a *self-admissible set (SAS)* if:

(i) each $s^a \in Q^a$ is admissible;
(ii) each $s^a \in Q^a$ is admissible with respect to $S^a \times Q^b$;
(iii) for any $s^a \in Q^a$, if $r^a \in \text{su}(s^a)$, then $r^a \in Q^a$;
and likewise for each $s^b \in Q^b$.

Definition 3.4. Set $S_0^i = S^i$ for $i = a, b$ and define inductively

$$S_{m+1}^i = \{s^i \in S_m^i : s^i \text{ is admissible with respect to } S_m^a \times S_m^b\}.$$

A strategy $s^i \in S_m^i$ is called *m-admissible*. A strategy $s^i \in \bigcap_{m=0}^{\infty} S_m^i$ is called *iteratively admissible (IA)*.

Note that there is an M such that $\bigcap_{m=0}^{\infty} S_m^i = S_M^i$ for $i = a, b$. Moreover, each set S_m^i is nonempty and, hence, the IA set is nonempty.

4. Lexicographic Probability Systems

Given a Polish space Ω, it will be helpful to fix a metric. (So "Polish" will mean complete separable metric.) Let $\mathcal{M}(\Omega)$ be the space of Borel probability measures on Ω with the Prohorov metric. Recall that $\mathcal{M}(\Omega)$ is again a Polish space and has the topology of weak convergence (Billingsley [1968, Appendix III]). Let $\mathcal{N}(\Omega)$ be the set of all finite sequences of Borel probability measures on Ω. That is, if $\sigma \in \mathcal{N}(\Omega)$, then there is some integer n with $\sigma = (\mu_0, \ldots, \mu_{n-1})$.

Define a metric on $\mathcal{N}(\Omega)$ as follows. The distance between two sequences of measures $(\mu_0, \ldots, \mu_{n-1})$ and $(\nu_0, \ldots, \nu_{n-1})$ of the same length is the maximum of the Prohorov distances between μ_i and ν_i for $i < n$. The distance between two sequences of measures of different lengths is 1. For each fixed n, this metric on the set of sequences in $N(\Omega)$ of length n is easily seen to be separable and complete, and thus Polish (this is the usual finite product metric). The whole space $N(\Omega)$ is thus a countable union of Polish spaces at uniform distance 1 from each other. This shows that $\mathcal{N}(\Omega)$ itself is a Polish space.

Definition 4.1. Fix $\sigma = (\mu_0, \ldots, \mu_{n-1}) \in \mathcal{N}(\Omega)$, for some integer n. Say σ is a *lexicographic probability system (LPS)* if σ is *mutually singular* — that is, for each $i = 0, \ldots, n-1$, there are Borel sets U_i in Ω with $\mu_i(U_i) = 1$

and $\mu_i(U_j) = 0$ for $i \neq j$. Write $\mathcal{L}(\Omega)$ for the set of LPS's and write $\overline{\mathcal{L}}(\Omega)$ for the closure of $\mathcal{L}(\Omega)$ in $\mathcal{N}(\Omega)$.

An LPS is a finite measure sequence where the measures are nonoverlapping (mutually singular). This has the usual interpretation: the player's primary hypothesis, secondary hypothesis, and so on, until an nth hypothesis.

In general, an LPS may have some null states which remain outside the support of each of its measures. We are also interested in the case that there are no such null states:

Definition 4.2. A *full-support sequence* is a sequence $\sigma = (\mu_0, \dots, \mu_{n-1}) \in \mathcal{N}(\Omega)$ such that $\Omega = \cup_{i<n}$ Supp μ_i. We write $\mathcal{N}^+(\Omega)$ for the set of full-support sequences and write $\mathcal{L}^+(\Omega)$ for the set of *full-support LPS's*.

Here, Supp μ_i denotes the support of μ_i, that is, the smallest closed set with μ_i-measure 1. The space $\overline{\mathcal{L}}(\Omega)$ is Polish, since it is a closed subspace of the Polish space $\mathcal{N}(\Omega)$. Also, the sets $\mathcal{N}^+(\Omega), \mathcal{L}(\Omega)$, and $\mathcal{L}^+(\Omega)$ are Borel (Corollary C.1).

Our definition of an LPS is a infinite version of the definition for finite spaces introduced by Blume *et al.* (1991a). Infinite spaces play a crucial role in this chapter — complete type structures (recall the discussion in Section 2.6) are infinite. (A note on terminology: Blume, Brandenburger, and Dekel [1991a] used the term LPS even when mutual singularity does not hold.)

5. Assumption

Here, we define formally the concept of assumption, which was introduced informally in Section 2.1. Fix a full-support LPS $\sigma = (\mu_0, \dots, \mu_{n-1})$ for Ann and fix an event E. Intuitively, Ann assumes E if she considers E infinitely more likely than not-E under σ. So, to define assumption, we first need to understand the idea of "infinitely more likely than."

Blume, Brandenburger, and Dekel (1991a, Definition 5.1) gave a definition of "infinitely more likely than" for the case of a finite space Ω and a full-support LPS $\sigma = (\mu_0, \dots, \mu_{n-1})$ (see their Axiom 5'). They say that a point ω_1 is infinitely more likely than a point ω_2 if ω_1 comes before ω_2 in the lexicographic ordering. For disjoint events F and G, they require that F is nonempty and each point in F is infinitely more likely than each point in G. Formally, the requirement is that: F is nonempty, and for each $\omega_1 \in F$ and $\omega_2 \in G$, $\mu_j(\omega_1) > 0$ and $\mu_k(\omega_2) > 0$ implies $j < k$. (The same idea of "infinitely more likely than" can be found in Battigalli [1996, p. 186] and Asheim and Dufwenberg [2003].)

We want a general (i.e., infinite) analog to this definition, so we work with open sets rather than just points. Call F_0 a *part* of F if $F_0 = U \cap F \neq \emptyset$ for some open U. Instead of asking that each point in F be infinitely more likely than each point in G, we require that each part of F be infinitely more likely than each part of G.

Definition 5.1. Fix a full-support LPS $\sigma = (\mu_0, \ldots, \mu_{n-1}) \in \mathcal{L}^+(\Omega)$ and disjoint events F and G. Then F is *infinitely more likely than* G under σ if F is nonempty and, for any part F_0 of F:

(a) $\mu_i(F_0) > 0$ for some i;
(b) if $\mu_j(F_0) > 0$ and there is a part G_0 of G with $\mu_k(G_0) > 0$, then $j < k$.

Note that for finite Ω, this is equivalent to the Blume, Brandenburger, and Dekel (1991a) definition. In particular, condition (a) is then automatically satisfied (since every point gets positive probability under some μ_j). In the general case, we need to require (a) explicitly. Without it, we could have that F is infinitely more likely than G, but at the same time G is infinitely more likely than F. This would not make sense. (See the online Supplemental material.)

The idea of assumption of an event E is simply that E is considered infinitely more likely than not-E:

Definition 5.2. Fix an event E and a full-support LPS $\sigma = (\mu_0, \ldots, \mu_{n-1}) \in \mathcal{L}^+(\Omega)$. Say E is *assumed* under σ if E is infinitely more likely than $\Omega \backslash E$ under σ.

We have the following characterization of assumption[8]:

Proposition 5.1. *Fix an event E and a full-support LPS $\sigma = (\mu_0, \ldots, \mu_{n-1}) \in \mathcal{L}^+(\Omega)$. An event E is assumed under σ if and only if there is a j such that:*

(i) $\mu_i(E) = 1$ *for all* $i \leq j$;
(ii) $\mu_i(E) = 0$ *for all* $i > j$;
(iii) *if* U *is open with* $U \cap E \neq \emptyset$, *then* $\mu_i(U \cap E) > 0$ *for some* i.

(We will sometimes say that E is *assumed at level j*. Also, we will refer to conditions (i)–(iii) of Proposition 5.1 as conditions (i)–(iii) of assumption.)

---

[8]Proofs not given in the main text can be found in the Appendices.

Note that if Ω is finite, conditions (i) and (ii) imply condition (iii). But this is not the case when Ω is infinite. (See the online Supplemental material.)

As with the usual notion of "belief" of an event E, assumption can be given an axiomatic treatment. Appendix A proposes two axioms: *Strict Determination* says that whenever Ann strictly prefers one act to another conditional on E, she has the same preference unconditionally. *Nontriviality* says that, conditional on any part of E, she can have a strict preference. In Appendix A, we show that Ann assumes E if and only if her preferences satisfy these axioms. We also relate this axiomatization to the axiomatization of "infinitely more likely than" in Blume, Brandenburger, and Dekel (1991a, Definition 5.1).

6. Properties of Assumption

We next mention some properties of assumption. (Again, we use an overbar to denote closure.)

Property 6.1 — Convexity: *If E and F are assumed under σ at level j, then any Borel set G lying between $E \cap F$ and $E \cup F$ is also assumed under σ at level j.*

Property 6.2 — Closure: *If E and F are assumed under σ at level j, then $\overline{E} = \overline{F}$. If E and F are assumed under σ, then either $\overline{E} \subseteq \overline{F}$ or $\overline{F} \subseteq \overline{E}$.*

The Convexity property refers to convexity in the sense of orderings (where the order is set inclusion), and is a two-sided monotonicity. The Closure property implies that, for a finite space, there is only one set that is assumed at each level. Also, in the finite case, if E and F are both assumed, then $E \subseteq F$ or $F \subseteq E$. Neither statement is true for an infinite space.

Overall, the mental picture we suggest for assumption is of rungs of a ladder, separated by gaps, where each rung is a convex family of sets with the same closure. (Each rung corresponds to the events assumed at the particular level.)

Next, notice that assumption is not monotonic. Here is an example: Set $\Omega = [0,1] \cup \{2,3\}$, and let $\sigma = (\mu_0, \mu_1)$ be a full-support LPS where μ_0 is uniform on $[0,1]$ and $\mu_1(\{2\}) = \mu_1(\{3\}) = \frac{1}{2}$. Then σ assumes $(0,1]$ but not $(0,1] \cup \{2\}$.

The best way to understand this nonmonotonicity is in terms of our axiomatic treatment.[9] Suppose Ann assumes $(0, 1]$ — that is, when she has a strict preference, she is willing to make a decision based solely on $(0, 1]$. (This is Strict Determination.) It does not seem natural to require that Ann also be willing to make a decision based only on $(0, 1] \cup \{2\}$. After all, she considers the possibility that 2 obtains. (Nontriviality implies that the state 2 must get positive weight under some measure — as it does under μ_1.) Once she considers this possibility, presumably she should also consider the possibility that 3 obtains. (To give 2 positive probability, she must look to her secondary hypothesis, which also gives 3 positive probability.) Of course, the state 3 may well matter for her preferences.

On the other hand, if Ann assumes $(0, 1]$, then certainly she should assume $[0, 1]$. Admitting the possibility of 0 does not force her to look to her secondary hypothesis — it does not force her to consider 2 or 3 possible. Formally, Ann assumes $[0, 1)$ and $(0, 1]$ at the same level. Convexity then requires her to assume $[0, 1]$ (at the same level).

Because of the nonmonotonicity, assumption fails one direction of conjunction. Returning to the example, Ann assumes $(0, 1] \cap ((0, 1] \cup \{2\})$ even though she does not assume $(0, 1] \cup \{2\}$. But the other direction of conjunction, and the analog for disjunction, are satisfied:

Property 6.3 — Conjunction and Disjunction: *Fix Borel sets $E_1, E_2, \ldots$ in Ω and suppose that, for each m, E_m is assumed under σ. Then $\bigcap_m E_m$ and $\bigcup_m E$ are assumed under σ.*

7. Type Structures

Fix again a two-player finite strategic-form game $\langle S^a, S^b, \pi^a, \pi^b \rangle$.

Definition 7.1. An (S^a, S^b)-*based type structure* is a structure

$$\langle S^a, S^b, T^a, T^b, \lambda^a, \lambda^b \rangle,$$

where T^a and T^b are nonempty Polish spaces, and $\lambda^a : T^a \to \overline{\mathcal{L}}(S^b \times T^b)$ and $\lambda^b : T^b \to \overline{\mathcal{L}}(S^a \times T^a)$ are Borel measurable. Members of T^a, T^b are called *types*. Members of $S^a \times T^a \times S^b \times T^b$ are called *states (of the world)*. A type structure is called *lexicographic* if $\lambda^a : T^a \to \mathcal{L}(S^b \times T^b)$ and $\lambda^b : T^b \to \mathcal{L}(S^a \times T^a)$.

---

[9] We thank a referee for this line of argument.

Definition 7.1 is based on a standard epistemic definition: A type structure enriches the basic description of a game by appending spaces of epistemic types for both players, where a type for a player is associated with a sequence of measures on the strategies and types for the other player. The difference from the standard definition is the use of a sequence of measures rather than one measure.

Our primary focus will be on lexicographic type structures, which have a natural interpretation in a game setting. Nonlexicographic type structures will play a useful role in the construction of lexicographic type structures. Note that lexicographic type structures can contain two different kinds of types — those associated with full-support LPS's and those associated with non-full-support LPS's. The reason for this was discussed in Section 2.4.

The following definitions apply to a given game and type structure. As before, they also have counterparts with a and b reversed. Write $\mathrm{marg}_{S^b}\,\mu_i$ for the marginal on S^b of the measure μ_i.

Definition 7.2. A strategy s^a is *optimal* under $\sigma = (\mu_0, \ldots, \mu_{n-1})$ if $\sigma \in \mathcal{L}(S^b \times T^b)$ and

$$(\pi^a(s^a, \mathrm{marg}_{S^b}\mu_i(s^b)))_{i=0}^{n-1} \geq^L (\pi^a(r^a, \mathrm{marg}_{S^b}\mu_i(s^b)))_{i=0}^{n-1}$$

for all $r^a \in S^a$.[10]

In words, Ann will prefer strategy s^a to strategy r^a if the associated sequence of expected payoffs under s^a is lexicographically greater than the sequence under r^a. (If σ is a length-1 LPS (μ_0), we will sometimes say that s^a is optimal under the measure μ_0 if it is optimal under (μ_0).)

Definition 7.3. A type $t^a \in T^a$ has *full support* if $\lambda^a(t^a)$ is a full-support LPS.

Definition 7.4. A strategy–type pair $(s^a, t^a) \in S^a \times T^a$ is *rational* if t^a has full support and s^a is optimal under $\lambda^a(t^a)$.

This is the usual definition of rationality, plus the full-support requirement, which is to capture our basic admissibility requirement. The following two lemmas say this formally:

Lemma 7.1. — *Blume, Brandenburger, and Dekel* (1991b): *Suppose s^a is optimal under a full-support LPS $(\mu_0, \ldots, \mu_{n-1}) \in \mathcal{L}^+(S^b \times T^b)$. Then there is a length-1 full-support LPS $(\nu_0) \in \mathcal{L}^+(S^b \times T^b)$, under which s^a is optimal.*

---

[10]If $x = (x_0, \ldots, x_{n-1})$ and $y = (y_0, \ldots, y_{n-1})$, then $x \geq^L y$ if and only if $y_j > x_j$ implies $x_k > y_k$ for some $k < j$.

Together with Lemma 3.1, this gives the following lemma:

Lemma 7.2. *If (s^a, t^a) is rational, then s^a is admissible.*

Fix an event $E \subseteq S^b \times T^b$ and write

$$A^a(E) = \{t^a \in T^a : \lambda^a(t^a) \text{ assumes } E\}.$$

The set $A^a(E)$ is Borel (Lemma C.3).

Let R_1^a be the set of rational strategy–type pairs (s^a, t^a). For finite m, define R_m^a inductively by

$$R_{m+1}^a = R_m^a \cap [S^a \times A^a(R_m^b)].$$

The sets R_m^a are Borel (Lemma C.4).

Definition 7.5. If $(s^a, t^a, s^b, t^b) \in R_{m+1}^a \times R_{m+1}^b$, say there is *rationality and mth-order assumption of rationality* (RmAR) at this state. If $(s^a, t^a, s^b, t^b) \in \bigcap_{m=1}^{\infty} R_m^a \times \bigcap_{m=1}^{\infty} R_m^b$, say there is *rationality and common assumption of rationality* (RCAR) at this state.

In words, there is RCAR at a state if Ann is rational, Ann assumes the event "Bob is rational," Ann assumes the event "Bob is rational and assumes Ann is rational," and so on, and similarly starting with Bob.

Note, we cannot replace this definition with $\widehat{R}_1^a = R_1^a$ and $\widehat{R}_{m+1}^a = \widehat{R}_1^a \cap [S^a \times A^a(\widehat{R}_m^b)]$. To clarify, suppose $(s^a, t^a) \in R_3^a$. Then $(s^a, t^a) \in R_1^a \cap [S^a \times A^a(R_1^b)] \cap [S^a \times A^a(R_1^b \cap [S^b \times A^b(R_1^a)])]$. In words, Ann is rational, she assumes the event "Bob is rational," and she assumes the event "Bob is rational and assumes Ann is rational." Now suppose $(s^a, t^a) \in \widehat{R}_3^a$. Then $(s^a, t^a) \in R_1^a \cap [S^a \times A^a(R_1^b \cap [S^b \times A^b(R_1^a)])]$. In words, Ann is rational, and she assumes the event "Bob is rational and assumes Ann is rational." But, because assumption is not monotonic, she might not assume the event "Bob is rational." We think that under a good definition of R2AR, Ann should assume this event.

Next is a notion of equivalence between type structures.

Definition 7.6. Two type structures $\langle S^a, S^b, T^a, T^b, \kappa^a, \kappa^b \rangle$ and $\langle S^a, S^b, T^a, T^b, \lambda^a, \lambda^b \rangle$ are *equivalent* if:

(i) they have the same strategy and type spaces;

(ii) for each $t^a \in T^a$, if either $\kappa^a(t^a)$ or $\lambda^a(t^a)$ belongs to $\mathcal{L}^+(S^b \times T^b)$, then $\kappa^a(t^a) = \lambda^a(t^a)$ (and likewise with a and b reversed).

Proposition 7.1.

(i) *For every type structure there is an equivalent lexicographic type structure.*

(ii) *If two type structures are equivalent, then for each m they have the same R_m^a and R_m^b sets.*

This proposition shows that any statement about rationality and mth-order assumption of rationality (for any m) that is true for every lexicographic type structure is true for every type structure. Conceptually, we are interested in type structures which satisfy the hypothesis of being lexicographic, but the proposition tells us that we will never need this hypothesis in our theorems. In practice, then, we will state and prove theorems for arbitrary type structures. By Proposition 7.1, in these proofs we can always assume without loss of generality that the type structure is lexicographic.

We conclude this section with the idea of a complete type structure (adapted from Brandenburger [2003]).

Definition 7.7. A type structure $\langle S^a, S^b, T^a, T^b, \lambda^a, \lambda^b \rangle$ is *complete* if $\mathcal{L}^+(S^b \times T^b) \subsetneq$ range λ^a and $\mathcal{L}^+(S^a \times T^a) \subsetneq$ range λ^b.

In words, a complete structure contains all full-support LPS's for Ann and Bob, and (at least) one non-full-support LPS.[11] (Refer back to Sections 2.4 and 2.6.) We see at once from the definition that any type structure which is equivalent to a complete type structure is complete.

Proposition 7.2. *For any finite sets S^a and S^b, there is a complete type structure $\langle S^a, S^b, T^a, T^b, \lambda^a, \lambda^b \rangle$ such that the maps λ^a and λ^b are continuous.*

Definition 7.8. A type structure $\langle S^a, S^b, T^a, T^b, \lambda^a, \lambda^b \rangle$ is *continuous* if it is equivalent to a type structure where the λ^a and λ^b maps are continuous.

Thus, in a continuous type structure, players associate neighboring full-support LPS's with neighboring full-support types. Propositions 7.1 and 7.2

---

[11] In the literature, the more common concept of a model of all possible types is the universal (or canonical) model. (See Armbruster and Böge [1979], Böge and Eisele [1979], Mertens and Zamir [1985], Brandenburger and Dekel [1993], Heifetz [1993], and Battigalli and Siniscalchi [1999], among others.) The completeness concept is well suited to our analysis.

immediately give the following corollary:

Corollary 7.1. *For any finite sets S^a and S^b, there exists a complete continuous lexicographic (S^a, S^b)-based type structure.*

8. Characterization of RCAR

Theorem 8.1.

(i) *Fix a type structure $\langle S^a, S^b, T^a, T^b, \lambda^a, \lambda^b \rangle$. Then $\mathrm{proj}_{S^a} \bigcap_{m=1}^{\infty} R_m^a \times \mathrm{proj}_{S^b} \bigcap_{m=1}^{\infty} R_m^b$ is an SAS.*

(ii) *Fix an SAS $Q^a \times Q^b$. There is a lexicographic type structure $\langle S^a, S^b, T^a, T^b, \lambda^a, \lambda^b \rangle$ with $Q^a \times Q^b = \mathrm{proj}_{S^a} \bigcap_{m=1}^{\infty} R_m^a \times \mathrm{proj}_{S^b} \bigcap_{m=1}^{\infty} R_m^b$.*

Proof. For part (i), if $\bigcap_m R_m^a \times \bigcap_m R_m^b = \emptyset$, then the conditions of an SAS are automatically satisfied. So we will suppose this set is nonempty.

Fix $S^a \in \mathrm{proj}_{S^a} \bigcap_m R_m^a$. Then $(s^a, t^a) \in \bigcap_m R_m^a$ for some $t^a \in T^a$. Certainly $(s^a, t^a) \in R_1^a$. Using Lemma 7.2, s^a is admissible, establishing condition (i) of an SAS. By Property 6.3, $t^a \in A^a(\bigcap_m R_m^b)$. We therefore get a picture like Figure 8.1 (for some $j < n$), and, as illustrated,

$$\bigcup_{i \leq j} \mathrm{Supp} \ \mathrm{marg}_{S^b} \mu_i = \mathrm{proj}_{S^b} \bigcap_m R_m^b.$$

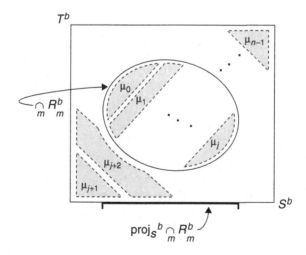

Figure 8.1.

(This is formally established as Lemma D.1 and uses condition (iii) of assumption.) As in Lemma 7.1, there is a length-1 LPS (ν_0) on S^b, with Supp $\nu_0 = \text{proj}_{S^b} \bigcap_m R_m^b$, under which s^a is optimal. Thus, s^a is admissible with respect to $S^a \times \text{proj}_{S^b} \bigcap_m R_m^b$, establishing condition (ii) of an SAS. Next suppose $r^a \in \text{su}(s^a)$. Then, for any t^a, $(s^a, t^a) \in R_1^a$ implies $(r^a, t^a) \in R_1^a$ (Lemma D.2), and so we have for all m, $(s^a, t^a) \in R_m^a$ implies $(r^a, t^a) \in R_m^a$. This establishes condition (iii) of an SAS.

For part (ii) of the theorem, fix an SAS $Q^a \times Q^b$. (Recall the convention that if $Q^a = \emptyset$, then $Q^b = \emptyset$ and vice versa.) By conditions (i) and (ii) of an SAS, for each $s^a \in Q^a$ there are measures $\nu_0, \nu_1 \in \mathcal{M}(S^b)$, with Supp $\nu_0 = S^b$ and Supp $\nu_1 = Q^b$, under which s^a is optimal. We can choose ν_0 so that r^a is optimal under ν_0 if and only if $r^a \in \text{su}(s^a)$. (This is Lemma D.4.)

Define type spaces $T^a = Q^a \cup \{t_*^a\}$ and $T^b = Q^b \cup \{t_*^b\}$, where t_*^a and t_*^b are arbitrary labels. For $t^a = s^a \in Q^a$, the associated $\lambda^a(t^a) \in \mathcal{L}^+(S^b \times T^b)$ will be a two-level full-support LPS (μ_0, μ_1), where $\text{marg}_{S^b} \mu_0 = \nu_1$ and $\text{marg}_{S^b} \mu_1 = \nu_0$.[12] (Further conditions are specified below.) Let $\lambda^a(t_*^a)$ be an element of $\mathcal{L}(S^b \times T^b) \backslash \mathcal{L}^+(S^b \times T^b)$. Define the map λ^b similarly.

Figure 8.2 shows the construction of $\lambda^a(t^a)$: Under the above specifications, points (s^b, s^b) on the diagonal are rational. That is, these points

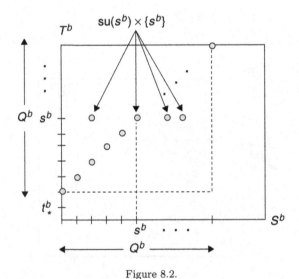

Figure 8.2.

---

[12] We reverse the indices for consistency with the proof of Theorem 9.1 below.

lie in R_1^b. Other points (r^b, s^b) are rational if and only if $r^b \in \mathrm{su}(s^b)$. By condition (iii) of an SAS, $\mathrm{su}(s^b) \subseteq Q^b$. So the set R_1^b contains the diagonal and is contained in the rectangle $Q^b \times Q^b$. Moreover, for each $s^b \in S^b$, $(s^b, t_*^b) \in (S^b \times T^b) \backslash R_1^b$. Thus, we can take the measures μ_0 and μ_1 to satisfy

$$\mathrm{marg}_{S^b} \mu_0 = \nu_1, \quad \mathrm{Supp}\, \mu_0 = R_1^b,$$
$$\mathrm{marg}_{S^b} \mu_1 = \nu_0, \quad \mathrm{Supp}\, \mu_1 = (S^b \times T^b) \backslash R_1^b.$$

Likewise for the map λ^b.

We now show that $\mathrm{proj}_{S^a} \bigcap_m R_m^a = Q^a$ and likewise for b. By the same argument as in the previous paragraph, $\mathrm{proj}_{S^a} R_1^a = Q^a$. Moreover, each $t^a \in Q^a$ assumes R_1^b. (Conditions (i) and (ii) are immediate for $j = 0$. Condition (iii) follows immediately from the fact that $S^b \times T^b$ is finite and each $t^a \in Q^a$ has full support.) So $R_2^a = R_1^a$. Likewise for b. Thus, $R_m^a = R_1^a$ and $R_m^b = R_1^b$ for all m, by induction. Certainly $\mathrm{proj}_{S^a} R_1^a \times \mathrm{proj}_{S^b} R_1^b = Q^a \times Q^a$. It follows that $\mathrm{proj}_{S^a} \bigcap_m R_m^a \times \mathrm{proj}_{S^b} \bigcap_m R_m^b = Q^a \times Q^b$, as required. $\qquad\square$

9. Characterization of R*m*AR in a Complete Structure

Theorem 9.1. *Fix a complete type structure* $\langle S^a, S^b, T^a, T^b, \lambda^a, \lambda^b \rangle$. *Then, for each* m,

$$\mathrm{proj}_{S^a} R_m^a \times \mathrm{proj}_{S^b} R_m^b = S_m^a \times S_m^b.$$

Proof. We may assume that the type structure is lexicographic. The proof is by induction on m. Begin by fixing some $(s^a, t^a) \in R_1^a$. By Lemma 7.2, $s^a \in S_1^a$. This shows that $\mathrm{proj}_{S^a} R_1^a \times \mathrm{proj}_{S^b} R_1^b \subseteq S_1^a \times S_1^b$.

Next fix some $s^a \in S_1^a$. By Lemma 3.1, there is an LPS $(\nu_0) \in \mathcal{L}^+(S^b)$ under which s^a is optimal. We want to construct an LPS $(\mu_0) \in \mathcal{L}^+(S^b \times T^b)$ with $\mathrm{marg}_{S^b} \mu_0 = \nu_0$. By completeness, there will then be a type t^a with $\lambda^a(t^a) = (\mu_0)$. By construction, the pair $(s^a, t^a) \in R_1^a$. This will establish that $\mathrm{proj}_{S^a} R_1^a \times \mathrm{proj}_{S^b} R_1^b = S_1^a \times S_1^b$.

To construct (μ_0), fix some $s^b \in S^b$ and set $X = \{s^b\} \times T^b$. Note that $\nu_0(s^b) > 0$. By rescaling and combining measures over different s^b, it is enough to find $(\xi_0) \in \mathcal{L}^+(X)$. By separability, X has a countable dense subset Y. So by assigning positive weight to each point in Y, we

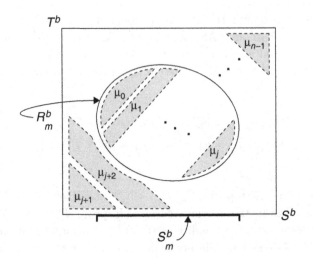

Figure 9.1.

get a measure ξ_0, where $\xi_0(Y) = 1$ and Supp ξ_0 is the closure of Y, as required.

Now assume the result for all $1 \leq i \leq m$. We will show it is also true for $i = m + 1$. Fix some $(s^a, t^a) \in R^a_{m+1}$, where $\lambda^a(t^a) = (\mu_0, \ldots, \mu_{n-1})$. Then $(s^a, t^a) \in R^a_m$ and so, by the induction hypothesis, $s^a \in S^a_m$. Also, $t^a \in A^a(R^b_m)$. Since $\text{proj}_{S^b} R^b_m = S^b_m$, by the induction hypothesis, we get a picture like Figure 9.1 (for some $j < n$). By the same argument as in the proof of Theorem 8.1, we conclude that s^a is admissible with respect to $S^a \times S^b_m$ (so certainly with respect to $S^a_m \times S^b_m$). Thus, $s^a \in S^a_{m+1}$.

Next fix some $s^a \in S^a_{m+1}$. It will be useful to set $S^b_0 = S^b$ and $R^b_0 = S^b \times T^b$. For each $0 \leq i \leq m$, there is a measure $\nu_i \in \mathcal{M}(S^b)$, with Supp $\nu_i = S^b_i$, under which s^a is optimal among all strategies in S^a. (This is Lemma E.1, which uses Lemma 3.1.) Thus, s^a is (lexicographically) optimal under the sequence of measures $(\nu_0, \ldots, \nu_m)$. Also, using the induction hypothesis, $S^b_i = \text{proj}_{S^a} R^b_i$ for all $0 \leq i \leq m$. We want to construct an LPS $(\mu_0, \ldots, \mu_m) \in \mathcal{L}^+(S^b \times T^b)$ where:

(i) $\text{marg}_{S^b} \mu_i = \nu_{m-i}$;

(ii) R^b_i is assumed at level $m - i$.

It will then follow from completeness that there is a t^a with $\lambda^a(t^a) = (\mu_0, \ldots, \mu_m)$, and hence $(s^a, t^a) \in R^a_{m+1}$. (Refer to Figure 9.2.)

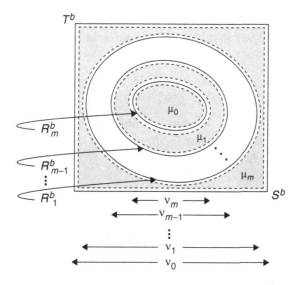

Figure 9.2.

Now fix some $s^b \in S^b$ and set $X = \{s^b\} \times T^b$ as above. Let h be the greatest $i \leq m$ such that $s^b \in S_i^b$. Note that for each $i \leq h$ we have $s^b \in S_i^b =$ Supp ν_i and so $\nu_i(s^b) > 0$.

By rescaling and combining the measures over different s^b, it is enough (using Lemma B.1) to find $(\xi_0, \ldots, \xi_h) \in \mathcal{L}^+(X)$ with:

(i) $\xi_0(X \cap R_h^b) = 1$;
(ii) $\xi_i(X \cap (R_{h-i}^b \backslash R_{h-i+1}^b)) = 1$ for each $1 \leq i \leq h$;
(v) $X \cap R_{h-i}^b \subseteq \bigcup_{j=0}^{i}$ Supp ξ_j for each $0 \leq i \leq h$.

Each R_{h-i}^b is Borel (Lemma C.4). We also have $\text{proj}_{S^b} R_{h-i}^b = \text{proj}_{S^b}(R_{h-i}^b \backslash R_{h-i+1}^b)$. (This is Lemma E.3. It is the place where we use the fact that a complete lexicographic type structure has a non-full-support LPS.) Since $s^b \in \text{proj}_{S^b} R_{h-i}^b$, for each $1 \leq i \leq h$ the set $X_i = X \cap (R_{h-i}^b \backslash R_{h-i+1}^b)$ is nonempty. The set $X_0 = X \cap R_h^b$ is also nonempty. The proof is finished by the same argument as in the base step above: By separability, each X_i has a countable dense subset Y_i. Assign positive probability to each point in Y_i to get a measure ξ_i, where $\xi_i(Y_i) = 1$ and Supp ξ_i is the closure of Y_i. Then $(\xi_0, \ldots, \xi_h) \in \mathcal{L}^+(X)$ and satisfies (iii)–(v), completing the induction. $\square$

10. A Negative Result

Definition 10.1. Say that player a is *indifferent* if $\pi^a(r^a, s^b) = \pi^a(s^a, s^b)$ for all r^a, s^a, s^b.

So if a player is not indifferent, then he has more than one "strategically distinct" strategy.

Theorem 10.1. *Fix a complete continuous type structure* $\langle S^a, S^b, T^a, T^b, \lambda^a, \lambda^b \rangle$. *If player a is not indifferent, then there is no state at which there is RCAR. In fact,*

$$\bigcap_{m=1}^{\infty} R_m^a = \bigcap_{m=1}^{\infty} R_m^b = \emptyset.$$

We come back to Theorem 10.1 in Sections 11D and 11E.

11. Discussion

Here, we discuss some open questions. The online Supplemental material contains other conceptual and technical discussion.

A. *LPS's*: We define an LPS to be a finite sequence of probability measures, not an infinite sequence. The main reason is that finite sequences suffice for what we do. But it would certainly be worth exploring extensions of our definition (see Halpern [2007]).

Would Theorem 10.1 go through with infinite sequences of measures? The intuition given in Section 2.7 appears to depend only on the condition that an LPS has a primary hypothesis, secondary hypothesis, and so forth. Given this, we will eventually hit the primary hypothesis, when trying to "count on" smaller and smaller events. In other words, it seems that the well-foundedness of an LPS is really what is responsible for the impossibility. The idea that a player has an initial hypothesis about a game seems very basic. That said, we do not know if Theorem 10.1 would be overturned if we used nonwell-founded LPS's.

B. *Assumption*: A weaker concept than assumption of an event E is to re quire only belief at level 0. That is, given an LPS $(\mu_0, \ldots, \mu_{n-1})$, we ask only that $\mu_0(E) = 1$. This is the concept used in Brandenburger (1992) and (effectively) in Börgers (1994). Ben Porath (1997) studied an extensive-form analog (which is, accordingly, weaker than strong belief).

All three papers obtain $S^\infty W$ strategies. (This is the Dekel and Fudenberg [1990] concept of one round of deletion of inadmissible strategies followed by iterated deletion of strongly dominated strategies.) Let us recast the analysis in the current epistemic framework.

Call a subset $Q^a \times Q^b$ of $S^a \times S^b$ a *weak best-response set* (WBRS) if (i) each $s^a \in Q^a$ is admissible, (ii) each $s^a \in Q^a$ is not strongly dominated with respect to $S^a \times Q^b$; and likewise with a and b interchanged. Every WBRS is contained in the $S^\infty W$ set, and the $S^\infty W$ set is a WBRS. We have the following analog to our Theorem 8.1:

> Let $Q^a \times Q^b$ be the projection into $S^a \times S^b$ of the states (s^a, t^a, s^b, t^b) satisfying rationality and common belief at level 0 of rationality. Then $Q^a \times Q^b$ is a WBRS. Conversely, given a WBRS $Q^a \times Q^b$, there is a type structure such that $Q^a \times Q^b$ is contained in the projection into $S^a \times S^b$ of the states (s^a, t^a, s^b, t^b) satisfying rationality and common belief at level 0 of rationality.

(Note that here the converse only has inclusion not equality.) We are not aware of an analog to Theorem 9.1.

C. *Irrationality*: We noted in Section 2.4 that there are two forms of irrationality in the chapter: strategy–type pairs (s^a, t^a) where s^a is not optimal under $\lambda^a(t^a)$, and strategy–type pairs (s^a, t^a) where $\lambda^a(t^a)$ is not full support. The presence of a non-full-support type is needed in the proofs of each of our three main theorems (Theorems 8.1, 9.1, and 10.1). In each case, the key fact is that there is a type t^a so that each (s^a, t^a) is irrational.

This raises the question: What would happen if we required all types to have full support — that is, if we ruled out the second form of irrationality? The strategies played under RCAR would still constitute an SAS (Theorem 8.1(i)). But, as the discussion of Figure 2.6 showed, not every SAS could now arise under RCAR. We do not know what subfamily of the SAS's would result and we leave this as an open question.

D. *Continuity*: In a continuous structure, players associate neighboring full-support LPS's with neighboring full-support types (Definition 7.8). Theorem 10.1 made use of this condition in addition to the condition that Ann is not indifferent. Under these hypotheses, $S^a \times T^a$ contains a nonempty open set of irrational pairs. This is used to get the first step of an induction (Lemma F.1). At each later step of the induction, continuity is again needed to guarantee that the pre-image of an open set is still open.

What happens to Theorem 10.1 if continuity is dropped? Alternatively put, does there exist a complete type structure in which the RCAR set is nonempty? We do not know.

E. *Infinite Games*: Finally, Theorem 10.1 may be suggestive of limitations to the analysis of infinite games.[13] For a fixed infinite game, it may

---

[13] We are grateful to Eddie Dekel for this observation.

be that one needs the full force of RCAR in a complete structure to obtain IA. Will this be possible? Of course, to answer this question, we have to rebuild all the ingredients of this chapters for infinite games. This must be left to future work.

Appendix A: Preference Basis

We begin with an axiomatic justification of assumption, that is, the conditions (i)–(iii) of Proposition 5.1.

Let Ω be a Polish space and let $\mathcal{A}$ be the set of all measurable functions from Ω to $[0,1]$. A particular function $x \in \mathcal{A}$ is an *act*, where $x(\omega)$ is the payoff to the player of choosing the act x if the true state is $\omega \in \Omega$. For $x, y \in \mathcal{A}$ and $0 \le \alpha \le 1$, write $\alpha x + (1-\alpha)y$ for the act that in state ω gives payoff $\alpha x(\omega) + (1-\alpha)y(\omega)$. For $c \in [0,1]$, write $\overrightarrow{c}$ for the constant act associated with c; that is, $\overrightarrow{c}(\omega) = c$ for all $\omega \in \Omega$. Also, given acts $x, z \in \mathcal{A}$ and a Borel subset E in Ω, write $(x_E, z_{\Omega \setminus E})$ for the act:

$$(x_E, z_{\Omega \setminus E})(\omega) = \begin{cases} x(\omega), & \text{if } \omega \in E, \\ z(\omega), & \text{if } \omega \notin E. \end{cases}$$

Let $\succsim$ be a preference relation on $\mathcal{A}$ and write $\succ$ (resp. $\sim$) for strict preference (resp. indifference). We maintain two axioms throughout:

A1 — Order: $\succsim$ is a complete, transitive, reflexive binary relation on $\mathcal{A}$.
A2 — Independence: For all $x, y, z \in \mathcal{A}$ and $0 < \alpha \le 1$,

$$x \succ y \text{ implies } \alpha x + (1-\alpha)z \succ \alpha y + (1-\alpha)z \quad \text{and}$$
$$x \sim y \text{ implies } \alpha x + (1-\alpha)z \sim \alpha y + (1-a)z.$$

Given a Borel set E, define conditional preference given E in the usual way:

Definition A.1. $x \succsim_E y$ if for some $z \in \mathcal{A}$, $(x_E, z_{\Omega \setminus E}) \succsim (y_E, z_{\Omega \setminus E})$.

(As is well known, under A1 and A2, $(x_E, z_{\Omega \setminus E}) \succsim (y_E, z_{\Omega \setminus E})$ holds for all z if it holds for some z.)

Given a full-support LPS $\sigma = (\mu_0, \ldots, \mu_{n-1}) \in \mathcal{L}^+(\Omega)$, define $\succsim^\sigma$ on $\mathcal{A}$ by

$$x \succsim^\sigma y \iff \left(\int_\Omega x(\omega) d\mu_i(\omega) \right)_{i=0}^{n-1} \ge^L \left(\int_\Omega y(\omega) d\mu_i(\omega) \right)_{i=0}^{n-1}.$$

Definition A.2. Say a set E is *believed* under $\succsim$ if E is Borel and, for all $x, y \in \mathcal{A}$, $x \sim_{\Omega \setminus E} y$.

This is just the statement that the event $\Omega\backslash E$ is Savage-null. We have the following characterization of belief.

Proposition A.1. *Fix* $\sigma = (\mu_0, \ldots, \mu_{n-1}) \in \mathcal{L}^+(\Omega)$ *and a Borel set E in Ω. The following statements are equivalent:*

(i) $\mu_i(E) = 1$ *for all i.*
(ii) E *is believed under $\succsim^\sigma$.*

Proof. Suppose (i) holds. Then $\mu_i(\Omega\backslash E) = 0$ for all i, and so for any $x, y \in \mathcal{A}, x \sim^\sigma_{\Omega\backslash E} y$. Thus, (ii) holds. Now suppose (ii) holds. Then $\overrightarrow{1} \sim^\sigma_{\Omega\backslash E} \overrightarrow{0}$. That is,

$$\left(\mu_i(\Omega\backslash E) + \int_E z(\omega)d\mu_i(\omega) \right)_{i=0}^{n-1} = \left(0 + \int_E z(\omega)d\mu_i(\omega) \right)_{i=0}^{n-1}$$

or $\mu_i(\Omega\backslash E) = 0$ for all i, as required. $\qquad\square$

Definition A.3. Say a set E is *assumed* under if $\succsim$ is Borel and satisfies the following conditions:

(i) *Nontriviality*: E is nonempty and, for each open set U with $E \cap U \neq \emptyset$, there are acts $x, y \in \mathcal{A}$ with $x \succ_{E \cap U} y$;
(ii) *Strict Determination*: For all acts $x, y \in \mathcal{A}$, $x \succ_E y$ implies $x \succ y$.

Proposition A.2. *Fix a full-support LPS $\sigma = (\mu_0, \ldots, \mu_{n-1}) \in \mathcal{L}^+(\Omega)$ and a Borel set E in Ω. Then E is assumed under σ if and only if E is assumed under $\succsim^\sigma$.*

Proof. First suppose that E is assumed under σ at level j. Fix an open set U with $E \cap U \neq \emptyset$. Then, by conditions (ii) and (iii) of assumption, there exists some $k \leq j$ with $\mu_k(E \cap U) > 0$. Let $x(\omega) = 1$ if $\omega \in E \cap U$ and let $x(\omega) = 0$ otherwise. Then the act $(x_{E \cap U}, \overrightarrow{0}_{\Omega\backslash(E \cap U)})$ is evaluated as $(\mu_0(E \cap U), \ldots, \mu_j(E \cap U), 0, \ldots, 0)$, where the kth entry is strictly positive. The act $(\overrightarrow{0}_{E \cap U}, \overrightarrow{0}_{\Omega\backslash(E \cap U)})$ is evaluated as $(0, \ldots, 0)$. Thus, $\overrightarrow{x} \succ^\sigma_{E \cap U} \overrightarrow{0}$, establishing Nontriviality. To establish Strict Determination, note that $x \succ^\sigma_E$ implies

$$\left(\int_E x\, d\mu_0, \ldots, \int_E x\, d\mu_j, \int_{\Omega\backslash E} z\, d\mu_{j+1}, \ldots, \int_{\Omega\backslash E} z\, d\mu_{n-1} \right)$$

$$>^L \left(\int_E y\, d\mu_0, \ldots, \int_E y\, d\mu_j, \int_{\Omega\backslash E} z\, d\mu_{j+1}, \ldots, \int_{\Omega\backslash E} z\, d\mu_{n-1} \right),$$

so that certainly

$$\left(\int_E x \, d\mu_0, \ldots, \int_E x \, d\mu_j, \int_{\Omega\backslash E} x \, d\mu_{j+1}, \ldots, \int_{\Omega\backslash E} x \, d\mu_{n-1} \right)$$

$$>^L \left(\int_E y \, d\mu_0, \ldots, \int_E y \, d\mu_j(\omega), \int_{\Omega\backslash E} y \, d\mu_{j+1}, \ldots, \int_{\Omega\backslash E} y \, d\mu_{n-1} \right).$$

Thus, $x \succ^\sigma y$, establishing Strict Determination.

Next, suppose E is assumed under $\succsim^\sigma$. We want to show that E is assumed under σ. Condition (iii) of assumption is immediate from Nontriviality, so we will show that σ satisfies conditions (i) and (ii).

Assume σ fails conditions (i) and (ii) of assumption. There are three cases to consider.

Case A.1 — $\mu_i(E) = 0$ *for all* i: This contradicts Nontriviality.

Case A.2 — $\mu_i(E) = 0$ *and* $\mu_h(E) = 1$ *where* $h > i$: Let U_i and U_h be Borel sets as in Definition 4.1 (i.e., with $\mu_i(U_i) = 1$ and, for $i \neq k$, $\mu_i(U_k) = 0$, and similarly for h). Define

$$x(\omega) = \begin{cases} 1, & \text{if } \omega \in E \cap U_h, \\ 0, & \text{otherwise}, \end{cases}$$

$$y(\omega) = \begin{cases} 1, & \text{if } \omega \in U_i \backslash E, \\ 0, & \text{otherwise}, \end{cases}$$

Acts x and $(x_E, \overrightarrow{0}_{\Omega\backslash E})$ are evaluated as $(0, \ldots, 0, 1, 0, \ldots, 0)$, where the 1 corresponds to μ_h. (Here, we use $\mu_k(U_h) = 0$ for all $k \neq h$.) Act y is evaluated as $(0, \ldots, 0, 1, 0, \ldots, 0)$, where the 1 corresponds to μ_i, while act $(y_E, \overrightarrow{0}_{\Omega\backslash E})$ is evaluated as $(0, \ldots, 0)$. Thus, $x \succ^\sigma_E y$. But since $h > i, y \succ^\sigma x$, contradicting Strict Determination.

Case A.3 — $0 < \mu_i(E) < 1$ *for some* i: Let U_i be a Borel set as in Definition 4.1 and define

$$x(\omega) = \begin{cases} \mu_i(U_i \backslash E), & \text{if } \omega \in E \cap U_i, \\ 0, & \text{otherwise}, \end{cases}$$

$$y(\omega) = \begin{cases} 1, & \text{if } \omega \in U_i \backslash E, \\ 0, & \text{otherwise}. \end{cases}$$

Acts x and $(x_E, \vec{0}_{\Omega \setminus E})$ are evaluated as

$$(0, \ldots, 0, \mu_i(U_i \setminus E)\mu_i(E \cap U_i), 0, \ldots, 0),$$

where the nonzero entry corresponds to μ_i. This entry is indeed nonzero, since $1 > \mu_i(E) > 0$ implies $\mu_i(U_i \setminus E) > 0$ and $\mu_i(E \cap U_i) > 0$. Act y is evaluated as

$$(0, \ldots, 0, \mu_i(U_i \setminus E), 0, \ldots, 0),$$

where the nonzero entry corresponds to μ_i. This entry is indeed nonzero, since $1 > \mu_i(E)$. The act $(y_E, \vec{0}_{\Omega \setminus E})$ is evaluated as $(0, \ldots, 0)$. Thus, $x \succ_E^\sigma y$. But since $1 > \mu_i(E \cap U_i)$, $y \succ^\sigma x$, contradicting Strict Determination. $\square$

Corollary A.1. *Fix a full-support LPS $\sigma = (\mu_0, \ldots, \mu_{n-1}) \in \mathcal{L}^+(\Omega)$ and a Borel set E in Ω. If E is believed under $\succsim^\sigma$, then E satisfies Nontriviality and Strict Determination.*

We conclude by mentioning the relationship between this axiomatization and the Blume, Brandenburger, and Dekel (1991a) axiomatization. Fix a finite state space and suppose $\succsim$ is represented by a full-support LPS. Impose Axiom 5$'$ in Blume, Brandenburger, and Dekel (1991a) (i.e., their full-support condition). They then say that E is infinitely more likely than not-E if E is nonempty and, for all acts $x, y, w, z, x \succ_E y$ implies $(x_E, w_{\Omega \setminus E}) \succ (y_E, z_{\Omega \setminus E})$. (See their Definition 5.1.) It is easily checked that E is infinitely more likely than $\Omega \setminus E$, in the sense of Blume *et al.*, if and only if Nontriviality and Strict Determination hold.

In Blume, Brandenburger, and Dekel (1991a), Axiom 5$'$ is needed to ensure that their Definition 5.1 carries the intended interpretation. (Without it, there might be no x, y with $x \succ_E y$, i.e., each measure in the LPS could assign zero probability to E.) Nontriviality plays an analogous role in our formulation.

Suppose $\succsim$ is represented by a full-support LPS σ. Fix an event E. In the context of a finite state space, Corollary 5.1 in Blume, Brandenburger, and Dekel (1991a) shows that $\succsim$ satisfies Nontriviality and Strict Determination if and only if σ satisfies conditions (i) and (ii) of assumption. For a finite state space and a full-support LPS, an event satisfies conditions (i) and (ii) of assumption if and only if it satisfies conditions (i)–(iii) of assumption. Proposition A.2 extends this result to infinite spaces.

Appendix B: Proofs for Sections 5 and 6

This appendix provides proofs that relate to the definition and properties of assumption.

Proof of Proposition 5.1. Suppose E is assumed under σ at level j. Condition (a) of Definition 5.1 follows immediately from condition (iii) of assumption. Next, suppose F is part of E and G is part of $\Omega \backslash E$. Suppose further that $\mu_i(F) > 0$ and $\mu_k(G) > 0$. Then, by conditions (i) and (ii) of assumption, $i \leq j < k$ as required.

For the converse, suppose conditions (a) and (b) of Definition 5.1 hold. By condition (b), whenever $\mu_i(E) > 0$ and $\mu_k(\Omega \backslash E) > 0$, we have that $i < k$. Moreover, by condition (a), there is some i with $\mu_i(E) > 0$. This establishes that there is some j satisfying conditions (i) and (ii) of assumption. Condition (iii) of assumption is immediate from condition (a) of Definition 5.1. □

It will be useful to have the following characterization of assumption.

Lemma B.1. *Fix a full-support LPS $\sigma \in \mathcal{L}^+(\Omega)$ and an event E. Then E is assumed under $\sigma = (\mu_0, \ldots, \mu_{n-1})$ at level j if and only if there is some j so that σ satisfies conditions (i) and (ii) plus the following condition:*

(iii') $E \subseteq \bigcup_{i \leq j} \text{Supp } \mu_i$.

Proof. First suppose that E is assumed under σ at level j. We will show that σ also satisfies (iii'). Consider the open set

$$U = \Omega \backslash \bigcup_{i \leq j} \text{Supp } \mu_i.$$

If $U \cap E \neq \emptyset$, then $\mu_i(U \cap E) > 0$ for some i. By condition (ii) of assumption, $i \leq j$. This implies that, for some $i \leq j, \mu_i(U) > 0$ and so $U \cap \text{Supp } \mu_i \neq \emptyset$, a contradiction. This says $U \cap E = \emptyset$ and so $E \subseteq \bigcup_{i \leq j} \text{Supp } \mu_i$, as required.

Next suppose that there is some j so that σ satisfies conditions (i) and (ii), and also (iii'). We will show that it satisfies condition (iii). Let U be an open set with $U \cap E \neq \emptyset$. By condition (iii'), for each $\omega \in U \cap E$, there is some $i \leq j$ with $\omega \in \text{Supp } \mu_i$. Since U is an open neighborhood of $\omega, \mu_i(U) > 0$. By condition (i) of assumption, $\mu_i(E \cap U) = \mu_i(U) > 0$, as required. □

We now turn to establish properties of the assumption operator.

Proof of Property 6.1 — Convexity: Let $\sigma = (\mu_0, \ldots, \mu_{n-1})$, and fix events E and F that are assumed under σ at level j. Fix also a Borel set

G with $E \cap F \subseteq G \subseteq E \cup F$. We will show that G is also assumed under σ at level j.

First fix $i \leq j$ and note that $\mu_i(E) = \mu_i(F) = 1$. So certainly $\mu_i(E \cap F) = 1$. Since $E \cap F \subseteq G$, $\mu_i(G) = 1$, establishing property (i) of assumption. Next fix $i > j$. Note that $\mu_i(E) = \mu_i(F) = 0$ and so $\mu_i(E \cup F) = 0$. Since $G \subseteq E \cup F$, $\mu_i(G) = 0$, establishing property (ii) of assumption. Finally, since E and F are assumed under σ at level j, Lemma B.1 says $E \cup F \subseteq \bigcup_{i \leq j} \operatorname{Supp} \mu_i$. So using the fact that $G \subseteq E \cup F$ and Lemma B.1, G is assumed under σ. □

Proof of Property 6.2 — Closure: Let $\sigma = (\mu_0, \ldots, \mu_{n-1})$ and suppose E is assumed under σ at level j. Then $\overline{E} = \bigcup_{i \leq j} \operatorname{Supp} \mu_i$. To see this, note that Lemma B.1 says that $E \subseteq \bigcup_{i \leq j} \operatorname{Supp} \mu_i$. Since $\bigcup_{i \leq j} \operatorname{Supp} \mu_i$ is closed, $\overline{E} \subseteq \bigcup_{i \leq j} \operatorname{Supp} \mu_i$. Moreover, for all $i \leq j, \mu_i(\overline{E}) = 1$ so that $\bigcup_{i \leq j} \operatorname{Supp} \mu_i \subseteq \overline{E}$.

If F is also assumed under σ at level j, then it is immediate that $\overline{E} = \overline{F}$. If F is assumed under σ at level $k > j$, then $\overline{E} \subseteq \overline{F}$, since $\bigcup_{i \leq j} \operatorname{Supp} \mu_i \subseteq \bigcup_{i \leq k} \operatorname{Supp} \mu_i$. □

Proof of Property 6.3 — Conjunction and Disjunction: We will only prove the Conjunction property. The proof of the Disjunction property is similar.

Let $\sigma = (\mu_0, \ldots, \mu_{n-1})$. For each m, E_m is assumed under σ at some level j_m. Let $j_M = \min\{j_m : m = 1, 2, \ldots\}$. Then, for each $m, \mu_i(E_m) = 1$ for all $i \leq j_M$. Thus, $\mu_i(\bigcap_m E_m) = 1$ for all $i \leq j_M$. Also, $\mu_i(E_M) = 0$ for all $i > j_M$. Then certainly $\mu_i(\bigcap_m E_m) = 0$ for all $i > j_M$. This establishes conditions (i) and (ii) of Proposition 5.1 (for $j = j_M$). Finally, using the fact that E_M is assumed at level j_M, and Lemma B.1,

$$\bigcap_m E_m \subseteq E_M \subseteq \bigcup_{i \leq j_M} \operatorname{Supp} \mu_i.$$

Again using Lemma B.1, this establishes condition (iii) of Proposition 5.1. □

Appendix C: Proofs for Section 7

In what follows, we will need to make use of the following characterizations of full support.

Lemma C.1. *A sequence* $\sigma = (\mu_0, \ldots, \mu_{n-1}) \in \mathcal{N}(\Omega)$ *has full support if and only if, for each nonempty open set U, there is an i with $\mu_i(U) > 0$.*

Proof. Fix a sequence $\sigma = (\mu_0, \ldots, \mu_{n-1}) \in \mathcal{N}(\Omega)$ which does not have full support. Then $U = \Omega \setminus \bigcup_{i<n} \operatorname{Supp} \mu_i$ is nonempty. The set U is open and $\mu_i(U) = 0$ for all i. For the converse, fix a full-support sequence $\sigma = (\mu_0, \ldots, \mu_{n-1}) \in \mathcal{N}(\Omega)$ and a nonempty open set U. Since σ has full support, $U \cap \operatorname{Supp} \mu_i \neq \emptyset$ for some i. Then $(\Omega \setminus U) \cap \operatorname{Supp} \mu_i$ is closed and strictly contained in $\operatorname{Supp} \mu_i$, so that $\mu_i((\Omega \setminus U) \cap \operatorname{Supp} \mu_i) < 1$. From this, $\mu_i(U) > 0$, as required.   □

In the next three lemmas, Borel without qualification means Borel in $\mathcal{N}(\Omega)$. We make repeated use of the following facts:

(i) There is a countable open basis $E_1, E_2, \ldots$ for Ω.

(ii) For each Borel set B in Ω and $r \in [0, 1]$, the set of μ such that $\mu(B) > r$ is Borel in $\mathcal{M}(\Omega)$.

(iii) For each Borel set Y in $\mathcal{M}(\Omega)$ and each k, the set of $\sigma = (\mu_0, \ldots, \mu_{n-1})$ in $\mathcal{N}(\Omega)$ such that $n > k$ and $\mu_k \in Y$ is Borel.

Fact (i) follows from the assumption that Ω is separable. Fact (ii) says that the function $\mu \mapsto \mu(B)$ is Borel, which follows from Kechris (1995, Theorem 17.24). Fact (iii) follows from the continuity of the projection function $\sigma \mapsto \mu_k$ from $\mathcal{N}(\Omega)$ to $\mathcal{M}(\Omega)$.

Let $\mathcal{N}_n(\Omega)$ be the set of all σ in $\mathcal{N}(\Omega)$ of length n, and define $\mathcal{N}_n^+(\Omega), \mathcal{L}_n(\Omega)$, and $\mathcal{L}_n^+(\Omega)$ analogously.

Lemma C.2. *Fix $n \in \mathbb{N}$. For any Polish space Ω, the sets $\mathcal{N}_n(\Omega), \mathcal{N}_n^+(\Omega)$, $\mathcal{L}_n(\Omega)$, and $\mathcal{L}_n^+(\Omega)$ are Borel.*

Proof. In this proof, $\sigma = (\mu_0, \ldots, \mu_{n-1})$ varies over $\mathcal{N}_n(\Omega)$. Recall that if $\sigma \in \mathcal{N}_n(\Omega)$ and $\tau \in \mathcal{N}(\Omega) \setminus \mathcal{N}_n(\Omega)$, then σ has distance 1 from τ. Thus, $\mathcal{N}_n(\Omega)$ is open and hence Borel.

By Lemma C.1 and fact (i), a sequence $\sigma \in \mathcal{N}_n(\Omega)$ has full support if and only if, for each basic open set E_i, there exists $j < n$ such that $\mu_j(E_i) > 0$. By facts (ii) and (iii), for each i and j the set of σ such that $\mu_j(E_i) > 0$ is Borel. Therefore, $\mathcal{N}_n^+(\Omega)$ is Borel.

Write $\mu \perp \nu$ if there is a Borel set $U \subseteq \Omega$ such that $\mu(U) = 1$ and $\nu(U) = 0$. It is easy to see that mutual singularity holds for an element $\sigma \in \mathcal{N}_n(\Omega)$ if and only if $\mu_i \perp \mu_j$ for all $i < j$. To prove that $\mathcal{L}_n(\Omega)$ is Borel, it suffices to prove that for each $i < j$, the set of σ such that $\mu_i \perp \mu_j$ is Borel. Note that $\mu_i \perp \mu_j$ if and only if for each m, there is an open set V such that $\mu_i(V) = 1$ and $\mu_j(V) < \frac{1}{m}$. By fact (i), this in turn holds if and only if for each m there exists k such that $\mu_i(E_k) > 1 - \frac{1}{m}$ and $\mu_j(E_k) < \frac{1}{m}$.

By facts (ii) and (iii), the set of σ such that $\mu_i(E_k) > 1 - \frac{1}{m}$ is Borel, and the set of σ such that $\mu_j(E_k) < \frac{1}{m}$ is Borel. The set of σ such that $\mu_i \perp \mu_j$ is a Borel combination of these sets, and hence is Borel, as required. Thus, $\mathcal{L}_n(\Omega)$ is Borel.

Since $\mathcal{L}_n^+(\Omega)$ is the intersection of the Borel sets $\mathcal{N}_n^+(\Omega)$ and $\mathcal{L}_n(\Omega)$, it is also Borel. $\qquad \square$

Corollary C.1. *For any Polish space Ω, the sets $\mathcal{N}^+(\Omega), \mathcal{L}(\Omega),$ and $\mathcal{L}^+(\Omega)$ are Borel.*

Proof. Each $\mathcal{N}_n^+(\Omega)$ is Borel and $\mathcal{N}^+(\Omega) = \bigcup_n \mathcal{N}_n^+(\Omega)$. Likewise for $\mathcal{L}(\Omega)$ and $\mathcal{L}^+(\Omega)$. $\qquad \square$

Lemma C.3. *For each Polish space Ω and Borel set E in Ω, the set of $\sigma \in \mathcal{L}^+(\Omega)$ such that E is assumed under σ is Borel.*

Proof. Fix n and $j < n$. By fact (ii), the sets of μ such that $\mu(E) = 1$ and such that $\mu(E) = 0$ are Borel in $\mathcal{M}(\Omega)$. Therefore, by fact (iii) and Corollary C.1, the set of $\sigma = (\mu_0, \ldots, \mu_{n-1}) \in \mathcal{L}_n^+(\Omega)$ such that conditions (i) and (ii) in Proposition 5.1 hold is Borel. Let $\{d_0, d_1, \ldots\}$ be a countable dense subset of E. For each k and $\mu \in M(\Omega)$, we have $d_k \in \text{Supp}\,\mu$ if and only if $\mu(B) > 0$ for every open ball B with center d_k and rational radius. Then by fact (ii), the set of μ such that $d_k \in \text{Supp}\,\mu$ is Borel in $\mathcal{M}(\Omega)$. We have $E \subseteq \bigcup_{i \leq j} \text{Supp}\,\mu_i$ if and only if $d_k \in \bigcup_{i \leq j} \text{Supp}\,\mu_i$ for all $k \in \mathbb{N}$. Therefore, the set of $\sigma \in \mathcal{L}_n^+(\Omega)$ with $E \subseteq \bigcup_{i \leq j} \text{Supp}\,\mu_i$ is Borel. By Lemma B.1, the set of $\sigma \in \mathcal{L}^+(\Omega)$ such that E is assumed under σ is Borel. $\qquad \square$

Lemma C.4. *For each m:*

(i) $R_m^a = R_1^a \cap \left[S^a \times \bigcap_{i < m} A^a(R_i^b) \right]$.
(ii) R_m^a *is Borel in $S^a \times T^a$.*

Proof. Part (i) is immediate.

Part (ii) is by induction. For $m = 1$, first note that since λ^a is Borel measurable, Lemma C.2 says that for each n the set $(\lambda^a)^{-1}(\mathcal{L}_n^+(S^b \times T^b))$ is Borel in T^a. From Definition 7.4, for each $s^a \in S^a$ there is a finite Boolean combination C of linear equations in $n \cdot |S^b|$ variables such that whenever $\lambda^a(t^a) = (\mu_0, \ldots, \mu_{n-1}) \in \mathcal{L}_n^+(S^b \times T^b)$, the pair (s^a, t^a) is rational if and only if C holds for $\{\text{marg}_{S^b} \mu_i(s^b) : i < n, s^b \in S^b\}$. Since S^a and S^b are finite, this shows that R_1^a is Borel in $S^a \times T^a$.

Assume the result holds for all $i \leq m$. Then, by Lemma C.3, for each $i \leq m, A^a(R_i^b)$ is Borel in T^a. So R_{m+1}^a is Borel. $\qquad \square$

Proof of Proposition 7.1. (i) Start with a type structure $\langle S^a, S^b,$ $T^a, T^b, \kappa^a, \kappa^b \rangle$. The case that $S^b \times T^b$ is a singleton is trivial, so we may assume that it is not. Pick any $\sigma \in \mathcal{L}(S^b \times T^b)$ which does not have full support. Define $\lambda^a(t^a) = \kappa^a(t^a)$ if $\kappa^a(t^a) \in \mathcal{L}^+(S^b \times T^b)$ and $\lambda^a(t^a) = \sigma$ otherwise. Since $\mathcal{L}^+(S^b \times T^b)$ is Borel, λ^a is a Borel map. Define λ^b similarly.

(ii) It is clear from the definitions that the two structures have the same rationality sets R_1^a and R_1^b. By induction, they also have the same sets R_m^a and R_m^b: Types associated with the R_m^a and R_m^b sets are all associated with full-support LPS's, so that only assumption by full-support LPS's is involved. $\qquad\square$

Proof of Proposition 7.2. Let T^a and T^b be the Baire space — that is, the metric space $\mathbb{N}^\mathbb{N}$ with the product metric, where $\mathbb{N}$ has the discrete metric. There is a continuous surjection λ^a (resp. λ^b) from T^a (resp. T^b) onto any Polish space, in particular onto $\overline{\mathcal{L}}(S^b \times T^b)$ (resp. $\overline{\mathcal{L}}(S^b \times T^b)$). (See Kechris, [1995, p. 13 and Theorem 7.9].) These maps give us a complete type structure. $\qquad\square$

Appendix D: Proofs for Section 8

Lemma D.1. *Suppose t^a assumes $E \subseteq S^b \times T^b$ at level j, where $\lambda^a(t^a) = (\mu_0, \ldots, \mu_{n-1})$. Then $\bigcup_{i \leq j} \operatorname{Supp} \operatorname{marg}_{S^b} \mu_i = \operatorname{proj}_{S^b} E$.*

Proof. Fix $s^b \in \operatorname{proj}_{S^b} E$, that is, $(s^b, t^b) \in E$ for some t^b. Then $\{s^b\} \times T^b$ is an open neighborhood of (s^b, t^b). So, by conditions (ii) and (iii) of Proposition 5.1, there is some $i \leq j$ with $\mu_i(E \cap (\{s^b\} \times T^b)) > 0$. Therefore, $0 < \mu_i(\{s^b\} \times T^b) = \operatorname{marg}_{S^b} \mu_i(s^b)$ and hence $s^b \in \operatorname{Supp} \operatorname{marg}_{S^b} \mu_i$. Next fix $s^b \notin \operatorname{proj}_{S^b} E$. Then $\{s^b\} \times T^b$ is disjoint from E. But for each $i \leq j$, we have $\mu_i(E) = 1$, so $\mu_i(\{s^b\} \times T^b) = \operatorname{marg}_{S^b} \mu_i(s^b) = 0$ and hence $s^b \notin \operatorname{Supp} \operatorname{marg}_{S^b} \mu_i$. $\qquad\square$

The next series of lemmas concerns the geometry of polytopes. We will first review some notions from geometry, then state the lemmas, then explain the connection between the geometric notions and games, then present some intuitive examples, and finally give the formal proofs of the lemmas.

Throughout this section, we will fix a finite set $X = \{x_1, \ldots, x_n\} \subseteq \mathbb{R}^d$. The *polytope* generated by X, denoted by P, is the closed convex hull of X — that is, the set of all sums $\sum_{i=1}^n \lambda_i x_i$, where $\lambda_i \geq 0$ for each i and

$\sum_{i=1}^{n} \lambda_i = 1$. The *affine hull* of P, denoted by aff(P), is the set of all affine combinations of finitely many points in P — that is, the set of all sums $\sum_{i=1}^{k} \lambda_i y_i$, where $y_1, \ldots, y_k \in P$ and $\sum_{i=1}^{k} \lambda_i = 1$. The *relative interior* of P, denoted by relint(P), is the set of all $x \in$ aff(P) such that there is an open ball $B(x)$ centered around x, with aff(P) $\cap B(x) \subseteq P$.

A *hyperplane* in $\mathbb{R}^d$ is a set of the form $H(u, \alpha) = \{x \in \mathbb{R}^d : \langle x, u \rangle = \alpha\}$ for some nonzero $u \in \mathbb{R}^d$. A hyperplane $H(u, \alpha)$ *supports* a polytope P if $\alpha = \sup\{\langle x, u \rangle : x \in P\}$. A *face* of P is either P itself or a set of the form $H \cap P$, where H is a hyperplane that supports P. If $F \neq P$ is a face of P, we say F is a *proper face*. A face $H \cap P$ is *strictly positive* if $H = H(u, \alpha)$ for some (u, α) such that each coordinate of u is strictly positive.

Given a point x in a polytope P, say the points $x_1, \ldots, x_k \in P$ each *support* $x \in P$ if there are $\lambda_1, \ldots, \lambda_k$, with $0 < \lambda_i \leq 1$ for each i, $\sum_{i=1}^{k} \lambda_i = 1$, and $x = \sum_{i=1}^{k} \lambda_i x_i$. Write su($x$) for the set of points that support $x \in P$. (Note the slight abuse of notation relative to that introduced before Definition 3.3.)

Here are the lemmas we will need:

Lemma D.2. *If F is a face of a polytope P and $x \in F$, then* su(x) $\subseteq F$.

Lemma D.3. *For each point x in a polytope P,* su(x) *is a face of P.*

Lemma D.4. *If x belongs to a strictly positive face of a polytope P, then* su(x) *is a strictly positive face of P.*

We now give the interpretation of the geometric notions in game theory. Let d be the cardinality of the finite strategy set S^b. Each strategy $s^a \in S^a$ corresponds to the point

$$\overrightarrow{\pi}^a(s^a) = (\pi^a(s^a, s^b) : s^b \in S^b) \in \mathbb{R}^d.$$

For any probability measure $\mu \in \mathcal{M}(S^a)$, $\overrightarrow{\pi}^a(\mu)$ is the point

$$\overrightarrow{\pi}^a(\mu) = \sum_{s^a \in S^a} \mu(s^a) \overrightarrow{\pi}^a(s^a).$$

Notice that $\overrightarrow{\pi}^a(\mu)$ is in the polytope P generated by the finite set $\{\overrightarrow{\pi}^a(s^a) : s^a \in S^a\}$.

Let us identify each probability measure $\nu \in \mathcal{M}(S^b)$ with the point $(\nu(s^b): s^b \in S^b) \in \mathbb{R}^d$. Then for each pair $(\mu, \nu) \in \mathcal{M}(S^a) \times \mathcal{M}(S^b)$, $\langle \overrightarrow{\pi}^a(\mu), \nu \rangle$ is the expected payoff to Ann. Thus, a pair (μ, ν) gives expected payoff α to Ann if and only if $\overrightarrow{\pi}^a(\mu)$ belongs to the hyperplane $H(\nu, \alpha)$. It follows that a set F is a strictly positive face of P if and only if there is a probability measure ν with support S^b such that

$$F = \{\overrightarrow{\pi}^a(\mu) : \mu \in \mathcal{M}(S^a) \text{ is optimal under } \nu\}.$$

Consider an admissible strategy s^a. By Lemma 3.1, $\overrightarrow{\pi}^a(s^a)$ is optimal under some measure ν with support S^b. That is, $\overrightarrow{\pi}^a(s^a)$ belongs to some strictly positive face of P. Lemma D.4 shows that $\mathrm{su}(\overrightarrow{\pi}^a(s^a))$ is a strictly positive face of P. So we can pick ν so that, for every $r^a \in S^a$, $\overrightarrow{\pi}^a(r^a)$ is optimal under ν if and only if $\overrightarrow{\pi}^a(r^a) \in \mathrm{su}(\overrightarrow{\pi}^a(s^a))$. This is the fact we use in the proof of Theorem 8.1(ii).

We next give some intuition for Lemmas D.2–D.4. Let P be a tetrahedron, as in Figure D.1. The point x^* is supported by the hyperplane H, and the corresponding face $H \cap P$ is the shaded region shown. The set of points that support x^*, that is, the set $\mathrm{su}(x^*)$, is the line segment from x_2 to x_4. Note that these points are also contained in the face $H \cap P$. The general counterpart of this is Lemma D.2.

Now a converse. In Figure D.1, the point x_3 lies in $H \cap P$, but does not support x^*. However, we can tilt the hyperplane H to get a new supporting hyperplane H' as in Figure D.2. Here, $H' \cap P$ is the line segment from x_2 to x_4, that is, exactly the set $\mathrm{su}(x^*)$. The general counterpart is Lemma D.3.

Consider another example in Figure D.3. Here P is the line segment from $(1,0)$ to $(1,1)$. Note that $\mathrm{su}((1,0)) = \{(1,0)\}$. The hyperplane H supports $(1,0)$, and $H \cap P = P$. We can tilt the hyperplane to get H',

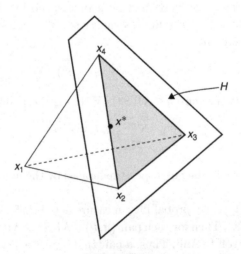

Figure D.1.

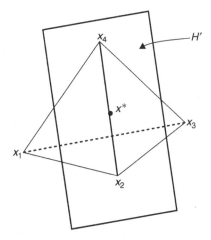

Figure D.2.

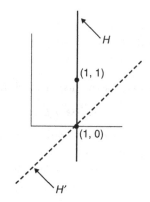

Figure D.3.

where $H' \cap P = \{(1,0)\}$ (in accordance with Lemma D.3). But note that we cannot do this if we require the hyperplane to be nonnegative. (Indeed, H is the unique nonnegative hyperplane supporting $(1,0)$.) Intuitively, though, we will have room to tilt the hyperplane and maintain nonnegativity — in fact, strict positivity — if the original hyperplane is strictly positive. This is Lemma D.4.

We now turn to the proofs of Lemmas D.2–D.4.

Proof of Lemma D.2. Fix a face F that contains x. If $F = P$, then certainly $\mathrm{su}(x) \subseteq F$. If $F \neq P$, there is a hyperplane $H = H(u, \alpha)$ that supports P with $F = H \cap P$. Fix $y \in \mathrm{su}(x)$. Then there are $x_1, \ldots, x_k \in P$ and $\lambda_1, \ldots, \lambda_k$ with $0 < \lambda_i \leq 1$ for each i, $\sum_{i=1}^{k} \lambda_i = 1$, $y = x_1$, and $x = \sum_{i=1}^{k} \lambda_i x_i$. Let $z = \sum_{i=2}^{k} (\lambda_i/(1-\lambda_1)) x_i$ and note that $z \in P$, since P is convex. Also note that $x = \lambda_1 y + (1 - \lambda_1) z$; that is, x lies on the line segment from y to z.

Since $x \in H$ and $y, z \in P$,

$$\langle x, u \rangle = \alpha, \quad \langle y, u \rangle \leq \alpha, \quad \langle z, u \rangle \leq \alpha.$$

Moreover, since x lies on the line segment from y to z,

$$\langle y, u \rangle \leq \langle x, u \rangle \leq \langle z, u \rangle.$$

It follows that $\langle y, u \rangle = \alpha$, so $y \in F$. $\square$

For the next proofs we need the following basic facts about a general polytope P (see Ziegler, 1998 [Chapter 2]):

P1: Every face of P is a polytope.
P2: Every face of a face of P is a face of P.
P3: If $x \in P$, either $x \in \mathrm{relint}(P)$ or x belongs to a proper face of P.
P4: P has finitely many faces.

We record an immediate consequence of P1–P4.

Lemma D.5. *If $x \in P$, then there exists a face F of P with $x \in \mathrm{relint}(F)$.*

Proof. If $x \in \mathrm{relint}(P)$, the result holds trivially. So suppose $x \notin \mathrm{relint}(P)$. By P3, x is contained in some proper face F of P. By P1, the face F is a polytope. Using P2 and P4, we can choose F so that there does not exist a proper face of F that contains x. P3 then implies $x \in \mathrm{relint}(F)$. $\square$

The next lemma establishes a fact about points in the relative interior of a face F of P.

Lemma D.6. *Let F be a face of P. If $x \in \mathrm{relint}(F)$, then $F \subseteq \mathrm{su}(x)$.*

Proof. Fix $x \in \mathrm{relint}(F)$ and some $x' \in F$. If $x' = x$, then certainly $x' \in \mathrm{su}(x)$. If not, consider the line going through both x and x', to be denoted by $L(x, x')$. Since $x \in \mathrm{relint}(F)$, there is some open ball $B(x)$ centered around x, with $\mathrm{aff}(F) \cap B(x) \subseteq F$. Then $\mathrm{aff}(F) \cap B(x)$ must meet $L(x, x')$. Certainly, we can find a point x'' both on $L(x, x')$ and in $\mathrm{aff}(F) \cap B(x)$, with $d(x', x) < d(x', x'')$ for the Euclidean metric d. Then there must exist $0 < \lambda < 1$ with $x = \lambda x' + (1 - \lambda) x''$. Since $x', x'' \in P$, this establishes $x' \in \mathrm{su}(x)$. $\square$

We now turn to the proofs of Lemmas D.3 and D.4.

Proof of Lemma D.3. Fix $x \in P$. By Lemma D.5, there exists a face F of P with $x \in \mathrm{relint}(F)$. We then have $\mathrm{su}(x) \subseteq F$ by Lemma D.2 and $F \subseteq \mathrm{su}(x)$ by Lemma D.6. $\qquad \square$

Proof of Lemma D.4. Let $H(u, \alpha) \cap P$ be a strictly positive face of P containing x. By Lemma D.3, $\mathrm{su}(x) = H(u', \alpha') \cap P$ is a face of P. Set

$$u'' = u' + \beta u, \quad \alpha'' = \alpha' + \beta \alpha$$

for some $\beta > 0$. If $y \in H(u', \alpha') \cap P$, we get

$$\langle y, u'' \rangle = \langle y, u' \rangle + \beta \langle y, u \rangle = \alpha' + \beta \alpha = \alpha'',$$

using $\mathrm{su}(x) \subseteq H(u, \alpha) \cap P$. If $y \in P \backslash H(u', \alpha')$, we get

$$\langle y, u'' \rangle = \langle y, u' \rangle + \beta \langle y, u \rangle < \alpha' + \beta \langle y, \mu \rangle \leq \alpha' + \beta \alpha = \alpha''.$$

Thus, $H(u'', \alpha'')$ is a supporting hyperplane with $\mathrm{su}(x) = H(u'', \alpha'') \cap P$. More over, we can choose $\beta > 0$ so that $u'' \gg 0$ as required. $\qquad \square$

Appendix E: Proofs for Section 9

Lemma E.1. *If $s^a \in S_m^a$, then there exists $\mu \in \mathcal{M}(S^b)$ with Supp $\mu = S_{m-1}^b$, such that $\pi^a(s^a, \mu) \geq \pi^a(r^a, \mu)$ for each $r^a \in S^a$.*

Proof. By Lemma 3.1, there exists $\mu \in \mathcal{M}(S^b)$ with Supp $\mu = S_{m-1}^b$, such that $\pi^a(s^a, \mu) \geq \pi^a(r^a, \mu)$ for all $r^a \in S_{m-1}^a$. Suppose there is an $r^a \in S^a \backslash S_{m-1}^a$ with

$$\pi^a(s^a, \mu) < \pi^a(r^a, \mu). \tag{E.1}$$

We have $r^a \in S_l^a \backslash S_{l+1}^a$, for some $l < m - 1$. Choose r^a (and l) so that there does not exist $q^a \in S_{l+1}^a$ with $\pi^a(s^a, \mu) < \pi^a(q^a, \mu)$.

Fix some $\nu \in \mathcal{M}(S^b)$ with Supp $\nu = S_l^b$ and define a sequence of measures $\mu^n \in \mathcal{M}(S^b)$, for each $n \in \mathbb{N}$, by $\mu^n = (1 - \frac{1}{n})\mu + \frac{1}{n}\nu$. Note that Supp $\mu^n = S_l^b$ for each n. Using $r^a \notin S_{l+1}^a$ and Lemma 3.1 applied to the $(l + 1)$-admissible strategies, it follows that for each n there is a $q^a \in S_l^a$ with

$$\pi^a(q^a, \mu^n) > \pi^a(r^a, \mu^n). \tag{E.2}$$

We can assume that $q^a \in S^a_{l+1}$. (Choose $q^a \in S^a_l$ to maximize the left-hand side of Equation (E.2) among all strategies in S^a_l.) Also, since S^a_{l+1} is finite, there is a $q^a \in S^a_{l+1}$ such that Equation (E.2) holds for infinitely many n. Letting $n \to \infty$ yields

$$\pi^a(q^a, \mu) \geq \pi^a(r^a, \mu). \tag{E.3}$$

From Equations (E.1) and (E.3), we get $\pi^a(q^a, \mu) > \pi^a(s^a, \mu)$, contradicting our choice of r^a. $\qquad\square$

The next lemma will guarantee that we will have enough room to build the measures we need to establish Lemma E.3. For $t^a, u^a \in T^a$, write $t^a \approx u^a$ if, for each i, the component measures $(\lambda^a(t^a))_i$ and $(\lambda^a(u^a))_i$ have the same marginals on S^b, and are mutually absolutely continuous (have the same null sets).

Lemma E.2. *In a complete type structure:*

(i) *If $\lambda^a(t^a) \in \mathcal{L}^+(S^b \times T^b)$ and $u^a \approx t^a$, then $\lambda^a(u^a) \in \mathcal{L}^+(S^b \times T^b)$.*

(ii) *If $\lambda^a(t^a) \in \mathcal{L}^+(S^b \times T^b)$, then there are continuum many u^a such that $u^a \approx t^a$.*

(iii) *For each set $E \subseteq S^b \times T^b$, the set $A^a(E)$ is closed under the relation $\approx$. In fact, for each j, if $t^a \approx u^a$ and E is assumed under $\lambda^a(t^a)$ at level j, then E is assumed under $\lambda^a(u^a)$ at level j.*

(iv) *If $t^a \approx u^a$, then for each m and $s^a \in S^a$, $(s^a, t^a) \in R^a_m$ if and only if $(s^a, u^a) \in R^a_m$.*

Proof. Part (i) follows from the fact that $\lambda^a(t^a) \in \mathcal{L}^+(S^b \times T^b)$ and the mutual absolute continuity of the component measures of $\lambda^a(t^a)$ and $\lambda^a(u^a)$. For part (ii), note that full support implies that $\mu_i = (\lambda^a(t^a))_i$ has infinite support for some i. Therefore, there are continuum many different measures ν_i with the same null sets and marginal on S^b as μ_i. The sequence of measures obtained by replacing μ_i by ν_i belongs to $\mathcal{L}^+(S^b \times T^b)$, and by completeness this sequence is equal to $\lambda^a(u^a)$ for some u^a. It follows that, for each such u^a, $u^a \approx t^a$. For part (iii), fix $\lambda^a(t^a)$ that assumes E at level j. It follows immediately from part (i) and the mutual absolute continuity of the component measures that if $u^a \approx t^a$, then $\lambda^a(u^a)$ also assumes E at level j. For part (iv), the case of $m = 1$ follows immediately from part (i). The case of $m > 1$ is proved by induction and makes use of part (iii). $\qquad\square$

Set $R^a_0 = S^a \times T^a$ and $R^b_0 = S^b \times T^b$.

Lemma E.3. *In a complete type structure,* $\text{proj}_{S^a} R_m^a = \text{proj}_{S^a} (R_m^a \backslash R_{m+1}^a)$ *for each* $m \geq 0$.

Proof. The proof is by induction on m.

$m = 0$: Choose t^a so that $\lambda^a(t^a) \notin \mathcal{L}^+(S^b \times T^b)$ and note that $S^a \times \{t^a\}$ is disjoint from R_1^a. So, $\text{proj}_{S^a}(R_0^a \backslash R_1^a) = S^a$.

$m = 1$: Fix $(s^a, t^a) \in R_1^a$. It suffices to show that there is a type $u^a \in T^a$ with $(s^a, u^a) \in R_1^a \backslash R_2^a$. To see this, first notice that there is a full-support LPS (μ) of length 1 such that s^a is optimal under (μ). (This is by Lemma 7.1.) By completeness, there is a type u^b such that $\lambda^b(u^b) \notin \mathcal{L}^+(S^a \times T^a)$. Construct a probability measure $\nu \in \mathcal{M}(S^b \times T^b)$ with $\text{marg}_{S^b}\mu = \text{marg}_{S^b}\nu$ and $\nu(S^b \times \{u^b\}) = 1$. Let ρ be the measure $(\mu + \nu)/2$. Then ρ is a full-support LPS, so by completeness there is a type $u^a \in T^a$ with $\lambda^a(u^a) = (\rho)$. Note that s^a is optimal under (ρ), so $(s^a, u^a) \in R_1^a$. But $\rho(R_1^b) \leq \frac{1}{2}$ because $\lambda^b(u^b) \notin \mathcal{L}^+(S^a \times T^a)$. So R_1^b is not assumed under (ρ) and therefore $(s^a, u^a) \notin R_2^a$.

$m \geq 2$: Assume the result holds for $m-1$. Let $(s^a, t^a) \in R_m^a$ and $\lambda^a(t^a) = \sigma = (\mu_0, \ldots, \mu_{n-1})$. Then $t^a \in A^a(R_i^b)$ for each $i < m$. We will find a type u^a such that $(s^a, u^a) \in R_m^a \backslash R_{m+1}^a$.

By the induction hypothesis and the fact that S^b is finite, there is a finite set $U \subseteq R_{m-1}^b \backslash R_m^b$ with $\text{proj}_{S^b} U = \text{proj}_{S^b} R_{m-1}^b$. Since $m \geq 2$, $U \subseteq R_1^b$, so $\lambda^b(t^b) \in \mathcal{L}^+(S^a \times T^a)$ for each $(s^b, t^b) \in U$. By Lemma E.2(ii), for each $(s^b, t^b) \in U$ there are continuum many u^b such that $u^b \approx t^b$, and hence there is a $u^b \approx t^b$ such that $\mu_i(\{s^b, u^b\}) = 0$ for all i. Form U' by replacing each $(s^b, t^b) \in U$ by a pair (s^b, u^b) with $u^b \approx t^b$ and $\mu_i(\{s^b, u^b\}) = 0$ for all i. Then U' is finite with $\mu_i(U') = 0$ for all i. By Lemma E.2(iv), $U' \subseteq R_{m-1}^b \backslash R_m^b$ and $\text{proj}_{S^b} U' = \text{proj}_{S^b} R_{m-1}^b$. It follows that the set U can be chosen so that $\mu_i(U) = 0$ for all i.

We will get a point $(s^a, u^a) \in R_m^a \backslash R_{m+1}^b$ by adding a measure to the beginning of the sequence σ. Since U is finite, $\text{proj}_{S^b} U = \text{proj}_{S^b} R_{m-1}^b$, and $\mu_0(R_{m-1}^b) = 1$, there is a probability measure ν such that $\nu(U) = 1$ and $\text{marg}_{S^b}\nu = \text{marg}_{S^b}\mu_0$. Let τ be the sequence $(\nu, \mu_0, \ldots, \mu_{n-1})$. Since $\sigma \in \mathcal{L}^+(S^b \times T^b)$ and $\mu_i(U) = 0$ for each i, we see that $\tau \in \mathcal{L}^+(S^b \times T^b)$. By completeness there is a $u^a \in T^a$ with $\lambda^a(u^a) = \tau$. Since ν has the same marginal on S^b as μ_0 and since $(s^a, t^a) \in R_1^a$, we have $(s^a, u^a) \in R_1^a$. Since $U \subseteq R_{m-1}^b$ and $t^a \in A^a(R_k^b)$ for each $k < m$, it follows that $u^a \in A^a(R_k^b)$ for each $k < m$. Then by Lemma C.4(i), we have $(s^a, u^a) \in R_m^a$. However, since U is disjoint from R_m^b we have $\nu(R_m^b) = 0$, so $u^a \notin A^a(R_m^b)$ and hence $(s^a, u^a) \notin R_{m+1}^a$. This completes the induction. $\quad\square$

Appendix F: Proofs for Section 10

For the following two lemmas we assume that $\langle S^a, S^b, T^a, T^b, \lambda^a, \lambda^b \rangle$ is a complete type structure in which the maps λ^a and λ^b are continuous.

Lemma F.1. *If player a is not indifferent, then $R_0^a \backslash \overline{R_1^a}$ is uncountable.*

Proof.   We have that $\pi^a(r^a, s^b) < \pi^a(s^a, s^b)$ for some r^a, s^a, s^b. Then S^a has more than one element and by completeness, T^b has more than one element. Therefore, using completeness again, there is a type $t^a \in T^a$ such that $\lambda^a(t^a) = (\mu_0, \mu_1)$ is a full-support LPS of length 2 and $\mu_0(\{s^b\} \times T^b) = 1$. Let U be the set of all $u^a \in T^a$ such that r^a is not optimal under $(\lambda^a(u^a))_0$, i.e., for some $q^a \in S^a$,

$$\sum_{s^b \in S^b} \pi^a(r^a, s^b) \text{marg}_{S^b}(\lambda^a(u^a))_0(s^b)$$

$$< \sum_{s^b \in S^b} \pi^a(q^a, s^b) \text{marg}_{S^b}(\lambda^a(u^a))_0(s^b).$$

We now show that $t^a \in U$. Note first that since $\mu_0(\{s^b\} \times T^b) = 1$, the function $\text{marg}_{S^b}(\lambda^a(t^a))_0$ has value 1 at s^b and 0 everywhere else in S^b. Therefore, for each $q^a \in S^a$,

$$\sum_{s^b \in S^b} \pi^a(q^a, s^b) \text{marg}_{S^b}(\lambda^a(t^a))_0(s^b) = \pi^a(q^a, s^a).$$

Since $\pi^a(r^a, s^b) < \pi^a(s^a, s^b)$, the inequality defining U holds with $(q^a, u^a) = (s^a, t^a)$ and hence $t^a \in U$.

We next show that U is open. Since λ^a is continuous, the function $u^a \mapsto (\lambda^a(u^a))_0$ is continuous. Convergence in the Prohorov metric is equivalent to weak convergence, so the function

$$u^a \mapsto \text{marg}_{S^b}(\lambda^a(u^a))_0(s^b) = \int 1(\{s^b\} \times T^b) d(\lambda^a(u^a))_0$$

is continuous. Thus, U is defined by a strict inequality between two continuous real functions of u^a and hence U is open.

Since $\{r^a\}$ is open in S^a, the set $\{r^a\} \times U$ is open in $S^a \times T^a$. By definition, the set $\{r^a\} \times U$ is disjoint from R_1^a. Now suppose $u^a \approx t^a$. Then $(\lambda^a(u^a))_0$ has the same marginals as $(\lambda^a(t^a))_0$, so $u^a \in U$ and hence $(r^a, u^a) \in \{r^a\} \times U$. Since $\{r^a\} \times U$ is open and disjoint from R_1^a, we have $(r^a, u^a) \notin \overline{R_1^a}$. By Lemma E.2, there are uncountably many u^a such that $u^a \approx t^a$, so $R_0^a \backslash \overline{R_1^a}$ is uncountable. $\square$

Lemma F.2. *Suppose that $m \geq 1$ and $R_{m-1}^b \backslash \overline{R_m^b}$ is uncountable. Then $R_m^a \backslash \overline{R_{m+1}^a}$ is uncountable.*

Proof. The proof is similar to the proof of Lemma E.3. Fix $(s^a, t^a) \in R_m^a$. By the proof of Theorem 9.1, we can choose t^a so that $\lambda^a(t^a) = \sigma = (\mu_0, \ldots, \mu_{m-1})$ and R_{m-1}^b is assumed at level 0. We will get uncountably many points $(s^a, u^a) \in R_m^a \backslash \overline{R_{m+1}^a}$ by adding one more measure to the beginning of the sequence σ and using Lemma E.2.

We claim that there is a finite set $U \subseteq R_{m-1}^b \backslash \overline{R_m^b}$ such that $\operatorname{proj}_{S^b} U = \operatorname{proj}_{S^b} R_{m-1}^b$ and $\mu_i(U) = 0$ for all $i < m$.

$m = 1$: Recall that, for each $(s^a, t^a) \in R_1^a$, there is a u^a such that $\lambda^a(u^a)$ is a full-support LPS and $(s^a, u^a) \in R_1^a \backslash R_2^a$. (This was shown in the proof of Lemma E.3.) The claim for $m = 1$ now follows from Lemma E.2 and the fact that S^a is finite.

$m \geq 2$: The claim was already established in the induction step of Lemma E.3.

Now, since $R_{m-1}^b \backslash \overline{R_m^b}$ is uncountable, there is a point $(s^b, t^b) \in R_{m-1}^b \backslash \overline{R_m^b}$ such that $\mu_i(s^b, t^b) = 0$ for all $i < m$. Therefore, we may also take U to contain such a point (s^b, t^b). Let ν be a probability measure such that $\nu(U) = 1$, $\operatorname{marg}_{S^b} \nu = \operatorname{marg}_{S^b} \mu_0$, and $\nu(s^b, t^b) = \operatorname{marg}_{S^b} \mu_0(s^b)$. Since R_{m-1}^b is assumed under σ at level 0, we have $(s^b, t^b) \in \operatorname{Supp} \mu_0$, and thus $\mu_0(\{s^b\} \times T^b) = \operatorname{marg}_{S^b} \mu_0(s^b) > 0$. Therefore, $\nu(s^b, t^b) > 0$.

Let τ be the sequence $(\nu, \mu_0, \ldots, \mu_{m-1})$. Since $(s^a, t^a) \in R_1^a, \lambda^a(t^a) = (\mu_0, \ldots, \mu_{m-1})$ is a full-support LPS. Also, $\mu_i(U) = 0$ for each i. Therefore, τ is mutually singular and so a full-support LPS. By completeness, there is a $v^a \in T^a$ with $\lambda^a(v^a) = \tau$. Then, $(\lambda^a(v^a))_0 = \nu$. As in Lemma E.3, we have $(s^a, v^a) \in R_m^a$. Given this, the proof of Lemma E.2(ii) shows that there are uncountably many $u^a \approx v^a$ such that $(\lambda^a(u^a))_0 = \nu$.

Suppose $u^a \approx v^a$ and $(\lambda^a(u^a))_0 = \nu$. Then, $\lambda^a(u^a)$ has length $m + 1$. By Lemma E.2, we have $(s^a, u^a) \in R_m^a$. However, since $(s^b, t^b) \notin \overline{R_m^b}$, the measure ν has an open neighborhood W, where, for each $\nu' \in W, \nu'(R_m^b) < 1$. (An example of such a neighborhood is the set $\{\nu' : \nu'(V) > \nu(s^b, t^b)/2\}$, where V is an open neighborhood of (s^b, t^b) which is disjoint from R_m^b.) Then the set

$$X = \{\xi \in \mathcal{N}_{m+1}(S^b \times T^b) : \xi_0 \in W\}$$

is an open neighborhood of $\lambda^a(u^a)$, and no LPS $\xi \in X$ can assume R_m^b at level 0. It follows that an LPS $\xi \in X$ cannot assume all of the $m + 1$ sets $R_k^b, k \leq m$, because by the inductive hypothesis all these sets have different closures, and hence by Property 6.2 at most one can be assumed at each level. By continuity of λ^a, the set $Y = (\lambda^a)^{-1}(X)$ is an open neighborhood of u^a. Then $\{s^a\} \times Y$ is an open neighborhood of (s^a, u^a) which is disjoint from R^a_{m+1}, so (s^a, u^a) is not in the closure of R^a_{m+1}. By Lemma E.2, there are uncountably many $u^a \approx v^a$ and, therefore, $R^a_m \backslash \overline{R^a_{m+1}}$ is uncountable. $\square$

Proof of Theorem 10.1. By Proposition 7.1(ii), it suffices to assume that λ^a and λ^b are continuous. As such, Lemma F.1 gives that the set $R^a_0 \backslash \overline{R^a_1}$ is uncountable. Then, by induction and Lemma F.2, for each m, the sets $R^a_{2m} \backslash \overline{R^a_{2m+1}}$ and $R^b_{2m+1} \backslash \overline{R^b_{2m+2}}$ are uncountable. Suppose that $(s^b, t^b) \in \bigcap_m R^b_m$. Then for each m, we have that R^a_m is assumed under $\lambda^b(t^b)$ at some level $j(m)$. Moreover, the sequence $j(m)$ is nonincreasing. Then by Property 6.2 and the fact that each $R^a_{2m} \backslash \overline{R^a_{2m+1}}$ is uncountable, we have that each $j(2m + 1) < j(2m)$. But this contradicts the fact that $\lambda^b(t^b)$ has finite length. $\square$

References

Armbruster, W and W Böge (1979). Bayesian game theory. In Möschlin, O and D Pallaschke (Eds.), *Game Thoery and Related Topics*, pp. 17–28. Amsterdam: North-Holland.

Asheim, G and M Dufwenberg (2003). Admissibility and common belief. *Games and Economic Behavior*, 42, 208–234.

Asheim, G and Y Søvik (2005). Preference-based belief operators. *Mathematical Social Sciences*, 50, 61–82.

Battigalli, P (1996). Strategic rationality orderings and the best rationalization principle. *Games and Economic Behavior*, 13, 178–200.

Battigalli, P (1997). On rationalizability in extensive games. *Journal of Economic Theory*, 74, 40–61.

Battigalli, P and M Siniscalchi (1999). Hierarchies of conditional beliefs and interactive epistemology in dynamic games. *Journal of Economic Thoery*, 88, 188–230.

Battigalli, P and M Siniscalchi (2002). Strong belief and forward-induction reasoning. *Journal of Economic Thoery*, 106, 356–391.

Ben Porath, E (1997). Rationality, Nash equilibrium, and backward induction in perfect information games. *Review of Economic Studies*, 64, 23–46.

Bernheim, D (1984). Rationalizable strategic behavior. *Econometrica*, 52, 1007–1028.

Billingsley, P (1968). *Convergence of Probability Measures*. New York, NY: Wiley.

Blume, L, A Brandenburger, and E Dekel (1991a). Lexicographic probabilities and choice under uncertainty. *Econometrica*, 59, 61–79.

Blume, L, A Brandenburger, and E Dekel (1991b). Lexicographic probabilities and equilibrium refinements. *Econometrica*, 59, 81–98.

Böge, W and Th Eisele (1979). On solutions of Bayesian games. *International Journal Game Theory*, 8, 193–215.

Börgers, T (1994). Weak dominance and approximate common knowledge. *Journal of Economic Theory*, 64, 265–276.

Brandenburger, A (1992). Lexicographic probabilities and iterated admissibility. In Dasgupta, P, D Gale, O Hart, and E Maskin (Eds.), *Economic Analysis of Markets and Games*, pp. 282–290. Cambridge, MA: MIT Press.

Brandenburger, A (2003). On the existence of 'complete' possibility structure. In Basili, M, N Dimitri, and I Gilboa (Eds.), *Cognitive Processes and Economic Behavior*, pp. 30–34. London, UK: Routledge.

Brandenburger, A and A Dekel (1993). Hierarachies of beliefs and common knowledge. *Journal of Economic Theory*, 59, 189–198.

Brandenburger, A and A Friedenberg (2003). The relationship between rationality on the matrix and the tree. Unpublished manuscript. Available at www.stern.nyu.edu/~abranden.

Brandenburger, A and A Friedenberg (2004). Self-admissible sets. Unpublished manuscript. Available at www.stern.nyu.edu/~abranden.

Brandenburger, A, A Friedenberg, and H J Keisler (2006). Notes on the relationship between strong belief and assumption. Unpublished manuscript. Available at www.stern.nyu.edu/~abranden.

Brandenburger, A, A Friedenberg, and H J Keisler (2008). Supplement to "Admissibility in games." *Econometrica Supplementary Material*, 76. Available at http://econometricsociety.org/ecta/Supmat/5602_extensions.pdf.

Dekel, E and D Fudenberg (1990). Rational behavior with payoff uncertainty. *Journal of Economic Theory*, 52, 243–267.

Halpern, J (2007). Lexicographic probability, conditional probability, and non-standard probability. Unpublished manuscript. Available at http://www.cs.cornell.edu/home/halpern.

Heifetz, A (1993). The Bayesian formulation of incomplete information — The non-compact case. *International Journal of Game Theory*, 21, 329–338.

Kechris, A (1995). *Classical Descriptive Set Theory*. New York, NY: Springer-Verlag.

Kohlberg, E and J-F Mertens (1986). On the strategic stability of equilibria. *Econometrica*, 54, 1003–1037.

Mailath, G, L Samuelson, and J Swinkels (1993). Extensive form reasoning in normal form games. *Econometrica*, 61, 273–302.

Marx, L and J Swinkels (1997). Order independence for iterated weak dominance. *Games and Economic Behavior*, 18, 219–245.

Mertens, J-F (1989). Stable equilibria — A reformulation. *Mathematics of Operations Research*, 14, 575–625.

Mertens, J-F and S Zamir (1985). Formulation of Bayesian analysis for games with incomplete information. *International Journal of Game Theory*, 14, 1–29.

Osborne, M and A Rubinstein (1994). *A Course in Game Theory*. Cambridge, MA: MIT Press.

Pearce, D (1984). Rational strategic behavior and the problem of perfection. *Econometrica*, 52, 1029–1050.

Rényi, A (1955). On a new axiomatic theory of probability. *Acta Mathematica Academiae Scientiarum Hungaricae*, 6, 285–335.

Samuelson, L (1992). Dominated stragies and common knowledge. *Games and Economic Behavior*, 4, 284–313.

Savage, L (1954). *The Foundations of Statistics*. New York, NY: Dover.

Shimoji, M (2004). On the equivalence of weak dominance and sequential best response. *Games and Economic Behavior*, 48, 385–402.

Ziegler, G (1998). *Lectures on Polytopes*, 2nd Edition. New York, NY: Springer.

Chapter 8

Self-Admissible Sets

Adam Brandenburger and Amanda Friedenberg

Best-response sets (Pearce [1984]) characterize the epistemic condition of "rationality and common belief of rationality." When rationality incorporates a weak-dominance (admissibility) requirement, the self-admissible set (SAS) concept (Brandenburger, Friedenberg, and Keisler [2008]) characterizes "rationality and common assumption of rationality." We analyze the behavior of SAS's in some games of interest — Centipede, the Finitely Repeated Prisoner's Dilemma, and Chain Store. We then establish some general properties of SAS's, including a characterization in perfect-information games.

1. Introduction

Consider the condition that Ann is rational, she thinks Bob is rational, she thinks he thinks she is rational, and so on. In this case, what strategies will Ann choose? This is a basic question in the epistemic approach to game theory. It has been asked when "rationality" means: (i) ordinary (subjective

Originally published in *Journal of Economic Theory*, 145, 785–811.

Keywords: admissibility; weak dominance; self-admissible sets; iterated admissibility; epistemic game theory; perfect-information games.

Note: Some of the results here were presented in two working papers, titled "When Does Common Assumption of Rationality Yield a Nash Equilibrium?" (July 2001) and "Common Assumption of Rationality in Games" (January 2002).

Acknowledgments: We owe much to joint work with H. Jerome Keisler. We are also indebted to Pierpaolo Battigalli, Drew Fudenberg, John Nachbar, Martin Osborne, Marciano Siniscalchi, Gus Stuart, the associate editor, and the referees for important comments. Geir Asheim, Konrad Grabiszewski, Elon Kohlberg, Alex Peysakhovich, and participants at presentations of this work provided valuable input. Brandenburger thanks Harvard Business School and the Stern School of Business for their support. Friedenberg thanks the Olin Business School and the W.P. Carey School of Business for their support.

expected utility) maximization on the matrix, (ii) admissibility (avoidance of weakly dominated strategies) on the matrix, and (iii) maximization at each information set in the tree.

There is a very intuitive answer for (i): Ann will choose an *iteratively undominated* (*IU*) strategy — i.e., a strategy which survives the iterated elimination of strongly dominated strategies. The idea goes back to Bernheim (1984) and Pearce (1984) (though they made an additional independence assumption). There is also a deeper answer, pioneered by Pearce. He introduced the concept of a *best-response set* (*BRS*): This is a subset $Q^a \times Q^b \subseteq S^a \times S^b$ (where S^a and S^b are Ann's and Bob's strategy sets) such that each of Ann's strategies in Q^a is undominated with respect to Q^b, and likewise with Ann and Bob interchanged.

Here is a sketch of how a formal epistemic analysis leads to a BRS. The first step is to add types (for Ann and Bob) to the description of the game. A particular type for Ann describes what she thinks about which strategy Bob chooses, what she thinks Bob thinks about which strategy she chooses, and so on. Likewise with Bob's types. With these ingredients, we can identify those strategy-type pairs for Ann which are rational, believe (i.e., assign probability 1 to the event) that Bob is rational, and so on. This is the set labelled RCBR ("rationality and common belief of rationality") in the left-hand panel in Figure 1.1. Applying a similar analysis for Bob leads to the right-hand panel.

The next step is to see what this analysis implies in terms of strategies that can be played. Formally, we project the RCBR sets into their respective strategy sets. This gives the middle panel in Figure 1.1. Can we say what the projections are? Yes, they constitute a BRS. This is the deeper answer to question (i).

Two observations. First, a given game can have several BRS's. A particular epistemic analysis will yield one such BRS — which one will

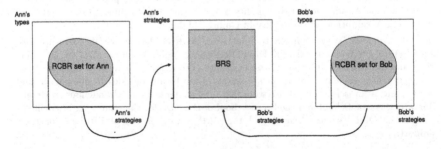

Figure 1.1.

depend on what type spaces we happen to begin with. Second, the greater precision in saying we get a BRS — rather than just saying that the strategies are IU — may seem a small matter. It is well known that every BRS is contained in the IU set, which is itself the largest BRS.

But, the greater precision becomes crucial in the case of admissibility — i.e., when we address question (ii). The obvious guess for the answer is that Ann will choose an *iteratively admissible* (*IA*) strategy — i.e., a strategy which survives the iterated elimination of inadmissible strategies. But this is wrong!

Here are the steps in order to see why. Brandenburger, Friedenberg, and Keisler (2008) formulate an analysis like that in Figure 1.1. Now, Ann is rational in the sense of admissibility, she assumes Bob is rational in the same sense, and so on. (Here, we say "assumes" rather than "believes." See Sections 3 and 8.2.) Likewise for Bob. The middle panel in Figure 1.1 now becomes a concept called a *self-admissible set* (*SAS*) in Brandenburger, Friedenberg, and Keisler (2008): This is a subset $Q^a \times Q^b \subseteq S^a \times S^b$ such that each of Ann's strategies in Q^a is admissible with respect to both S^b and Q^b, and, in addition, Q^a satisfies a maximality condition. Likewise with Ann and Bob interchanged. (Details will be given later.)

Note the analogy to BRS. (Obviously, if one of Ann's strategies in Q^a is not strongly dominated with respect to Q^b, it is also undominated with respect to S^b. With admissibility, the two conditions must be explicitly stated. The maximality condition is not a point of difference — again, details later.)

But — and this is the key point — unlike with BRS's and the IU set, an SAS need not be contained in the IA set. Indeed, while the IA set of a game constitutes one SAS of that game, there can be other SAS's which are even disjoint from it.

This is why a separate analysis of SAS's is in order. Much is known about the behavior of IA in various games, but not about SAS. This chapter aims to fill in the picture. We preview our investigation in Section 2.

Before that, a little more on the foundation of the concept. In particular, why is the SAS-IA relationship different from the BRS-IU relationship? The reason is a basic non-monotonicity in admissibility: The component-by-component union of two SAS's need not be an SAS. Consider the game in Figure 1.2.[1] There are five SAS's: $\{(U, L)\}, \{(U, C)\}, \{U$

---

[1]We thank a referee for this example. It is similar to examples in Asheim and Dufwenberg (2003).

Bob

	L	C	R
U	1, 1	1, 1	0, 0
M	1, 1	0, 0	1, 0
D	0, 0	0, 1	0, 0

Ann

Figure 1.2.

$\times \{L, C\}, \{(M, L)\}$, and $\{U, M\} \times \{L\}$. But $\{U, M\} \times \{L, C\}$ is not an SAS. Kohlberg and Mertens (1986, p. 1017) gave the 'philosophical' explanation (albeit in a different context): Under admissibility, adding new possibilities can change previously good strategies into bad ones. While M (resp. C) is admissible with respect to $\{L\}$ (resp. $\{U\}$), it becomes inadmissible once C (resp. M) is added. This appears to be the fundamental reason for the greater complexity of SAS vs. BRS theory. We expand on this point in Section 8.2.

Finally, let us return to the third version of the question we posed at the outset — concerning epistemic analysis of rationality in the tree. This is not the subject of this chapter, but there is a parallel. Two concepts arise in this case: extensive-form rationalizability (Pearce [1984], Battigalli and Siniscalchi [2002]) and extensive-form best-response sets (Battigalli and Friedenberg [2009]). They are the analogs to IU-BRS and IA-SAS. But, unlike IU-BRS and again like IA-SAS, there is a non-monotonicity: extensive-form best-response sets are not necessarily contained in the extensive-form rationalizable set.

2. Preview

In Section 3, we give the formal definition of an SAS and review epistemic foundations of the concept. Section 4 studies how SAS behaves in some of the most commonly studied games — Centipede, the Finitely Repeated Prisoner's Dilemma, and Chain Store. In Sections 5–7 we develop some general properties, first in the strategic form and second in the extensive form — including a characterization in perfect-information games. Section 8 contains some conceptual discussion and also covers related work.

Here, we make two preliminary comments on the philosophy underlying our approach. First, SAS is a strategic-form concept, yet the bulk of the chapter investigates its behavior in extensive-form games. Why? The easy answer is that we are following the Kohlberg and Mertens (1986, Sections 2.4–2.5) invariance doctrine: Our analysis should depend only on the strategic form, even if our primary interest is in behavior in game trees.

Recall that Dalkey (1953), Thompson (1952), and Elmes and Reny (1994) showed that two trees have the same reduced strategic form — i.e., the strategic form after elimination of duplicate rows and columns — if and only if they differ by a sequence of four elementary transformations.[2] Kohlberg and Mertens added a fifth convex-combination transformation. If all five transformations are viewed as "irrelevant for correct decision making" (Kohlberg and Mertens [1986, p. 1011]), then we get a requirement of invariance to the fully reduced strategic form — i.e., invariance to the strategic form after elimination of convex combinations. We show that SAS satisfies this full invariance requirement (Proposition 5.2).

In fact, there is a deeper level to the invariance issue. Epistemics are a tool to formalize reasoning in games. If the idea of invariance is that reasoning should not change across equivalent games, then should not invariance be stated at the level of the epistemics?[3] The answer, surely, is yes. We would like to have a full-blown principle of "epistemic invariance" and to be able to investigate whether various solution concepts — SAS included — satisfy it. See Section 8.6 for further discussion.

Our second preliminary comment concerns the broader question of what we hope to learn by investigation of SAS — or, indeed, some other epistemically derived concept. Our view is that a primary purpose of epistemic game theory is to formalize intuitive notions which one has about strategic situations — notions such as that of a "best (or rational?) course of action," or that of the importance of "thinking about what the other player is thinking," and so on. The goal is that, by carrying out such formalization and by investigating properties of the resulting concepts, we improve our understanding of the underlying strategic situations. In this chapter we take as given that the SAS concept is an embodiment of certain intuitions (following Brandenburger, Friedenberg, and Keisler [2008]). Our goal here is the next step — to uncover properties of the concept.

3. Self-Admissible Sets

We now give a formal definition of the SAS concept and sketch its epistemic basis. To begin, fix a two-player strategic-form game $\langle S^a, S^b; \pi^a, \pi^b \rangle$, where

---

[2]The Elmes–Reny transformations differ from the Dalkey–Thompson ones by preserving perfect recall throughout.

[3]We are grateful to a referee for raising this point.

S^a, S^b are the (finite) strategy sets and π^a, π^b are payoff functions for Ann and Bob, respectively. (We focus on two players, but our analysis extends readily to n players.) The definitions to come all have counterparts with a and b reversed.

Given a finite set X, let $\mathcal{M}(X)$ denote the set of all probability measures on X. We extend π^a to $\mathcal{M}(S^a) \times \mathcal{M}(S^b)$ in the usual way, i.e., $\pi^a(\sigma^a, \sigma^b) = \sum_{s^a \in S^a} \sum_{s^b \in S^b} \sigma^a(s^a)\sigma^b(s^b)\pi^a(s^a, s^b)$. Throughout, we adopt the convention that in a product $X \times Y$, if $X = \emptyset$ then $Y = \emptyset$ (and vice versa).

Definition 3.1. Fix $X \times Y \subseteq S^a \times S^b$ and some $s^a \in X$. Say s^a is *weakly dominated with respect to* $X \times Y$ if there exists $\sigma^a \in \mathcal{M}(S^a)$, with $\sigma^a(X) = 1$, such that $\pi^a(\sigma^a, s^b) \geq \pi^a(s^a, s^b)$ for every $s^b \in Y$, and $\pi^a(\sigma^a, s^b) > \pi^a(s^a, s^b)$ for some $s^b \in Y$. Otherwise, say s^a is *admissible with respect to* $X \times Y$. If s^a is admissible with respect to $S^a \times S^b$, simply say that s^a is *admissible*.

Definition 3.2. Fix $X \subseteq S^a$. Say $s^a \in X$ is *optimal under* $\mu^b \in \mathcal{M}(S^b)$ *given* X if $\pi^a(s^a, \mu^b) \geq \pi^a(r^a, \mu^b)$ for all $r^a \in X$. If $s^a \in S^a$ is optimal under μ^b given S^a, simply say that s^a is *optimal under* μ^b.

Remark 3.1. A strategy s^a is admissible with respect to $X \times Y$ if and only if there exists $\mu^b \in \mathcal{M}(S^b)$, with $\operatorname{Supp} \mu^b = Y$, such that s^a is optimal under μ^b given X.

Next, we need:

Definition 3.3. Fix some $s^a \in S^a$ and suppose there is $\varphi^a \in \mathcal{M}(S^a)$ with $\pi^a(\varphi^a, s^b) = \pi^a(s^a, s^b)$ for all $s^b \in S^b$. Then if $r^a \in \operatorname{Supp} \varphi^a$, say r^a *supports* s^a (*via* φ^a). Write $\operatorname{su}(s^a)$ for the set of $r^a \in S^a$ that support s^a.

In words, this says that $\operatorname{su}(s^a)$ consists of those strategies for Ann that are part of some convex combination equivalent (for her) to s^a. With this, we can give the definition of a self-admissible set.

Definition 3.4. Fix $Q^a \times Q^b \subseteq S^a \times S^b$. The set $Q^a \times Q^b$ is a *self-admissible set* (*SAS*) if:

(i) each $s^a \in Q^a$ is admissible with respect to $S^a \times S^b$;
(ii) each $s^a \in Q^a$ is admissible with respect to $S^a \times Q^b$;
(iii) for any $s^a \in Q^a$, if $r^a \in \operatorname{su}(s^a)$ then $r^a \in Q^a$;

and likewise for each $s^b \in Q^b$.

Definition 3.4 brings out the analogy to best-response sets (BRS's). To repeat from earlier, a strong-dominance version of condition (i) is implied by a strong-dominance version of condition (ii). For weak dominance, we need to stipulate the additional condition. Condition (iii) could be added to the definition of a BRS. It is without loss of generality in the following sense: Any BRS $Q^a \times Q^b$ not satisfying condition (iii) is contained in a larger BRS that does satisfy condition (iii).[4] (This is a consequence of Lemma A.1 below.) By contrast, a set $Q^a \times Q^b$ satisfying only conditions (i) and (ii) of Definition 3.4 may not be contained in any SAS. (See Brandenburger, Friedenberg, and Keisler [2008, Section 2.3] for an example.)

Let us briefly review how an epistemic analysis of admissibility leads to the SAS concept. The analysis starts from a basic challenge for admissibility in games, identified by Samuelson (1992). On the one hand, admissibility requires that, if Ann is rational, she should not rule out any of Bob's strategies (per Remark 3.1). On the other hand, if Ann thinks Bob is rational, then she should rule out the possibility that Bob plays an inadmissible strategy. There appears to be a conflict between the requirements that: (i) Ann is rational, and (ii) Ann thinks that Bob is rational. An epistemic analysis of admissibility must face this tension.

Brandenburger, Friedenberg, and Keisler (2008) resolve the tension by asking that Ann consider it infinitely less likely — but not impossible — that Bob is irrational vs. rational. Ann is equipped with a lexicographic probability system (LPS) on Bob's strategies and types — i.e., on $S^b \times T^b$. This is a sequence of measures $(\mu_1, \ldots, \mu_n)$, where μ_1 represents Ann's primary hypothesis, μ_2 represents her secondary hypothesis, and so on. (See Blume, Brandenburger, and Dekel [1991].) Now, Ann can consider one strategy-type pair (s^b, t^b) for Bob infinitely more likely than another pair (r^b, u^b) — e.g., if μ_1 assigns probability 1 to (s^b, t^b), while μ_2 assigns positive probability to (r^b, u^b). More generally, say Ann *assumes* an event $E \subseteq S^b \times T^b$ if, under her LPS, all of E is infinitely more likely than all of not-E.

With these ingredients, the epistemic condition of rationality and common assumption of rationality $(RCAR)$ is expressible. (A strategy-type pair (s^a, t^a) is *rational* if t^a is associated with a full-support LPS and s^a lexicographically maximizes the sequence of expected payoffs Ann

---

[4]In private communication, David Pearce told us that he was aware of the maximality condition, but, given this property, did not include it in his definition. In fact, if the definition of BRS is derived epistemically, per Figure 1.1, maximality would be automatically incorporated.

gets under the LPS associated with t^a.) Theorem 8.1 in Brandenburger, Friedenberg, and Keisler (2008) says that RCAR is characterized by the SAS concept. That is, fixing a game and a type structure (analogous to the one in Figure 1.1), the strategies that are played under RCAR constitute an SAS of the game. Conversely, fixing a game and an SAS of the game, there is a type structure so that the strategies played under RCAR are those in the SAS. In Section 8.3, we will discuss other epistemic analyses of admissibility.

As we proceed to investigate properties of SAS's, we will compare what we find with properties of the IA set. (We take IA to mean simultaneous maximal deletion.) This is a natural comparison since the two concepts are related at the epistemic level. (Again, see Brandenburger, Friedenberg, and Keisler [2008].) Here is the formal definition:

Definition 3.5. Set $S_0^i = S^i$ for $i = a, b$, and define inductively

$$S_{m+1}^i = \{s^i \in S_m^i : s^i \text{ is admissible with respect to } S_m^a \times S_m^b\}.$$

A strategy $s^i \in S_m^i$ is called *m-admissible*. A strategy $s^i \in \bigcap_{m=0}^{\infty} S_m^i$ is called *iteratively admissible* (*IA*).

Since the game is finite, there is an M such that $S_M^i = \bigcap_{m=0}^{\infty} S_m^i$ for $i = a, b$. Moreover, each S_M^i is nonempty and therefore the IA set is nonempty.

4. Applications

We now begin our investigation of properties of SAS's. In this section, we look at what the SAS concept gives in three canonical examples — Centipede (Rosenthal [1981]), the Finitely Repeated Prisoner's Dilemma, and Chain Store (Selten [1978]). This will give some pointers to general properties to investigate in the following sections.

Example 4.1. (*Centipede*) Consider n-legged Centipede, as in Figure 4.1. If $Q^a \times Q^b$ is an SAS of Centipede, and $(s^a, s^b) \in Q^a \times Q^b$, then s^a is Ann's strategy of playing *Out* at the first node.[5]

Indeed, suppose, to the contrary, that there is an $(s^a, s^b) \in Q^a \times Q^b$ where s^a involves Ann's playing *In* at the first node. In particular, pick a profile (s^a, s^b) which yields the longest path of play (before Ann or Bob

---

[5]We consider the reduced strategic form of the game. This will suffice, given Proposition 5.2 to come.

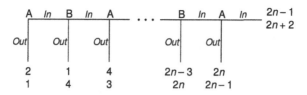

Figure 4.1.

plays *Out*). (The fact that some player does play *Out* on this path follows from admissibility — i.e., condition (i) of the definition of an SAS.) Let h be the node on this path at which *Out* is played. Suppose Bob moves at h. (A similar argument applies if Ann moves at h.) Then, by condition (ii) of the definition of an SAS, and Remark 3.1, Ann's strategy s^a must be optimal under a measure that assigns: (i) probability 1 to Bob's playing *Out* at node h or earlier; and (ii) positive probability to Bob's playing *In* until node h and *Out* at h. Now consider the strategy r^a for Ann that plays *In* until node h' (where h' is the immediate predecessor of h) and plays *Out* at h'. Then r^a does strictly better than s^a under any such measure — a contradiction.

The IA set for Centipede is easily verified to be $\{(Out,\ Out)\}$, and $\{(Out,\ Out)\}$ is also an SAS of the game. Thus, in particular, the set of SAS's is nonempty.

This analysis of Centipede seems very intuitive. It starts at the root of the tree and works forwards to reach a contradiction: If it is Bob who ends the game (by playing *Out* at node h), then Ann should have ended the game earlier.[6] (Of course, our goal is not to defend *Out* as the only rational choice. Indeed, the analysis says that RCAR — a much stronger condition — yields this conclusion.) The next example also involves a forward-looking argument.[7]

Example 4.2. (*Finitely Repeated Prisoner's Dilemma*) Consider the Prisoner's Dilemma, as in Figure 4.2. Fix an SAS of the game played T times (for some integer T). Any strategy profile in the SAS yields the Defect-Defect path throughout.

---

[6]The same proof applies to Nash equilibrium. If (σ^a, σ^b) is an equilibrium of Centipede, then σ^a puts probability 1 on Ann's playing *Out* at the first node. Just apply our argument to Supp σ^a × Supp σ^b.

[7]Other papers — albeit with different epistemics — employ forward-looking arguments. For Centipede, see Aumann (1998). For FRPD, see Stuart (1997).

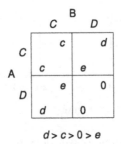

$$d > c > 0 > e$$

Figure 4.2.

The proof will use a projection property of SAS's established later (Proposition 6.2). We argue by induction on the number of rounds. For $T = 1$, the result is immediate from the fact that any strategy in an SAS is admissible (a fortiori, not strongly dominated). Now, assume the result for T, and fix an SAS $Q^a \times Q^b$ of the $(T+1)$-fold game. Suppose $s^a \in Q^a$ involves Ann's playing C on the first round. Then, for any $s^b \in Q^b$, Ann gets a first-round payoff of c if s^b involves Bob's playing C on the first round, and e if s^b involves Bob's playing D on the first round. These are also Ann's total payoffs from the game when (s^a, s^b) is played, since the induction hypothesis, together with the fact that SAS's induce SAS's on subtrees (see Proposition 6.2 for a precise statement), implies that the profile (s^a, s^b) must yield the Defect-Defect path on rounds $2, \ldots, T+1$. Suppose instead that Ann chooses the "Defect always" strategy. Then she gets a first-round payoff of d if s^b involves Bob's playing C on the first round, and 0 if s^b involves Bob's playing D on the first round. On subsequent rounds Ann gets at least 0. But then the "Defect always" strategy does strictly better than s^a against every $s^b \in Q^b$, contradicting the definition of an SAS.

The IA set for FRPD consists of a unique strategy pair, where each player chooses Defect regardless of history.

Example 4.3. (*Chain Store*) Consider the version of Chain Store in Figure 4.3. By admissibility, the unique SAS is $\{(In, \ Cede)\}$.

Now consider twice-repeated Chain Store. The tree is given in Figure 4.4 and the (reduced) strategic form in Figure 4.5.

For the entrant E, all strategies are admissible. For the incumbent I, only strategies ac and ae are admissible. From this, we can see that if

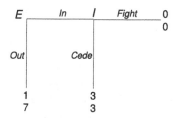

Figure 4.3.

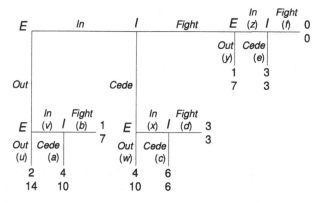

Figure 4.4.

	ac	ad	bc	bd	ae	af	be	bf
u	2, 14	2, 14	2, 14	2, 14	2, 14	2, 14	2, 14	2, 14
v	4, 10	4, 10	1, 7	1, 7	4, 10	4, 10	1, 7	1, 7
wy	4, 10	4, 10	4, 10	4, 10	1, 7	1, 7	1, 7	1, 7
wz	4, 10	4, 10	4, 10	4, 10	3, 3	0, 0	3, 3	0, 0
xy	6, 6	3, 3	6, 6	3, 3	1, 7	1, 7	1, 7	1, 7
xz	6, 6	3, 3	6, 6	3, 3	3, 3	0, 0	3, 3	0, 0

Figure 4.5.

$Q^E \times Q^I$ is an SAS either: (i) $Q^I = \{ac\}$ and $Q^E = \{xy, xz\}$ or $\{xz\}$; (ii) $Q^I = \{ae\}$ and $Q^E = \{v\}$; (iii) $Q^I = \{ac, ae\}$ and $Q^E = \{v\}$. Two paths of play are possible under the SAS concept. In one, E enters in both periods and I cedes in both periods. In the other, E stays out in the first period (because there is sufficient chance the incumbent would fight) and enters only in the second period, at which point I cedes.

The IA set for twice-repeated Chain Store is the singleton $\{(xz, ac)\}$, corresponding to enter regardless for I and cede regardless for E.

What do these three examples tell us about the behavior of SAS's?

First, we see that, in each case, the IA set is one SAS of the game. We will show this is true in general. A consequence is that any game possesses at least one SAS.

We also see that there are SAS's distinct from the IA set. For example, $\{Out\} \times \{Out, In\}$ is an SAS of three-legged Centipede, while the IA set is the singleton $\{(Out, Out)\}$. The difference between SAS and IA is more stark in twice-repeated Chain Store. In this game, an SAS may even give a different outcome from the IA set. There are SAS's in which the entrant stays out in the first period — not so in the IA set.

In particular, in Centipede, SAS yields the backward-induction (BI) outcome, while, in twice-repeated Chain Store, SAS allows non-BI outcomes. This is different from IA, and prompts the questions: Why does SAS yield the BI outcome in some games but not others? When does SAS yield the BI outcome? Can we characterize SAS's in perfect-information games?

In the following sections we will give answers to these and other questions about the SAS concept.

5. Strategic-Form Properties of SAS's

Here we record two basic strategic-form properties: existence and invariance. The proofs are relegated to Appendix A.

Proposition 5.1. *Any finite game possesses a nonempty SAS. In particular, the IA set is an SAS.*

Next, invariance. Referring back to our discussion in Section 2, we see that, since SAS is defined on the strategic form, we have only to show invariance with respect to the addition of strategies that are convex combinations (for all players) of existing strategies. (Of course, this covers addition of duplicate strategies.) To establish that this is true, consider games $G = \langle S^a, S^b, \pi^a, \pi^b \rangle$ and $\bar{G} = \langle S^a \cup \{q^a\}, S^b, \bar{\pi}^a, \bar{\pi}^b \rangle$, where $q^a \notin S^a$, and:

(i) $\bar{\pi}^a | S^a \times S^b = \pi^a$ and $\bar{\pi}^b | S^b \times S^a = \pi^b$ (where "|" denotes the restriction);
(ii) there is a $\varphi^a \in \mathcal{M}(S^a)$ such that $\bar{\pi}^a(q^a, s^b) = \pi^a(\varphi^a, s^b)$ and $\bar{\pi}^b(s^b, q^a) = \pi^b(s^b, \varphi^a)$ for each $s^b \in S^b$.

Proposition 5.2.

(a) *Let $\bar{Q}^a \times \bar{Q}^b$ be an SAS of $\bar{G}$. Then $(\bar{Q}^a \backslash \{q^a\}) \times \bar{Q}^b$ is an SAS of G.*
(b) *Let $Q^a \times Q^b$ be an SAS of G. If q^a does not support any strategy in Q^a, then $Q^a \times Q^b$ is an SAS of $\bar{G}$. Otherwise, $(Q^a \cup \{q^a\}) \times Q^b$ is an SAS of $\bar{G}$.*

Proposition 5.1 states that the IA set is one SAS. So, Proposition 5.2 says that the IA set remains an SAS after the addition or deletion of convex combinations. Indeed, we can go further — the IA set is also invariant to the fully reduced strategic form. (See Proposition A.1 in Appendix A.)

6. Extensive-Form Properties of SAS's

We will consider extensive-form games with perfect recall (Kuhn [1950, 1953]). Let S^a, S^b be the strategy sets associated with an extensive form Γ.[8] Write H^a (resp. H^b) for the information sets at which Ann (resp. Bob) moves, and $S^a(h)$ (resp. $S^b(h)$) for the subset of S^a (resp. S^b) that allows information set h. Let Z be the set of terminal nodes. Let $\zeta : S^a \times S^b \to Z$ be the map from strategy profiles to terminal nodes. Extensive-form payoff functions are maps $\Pi^a : Z \to \mathbb{R}$ and $\Pi^b : Z \to \mathbb{R}$. The strategic form induced by Γ is then $G = \langle S^a, S^b, \pi^a, \pi^b \rangle$, where $\pi^a = \Pi^a \circ \zeta$ and $\pi^b = \Pi^b \circ \zeta$. (Note the abuse of notation: ζ maps $S^b \times S^a$ to Z in the definition of π^b.)

A very basic requirement of a solution concept defined on the matrix is that it should induce optimal behavior in the tree.

Definition 6.1. A strategy $s^a \in S^a$ is *(extensive-form) rational* if, for each information set $h \in H^a$ allowed by s^a, there is some $\mu^b \in \mathcal{M}(S^b)$, with $\mu^b(S^b(h)) = 1$, under which s^a is optimal among all strategies in $S^a(h)$.

---

[8]In light of Proposition 5.2, we can (and often do) conflate strategies with plans of action. No confusion should result.

It is a standard argument that a strategy which is admissible in the matrix is extensive-form rational in every associated tree with perfect recall. (Use the full-support measure in Remark 3.1 to build a measure μ^b at each information set h.) So, we certainly get:

Proposition 6.1. *Fix an extensive-form game Γ with induced strategic form G. If $Q^a \times Q^b$ is an SAS of G, then any $s^a \in Q^a$ (resp. $s^b \in Q^b$) is extensive-form rational in Γ.*

More subtle extensive-form properties involve relating what a solution concept gives on the whole tree to what it gives on parts of the tree. One such property (introduced by Kohlberg and Mertens [1986, p. 1012]) is projection: If a strategy profile lies in a solution for the whole tree, then it should also lie in a solution for any part of the tree that it reaches.

SAS's satisfy projection: Any SAS of game G that allows subtree Δ induces an SAS on the strategic form of Δ. We will need some notation to prove this. Fix an extensive form Γ and associated strategic form G. Let Δ be a proper subtree of Γ, with strategic form D. Let S_Δ^a (resp. S_Δ^b) be the subset of S^a (resp. S^b) consisting of those strategies that allow subtree Δ. Note that, up to duplication of strategies, we identify S_Δ^a with the set of Ann's strategies on subtree Δ. (Likewise for Bob.) Since SAS's are invariant to the addition or deletion of duplicate strategies (Proposition 5.2), we can indeed make this identification. Now the formal statement:

Proposition 6.2. *Let $Q^a \times Q^b$ be an SAS of G, and suppose that $(Q^a \cap S_\Delta^a) \times (Q^b \cap S_\Delta^b) \neq \emptyset$. Then, $(Q^a \cap S_\Delta^a) \times (Q^b \cap S_\Delta^b)$ is an SAS of D, up to the addition of strategies that are duplicates on Δ.*

Proof. Each $s^a \in Q^a \cap S_\Delta^a$ is optimal under some $\mu^b \in \mathcal{M}(S^b)$ with Supp $\mu^b = S^b$ (condition (i) of the definition of an SAS). We have $\mu^b(S_\Delta^b) > 0$, so $\mu^b(\cdot|S_\Delta^b)$ is well defined with Supp $\mu^b(\cdot|S_\Delta^b) = S_\Delta^b$. Suppose s^a is not optimal under $\mu^b(\cdot|S_\Delta^b)$, i.e., there is some $r^a \in S_\Delta^a$ with

$$\sum_{s^b \in S_\Delta^b} \pi^a(r^a, s^b)\mu^b(s^b|S_\Delta^b) > \sum_{s^b \in S_\Delta^b} \pi^a(s^a, s^b)\mu^b(s^b|S_\Delta^b).$$

Define a new strategy $q^a \in S_\Delta^a$ that agrees with r^a at each information set (for a) in Δ, and agrees with s^a at information sets not in Δ. Then, using the fact that $\mu^b(S_\Delta^b) > 0$,

$$\sum_{s^b \in S^b} \pi^a(q^a, s^b)\mu^b(s^b) > \sum_{s^b \in S^b} \pi^a(s^a, s^b)\mu^b(s^b),$$

a contradiction. Thus, s^a is admissible with respect to $S_\Delta^a \times S_\Delta^b$.

The argument for condition (ii) is very similar. Each $s^a \in Q^a \cap S_\Delta^a$ is optimal under some $\nu^b \in \mathcal{M}(S^b)$ with $\operatorname{Supp} \nu^b = Q^b$. Since $Q^b \cap S_\Delta^b \neq \emptyset$, we have $\nu^b(Q^b \cap S_\Delta^b) > 0$, so that $\nu^b(\cdot|Q^b \cap S_\Delta^b)$ is well defined with $\operatorname{Supp} \nu^b(\cdot|Q^b \cap S_\Delta^b) = Q^b \cap S_\Delta^b$. Suppose there is some $r^a \in S_\Delta^a$ with

$$\sum_{s^b \in S_\Delta^b} \pi^a(r^a, s^b)\nu^b(s^b|Q^b \cap S_\Delta^b) > \sum_{s^b \in S_\Delta^b} \pi^a(s^a, s^b)\nu^b(s^b|Q^b \cap S_\Delta^b).$$

Define a new strategy $q^a \in S_\Delta^a$ that agrees with r^a at each information set (for a) in Δ, and agrees with s^a at information sets not in Δ. Then,

$$\sum_{s^b \in S^b} \pi^a(q^a, s^b)\nu^b(s^b) > \sum_{s^b \in S^b} \pi^a(s^a, s^b)\nu^b(s^b),$$

a contradiction.

Finally, fix $s^a \in Q^a \cap S_\Delta^a$, and suppose there is $\varphi^a \in \mathcal{M}(S_\Delta^a)$ such that $\pi^a(\varphi^a, s^b) = \pi^a(s^a, s^b)$ for all $s^b \in S_\Delta^b$. Given each $r^a \in \operatorname{Supp} \varphi^a$, let $f(r^a)$ be the strategy that agrees with r^a at each information set (of a) in Δ, and agrees with s^a at all other information sets. Define a measure $\rho^a \in \mathcal{M}(S^a)$ where $\rho^a(f(r^a)) = \varphi^a(r^a)$, for $r^a \in \operatorname{Supp} \varphi^a$. Then $\pi^a(\rho^a, s^b) = \pi^a(s^a, s^b)$ for all $s^b \in S^b$. Thus, condition (iii) applied to $Q^a \times Q^b$ implies that each $q^a \in \operatorname{Supp} \rho^a$ is in Q^a. But each $r^a \in \operatorname{Supp} \varphi^a$ agrees with $f(r^a) \in \operatorname{Supp} \rho^a$ on subtree Δ, by construction. So, by our identification of strategies, we have that condition (iii) is satisfied. □

By Proposition 5.1, the IA set is one SAS, so, by the projection property, the projection of the IA set into a reached subtree constitutes an SAS of the subtree. Notice this does not say that IA projects to the IA set of the subtree — only to some SAS of the subtree. The game in Figure 6.1 is an example.[9] The IA set is $\{(In\text{-}D, L)\}$. But the IA set of the subtree following In consists of the entire set $\{U, D\} \times \{L, R\}$ — different from the projection $\{(D, L)\}$. Of course, the projection $\{(D, L)\}$ does form

---

[9]We thank a referee for the example.

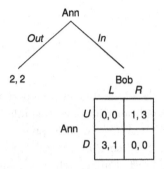

Figure 6.1.

an SAS, as it must. Is there a game where the projection of the IA set of the whole tree is even disjoint from the IA set on a reached subtree? We do not know and leave this as an open question.

It is also important to note that a solution concept may satisfy the projection property, but, nonetheless, may not yield the BI outcome in perfect-information (PI) games. SAS is a case in point. Refer back to the twice-repeated Chain Store game (Example 4.3 in Section 4). Pure-strategy Nash equilibrium is another example: Let (s^a, s^b) be a pure-strategy profile for Γ that reaches subtree Δ. Then, if the restrictions of s^a and s^b to Δ fail to constitute a Nash equilibrium of Δ, the profile (s^a, s^b) must fail to be a Nash equilibrium of Γ.

We can go further: A solution concept may satisfy the (extensive-form) rationality and projection properties, but may not give the BI outcome. Again, SAS is a case in point. So is pure-strategy Nash equilibrium in extensive-form rational strategies.

If not BI, what does SAS yield in general PI games? The parallel between SAS and Nash equilibrium gives a clue. So do the analyses of the Centipede and Chain Store games (Examples 4.1 and 4.3). In any SAS of Centipede, Ann plays *Out* immediately. Footnote 6 pointed out that the same is true of any Nash equilibrium of Centipede (and that the proof is the same). In the once-repeated Chain Store game, there is a Nash equilibrium in which the entrant stays out and the incumbent fights. But, this involves an inadmissible strategy for the incumbent. In the unique admissible equilibrium, the entrant enters and incumbent cedes — the same as in the unique SAS. In twice-repeated Chain Store, there is an SAS in which the entrant stays out in the first period. This is not possible under BI, but is possible if we look at admissible Nash equilibria.

The conjecture, then, is that SAS yields an admissible Nash equilibrium in PI games. This is almost correct. The next section gives an exact statement and proof.

7. Perfect-Information Games

We now come to a characterization of SAS in PI games. We impose a no-ties condition on payoffs. For this, we need some definitions: An *outcome* is a payoff vector $\Pi(z) = (\Pi^a(z), \Pi^b(z))$. Two terminal nodes z and z' are *outcome equivalent* if $\Pi(z) = \Pi(z')$. We also say that two strategy profiles (s^a, s^b) and (r^a, r^b) are outcome equivalent if $\zeta(s^a, s^b)$ and $\zeta(r^a, r^b)$ are outcome equivalent.

Definition 7.1. A tree Γ satisfies the *Single Payoff Condition (SPC)* if for all z, z' in Z, if Ann (resp. Bob) moves at the last common predecessor of z and z', then $\Pi^a(z) = \Pi^a(z')$ implies $\Pi^b(z) = \Pi^b(z')$ (resp. $\Pi^b(z) = \Pi^b(z')$ implies $\Pi^a(z) = \Pi^a(z')$).

In words, Ann is indifferent between two terminal nodes over which she is decisive, only if those two terminal nodes are outcome equivalent. SPC appears to be a minimal no-ties condition if BI is to select a unique outcome in the tree. A generic tree satisfies SPC. Nongeneric trees can also satisfy SPC and many games of interest are nongeneric. (Zero-sum games are one example. The twice-repeated Chain Store game, as given in Figure 4.4, satisfies SPC. For additional examples, see the discussions of nongenericity in Mertens [1989, p. 582] and Marx and Swinkels [1997, pp. 224, 225].) A tree that satisfies "no relevant ties" (Battigalli [1997, p. 48] satisfies SPC, but the converse need not hold. In a PI tree, SPC is equivalent to the "transfer of decisionmaker indifference" condition (Marx and Swinkels [1997]).[10]

Now the characterization result:

Proposition 7.1. *Fix a PI tree Γ satisfying SPC, and let G be the strategic form of Γ.*

(a) *Fix an SAS $Q^a \times Q^b$. Then there is a pure Nash equilibrium of G such that each profile $(s^a, s^b) \in Q^a \times Q^b$ is outcome equivalent to this equilibrium.*

---

[10]This latter is a condition on the strategic form. In the proof to come, we need to make use of properties of the extensive form — hence our use of SPC.

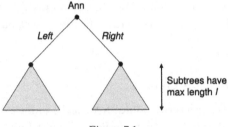

Figure 7.1.

(b) *Fix an admissible pure Nash equilibrium (s^a, s^b) of G. Then there is an SAS of G that contains (s^a, s^b).*

Before the formal argument, we give a sketch of the proof. For part (a), the argument is forward looking, by induction on the length of the tree. Suppose the result is true for a tree of length l. Here is why it is then true for a tree of length $(l + 1)$. Refer to Figure 7.1.

Fix an SAS $Q^a \times Q^b$ and suppose there is a strategy in Q^a where Ann plays *Left*. Then, the projection of $Q^a \times Q^b$ into the *Left* subtree — which we will denote $Q_L^a \times Q_L^b$ — forms an SAS of this subtree. By the induction hypothesis, we can find a Nash equilibrium (s_L^a, s_L^b) of the *Left* subtree, so that each profile in $Q_L^a \times Q_L^b$ is outcome equivalent to this Nash equilibrium. We want to show that (s_L^a, s_L^b) can be made into a Nash equilibrium of the whole tree. To do this, we give Ann the strategy that selects *Left* and then follows the choices prescribed by s_L^a. Denote this strategy by s^a. On the *Left* subtree, Bob will follow the choices prescribed by s_L^b. When Ann plays s^a, Bob then has no incentive to deviate. When Bob plays s_L^b, Ann has no incentive to deviate to strategies that lead to the *Left* subtree. It remains to specify choices for Bob on the *Right* subtree, so that Ann also has no incentive to deviate to the *Right* subtree. This step is achieved via the Minimax Theorem for PI games.

This sketch omits some important details. One is the role of SPC, which is, in fact, necessary for both parts (a) and (b) of Proposition 7.1.

Figures 7.2 and 7.3 are two trees that fail SPC. In Figure 7.2, we see that part (a) is false without SPC.[11] Here, the set $\{Out, In\} \times \{Down, Across\}$ is an SAS, but $(In, Down)$ is not outcome equivalent to a Nash equilibrium. In Figure 7.3, we see that part (b) is false without SPC. Here, $(Out, Across)$ is an admissible Nash equilibrium (is even outcome equivalent to a BI strategy

---

[11] Drew Fudenberg kindly provided this example.

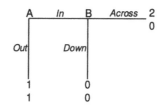

Figure 7.2.

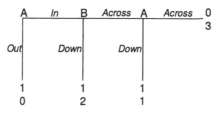

Figure 7.3.

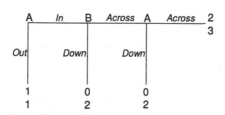

Figure 7.4.

profile), but is not contained in any SAS. In particular, the unique SAS (and so the IA set) is $\{Out, In\text{-}Down\} \times \{Down\}$. (To see this, fix an SAS $Q^a \times Q^b$. For Ann, the strategy $In\text{-}Across$ is inadmissible, so cannot be part of an SAS. Using this and condition (iii) of an SAS, $Q^a = \{Out, In\text{-}Down\}$. For Bob, only $Down$ is admissible with respect to $\{Out, In\text{-}Down\}$.)

Let us also comment on the gap between parts (a) and (b) of Proposition 7.1. Part (a) involves a strategy profile which is admissible (condition (i) of an SAS) and outcome equivalent to a Nash equilibrium. Part (b) starts with an admissible Nash profile and says that there is an SAS containing this profile. Can the gap between the two directions be closed?

Part (b) cannot be improved to read: *Fix an admissible pure profile* (s^a, s^b) *of G that is outcome equivalent to a Nash equilibrium of G. Then* (s^a, s^b) *is contained in some SAS.* The tree in Figure 7.4 satisfies SPC, and

(*Out, Across*) is an admissible profile that is outcome equivalent to the Nash equilibrium (*Out, Down*). But the unique SAS is $\{(In\text{-}Across, Across)\}$. Suppose we try instead to improve part (a) to read: *Fix an SAS $Q^a \times Q^b$. Then there is an admissible pure Nash equilibrium of G such that each profile $(s^a, s^b) \in Q^a \times Q^b$ is outcome equivalent to this equilibrium.* We do not know if this stronger statement is true.

Proof of Proposition 7.1(a). By induction on the length of the tree.

Consider a tree of length 1, and assume that Ann moves at the initial node. (We then define π^a and π^b on S^a alone.) If $s^a \in Q^a$, then $\pi^a(s^a) \geq \pi^a(r^a)$ for all $r^a \in S^a$. Thus, (s^a) is an admissible Nash equilibrium. If $r^a \in Q^a$, then we must have $\pi^a(r^a) = \pi^a(s^a)$. Since Ann moves at the last common predecessor of $\zeta(r^a)$ and $\zeta(s^a)$, SPC implies $\pi^b(r^a) = \pi^b(s^a)$, establishing that (r^a) is outcome equivalent to (s^a).

Now suppose the result is true for any tree of length l or less, and consider a tree of length $l + 1$. We can again assume that Ann moves at the initial node. Denote the subtrees that follow Ann's initial move by $\Delta_1, \ldots, \Delta_K$. Also, let S_k^a denote the strategies for Ann that allow subtree Δ_k. (Up to duplication, this subset can be identified with Ann's strategies for Δ_k.) Let S_k^b be the corresponding set for Bob. We have $S^b = \times_{k=1}^K S_k^b$.[12] (Thus, a strategy for Bob on the whole tree is some $s^b = (s_1^b, \ldots, s_K^b)$, where each s_k^b specifies Bob's choices at the information sets in subtree Δ_k.)

By the projection property (Proposition 6.2), for each k such that $Q^a \cap S_k^a \neq \emptyset$, we have that $(Q^a \cap S_k^a) \times \mathrm{proj}_{S_k^b} Q^b$ is an SAS of (the strategic form of) Δ_k. By the induction hypothesis, for each such k, each profile in $(Q^a \cap S_k^a) \times \mathrm{proj}_{S_k^b} Q^b$ is outcome equivalent.

Suppose there are strategies $s^a, r^a \in Q^a$ that reach two distinct subtrees, say Δ_1 and Δ_2, respectively. By condition (ii) of an SAS, there is a $\sigma^b \in \mathcal{M}(S^b)$, with $\mathrm{Supp}\, \sigma^b = Q^b$, such that s^a is optimal under σ^b. Since s^a reaches Δ_1 and r^a reaches Δ_2, this implies

$$\sum_{s_1^b \in \mathrm{proj}_{S_1^b} Q^b} \mathrm{marg}_{S_1^b} \sigma^b(s_1^b) \pi^a(s^a, s_1^b)$$

$$\geq \sum_{s_2^b \in \mathrm{proj}_{S_2^b} Q^b} \mathrm{marg}_{S_2^b} \sigma^b(s_2^b) \pi^a(r^a, s_2^b).$$

---

[12]Do notice, our approach here differs slightly from that in the proof of Proposition 6.2.

(We use π^a for the induced payoff functions on $S_1^a \times S_1^b$ and $S_2^a \times S_2^b$; no confusion should result.)

The induction hypothesis gives that $\pi^a(s^a, s_1^b)$ is constant for all $s_1^b \in \mathrm{proj}_{S_1^b} Q^b$, and, likewise, that $\pi^a(r^a, s_2^b)$ is constant for all $s_2^b \in \mathrm{proj}_{S_2^b} Q^b$. Thus, we have $\pi^a(s^a, s_1^b) \geq \pi^a(r^a, s_2^b)$ for all such s_1^b and s_2^b. But, symmetrically, we can apply condition (ii) of an SAS to r^a, to get the opposite inequality. Thus, $\pi^a(s^a, s_1^b) = \pi^a(r^a, s_2^b)$ for all $s_1^b \in \mathrm{proj}_{S_1^b} Q^b$ and $s_2^b \in \mathrm{proj}_{S_2^b} Q^b$. Also, since Ann moves at the last common predecessor of $\zeta(s^a, s_1^b)$ and $\zeta(r^a, s_2^b)$, SPC implies that $\pi^b(s_1^b, s^a) = \pi^b(s_2^b, r^a)$.

We have now shown that every profile in $Q^a \times Q^b$ is outcome equivalent.

The final step is to construct a Nash equilibrium to which (s^a, s_1^b) is outcome equivalent. By the induction hypothesis, there is a Nash equilibrium of Δ_1 to which (s^a, s_1^b) is outcome equivalent. Denote it as (q_1^a, q_1^b). We show that for each $k = 2, \ldots, K$, there is a strategy $q_k^b \in S_k^b$ such that $\pi^a(q_1^a, q_1^b) \geq \pi^a(r^a, q_k^b)$ for all $r^a \in S_k^a$. The profile $(q_1^a, (q_1^b, \ldots, q_K^b))$ will then be the desired equilibrium.

Since s^a is optimal under σ^b, as above, we have for each $k = 2, \ldots, K$,

$$\pi^a\left(q_1^a, q_1^b\right) = \pi^a\left(s^a, s_1^b\right) \geq \sum_{s_k^b \in \mathrm{proj}_{S_k^b} Q^b} \mathrm{marg}_{S_k^b} \sigma^b(s_k^b) \pi^a(r^a, s_k^b)$$

for all r^a that allow Δ_k. Letting $(\bar{r}_k^a, \bar{r}_k^b) \in \arg\max_{S_k^a} \min_{S_k^b} \pi^a(\cdot, \cdot)$, we have in particular

$$\pi^a(q_1^a, q_1^b) \geq \sum_{s_k^b \in \mathrm{proj}_{S_k^b} Q^b} \mathrm{marg}_{S_k^b} \sigma^b(s_k^b) \pi^a(\bar{r}_k^a, s_k^b).$$

But $\pi^a(\bar{r}_k^a, r_k^b) \geq \pi^a(\bar{r}_k^a, \bar{r}_k^b)$ for any $r_k^b \in S_k^b$, by definition. So

$$\pi^a\left(q_1^a, q_1^b\right) \geq \sum_{s_k^b \in \mathrm{proj}_{S_k^b} Q^b} \mathrm{marg}_{S_k^b} \sigma^b(s_k^b) \pi^a\left(\bar{r}_k^a, \bar{r}_k^b\right) = \pi^a\left(\bar{r}_k^a, \bar{r}_k^b\right).$$

Set $(\underline{r}_k^a, \underline{r}_k^b) \in \arg\min_{S_k^b} \max_{S_k^a} \pi^a(\cdot, \cdot)$. By the Minimax Theorem for PI games (see, e.g., Ben Porath [1997]), $\pi^a(\underline{r}_k^a, \underline{r}_k^b) = \pi^a(\bar{r}_k^a, \bar{r}_k^b)$. It follows that $\pi^a(q_1^a, q_1^b) \geq \pi^a(\bar{r}_k^a, \bar{r}_k^b) = \pi^a(\underline{r}_k^a, \underline{r}_k^b)$. But $\pi^a(\underline{r}_k^a, \underline{r}_k^b) \geq \pi^a(r_k^a, \underline{r}_k^b)$ for any $r_k^a \in S_k^a$, by definition. So $\pi^a(q_1^a, q_1^b) \geq \pi^a(r_k^a, \underline{r}_k^b)$. Setting $q_k^b = \underline{r}_k^b$ gives the desired profile. $\qquad\square$

For the proof of Proposition 7.1 (b), we will need a preliminary lemma, which is proved in Appendix A.

Lemma 7.1. *Fix a PI tree. If s^a is admissible, then $\pi^a(s^a, s^b) = \pi^a(r^a, s^b)$ for each $r^a \in \mathrm{su}(s^a)$ and $s^b \in S^b$.*

Proof of Proposition 7.1(b). Fix a PI tree Γ satisfying SPC. Let G be the associated strategic form. We show that if (s^a, s^b) is an admissible Nash equilibrium of G, then $\mathrm{su}(s^a) \times \mathrm{su}(s^b)$ is an SAS of G.

Each $r^a \in \mathrm{su}(s^a)$ is admissible, since s^a is admissible (Corollary A.1). So condition (i) of an SAS is satisfied.

Next, condition (ii). It suffices to show that, for each $q^a \in S^a$ and $r^b \in \mathrm{su}(s^b)$, $\pi^a(q^a, s^b) = \pi^a(q^a, r^b)$. If so, s^a must be admissible with respect to $\mathrm{su}(s^b)$. To see this last statement: Fix $r^b \in \mathrm{su}(s^b)$ and notice that if the claim holds then

$$\pi^a\left(s^a, r^b\right) = \pi^a\left(s^a, s^b\right) \geq \pi^a\left(q^a, s^b\right) = \pi^a\left(q^a, r^b\right),$$

where the two equalities follow from the claim and the inequality follows from the fact that (s^a, s^b) is a Nash equilibrium. With this, s^a is optimal under any measure $\sigma^b \in \mathcal{M}(S^b)$ with $\mathrm{Supp}\ \sigma^b = \mathrm{su}(s^b)$, and so s^a is admissible with respect to $\mathrm{su}(s^b)$.

Fix $q^a \in S^a$ and $r^b \in \mathrm{su}(s^b)$. By Lemma 7.1 applied to s^b (which is admissible), we have $\pi^b(s^b, q^a) = \pi^b(r^b, q^a)$. If (q^a, s^b) and (q^a, r^b) reach the same terminal node, then certainly $\pi^a(q^a, s^b) = \pi^a(q^a, r^b)$. If not, Bob moves at the last common predecessor of $\zeta(q^a, s^b)$ and $\zeta(q^a, r^b)$, so that SPC establishes the desired result.

Finally, condition (iii) follows immediately from Lemma A.3. $\quad\square$

To sum up: In a PI tree satisfying SPC, each SAS is outcome equivalent to some Nash equilibrium, and each admissible Nash equilibrium is contained in some SAS. In particular, then, an SAS need not yield the (unique) BI outcome in such a game. However, a converse does hold.

Proposition 7.2. *Fix a PI tree satisfying SPC. There is an SAS that is outcome equivalent to the BI outcome.*

Proof. Proposition A.2 in Appendix A shows that in a PI tree satisfying SPC, there is an admissible BI strategy profile. Of course, this profile is a Nash equilibrium, so, by Proposition 7.1(b), there is an SAS that contains it. $\quad\square$

Note that this is only a partial converse. A BI strategy profile need not be admissible, and, therefore, need not be contained in any SAS. (In particular, then, a BI profile need not be contained in the IA set.)

8. Discussion

We conclude with some comments on conceptual matters and related literature.

8.1. Player-specific type structures

In Figure 1.1, we presented the basic 'architecture' of an epistemic analysis. The starting point is to add types to the description of the game. With this, we can analyze the conditions of RCBR or RCAR. The projection of the RCBR (resp. RCAR) set into the strategy sets forms a BRS (resp. SAS) of the game. As for which BRS or SAS obtains, this depends on which particular types we add.

How should we think about the choice of one vs. another type structure? In any particular structure, certain beliefs, beliefs about beliefs,..., will be present (i.e., will be induced by a type) and others will not be. So, there is an important implicit assumption behind the choice of a structure. This is that it is "transparent" to the players that the beliefs in the type structure — and only those beliefs — are possible. (See Battigalli and Friedenberg [2009, Appendix A] for a formal treatment of this point, for the RCBR case.) Why would there be such a "transparent" restriction on beliefs? The idea is that there is a "context" (Brandenburger, Friedenberg, and Keisler [2008, Section 2.8]) to the strategic situation (e.g., history, conventions, etc.) and this "context" causes the players to rule out certain beliefs.

Notice what is involved: Ann and Bob think the same way about which beliefs are — and are not — possible. This is a substantive (if again implicit) assumption. While the assumption is standard in epistemic game theory, it is clearly important to investigate the implications of dropping it.[13]

Presumably, the new ingredient would be the concept of a "player-specific type structure," where we specify a (potentially different) type structure for each player. Now, a type t^a for Ann, in Ann's type structure, could consider possible a certain type t^b for Bob, even though, in Bob's type structure, there is no type for Bob with the same hierarchy of beliefs as t^b. Of course, a special case will be when the player-specific type structures coincide. Call such a (common) type structure a "player-independent type structure."

The BRS concept characterizes RCBR across all player-independent type structures. If we now characterize RCBR across all player-specific type

---

[13]We are grateful to a referee for very fruitful exchanges on this issue.

structures, we will get all the BRS's and also some new sets. It can be shown these new sets need not be BRS's. We conjecture that they will be contained in the IU set. If so, the extra precision we get in identifying these sets above and beyond BRS's may seem a small matter — much as the extra precision in identifying BRS's above and beyond the IU set may seem a small matter.

Contrast with the case of admissibility. In the Introduction, we pointed to a basic nonmonotonicity in admissibility, and explained that, precisely because of this nonmonotonicity, SAS's needed to be understood separately from the IA set. For the same reason, it seems that a full characterization of RCAR across all type structures, including the player-specific type structures, will be needed. This is left for future work.

8.2. *Nonmonotonicity continued*

We began with the observation that there is a basic nonmonotonicity in admissibility: Adding new possibilities can change previously good strategies into bad ones. Refer back to Figure 1.2. There, we said that when we add the strategy C to the set $\{U, M\} \times \{L\}$, we introduce a new possibility and, so, the strategy M may now turn into a bad strategy. But this argument was incomplete. After all, the strategy C is already in the matrix. So, presumably, Ann should have already considered this possibility.

The answer brings us back to Samuelson's (1992) basic tension between admissibility and strategic reasoning (mentioned in Section 3). On the one hand, admissibility requires that Ann includes all of Bob's strategies. On the other hand, strategic reasoning requires that Ann excludes Bob's irrational strategies. Back to Figure 1.2, and consider a solution concept with $Q^a \times Q^b = \{U, M\} \times \{L\}$. Yes, Ann should include C, since she should include all possibilities. But, she should also exclude C, since it is inconsistent with the solution. So, if C is now added to Q^b, it does 'make a difference.' Ann should no longer exclude C, which turns the previously good strategy M into a bad one. We conclude that nonmonotonicity of the solution concept is part and parcel of any epistemic analysis that addresses the inclusion – exclusion problem.

How does this verbal argument play out at a more formal level? Let us specialize to the case of an LPS $(\mu_1, \ldots, \mu_n)$. Assumption lies strictly between the concepts of "belief at the 1st level" and "belief." (See Brandenburger, Friedenberg, and Keisler [2008, Proposition 5.1].) In particular, Ann believes at the 1st level (resp. believes) the event $E =$ "Bob is rational" if $\mu_1(E) = 1$ (resp. $\mu_i(E) = 1$ for all i). Back again to

Figure 1.2. We can construct an LPS $(\mu_1, \ldots, \mu_n)$ which has full support and is such that Ann believes at the 1st level the event $\{L, C\}$.[14] But, we also want M to be optimal under this LPS — and, for this, it must be that the irrational R is considered at least as likely as C. That is, if $\mu_i(C) > 0$ then there is some $j \leq i$ with $\mu_j(R) > 0$. Of course, by inclusion, there is some i with $\mu_i(C) > 0$. Thus, by asking for inclusion, we forgo exclusion. Under belief, we get exclusion, but lose inclusion. Ann can believe the event "Bob is rational" only if each measure μ_i assigns probability 0 to one of Bob's strategies — viz., the irrational R.

Assumption gives both inclusion and exclusion. The 'cost' is nonmonotonicity: An LPS may assume $\{L\}$ but not $\{L, C\}$. In particular, then, the strategy M can be optimal for Ann if she assumes L. Yet M cannot be optimal if she assumes $\{L, C\}$. Thus, the nonmonotonicity of assumption yields a nonmonotonicity of the solution concept: $Q^a \times Q^b$ is an SAS but $Q^a \times (Q^b \cup \{s^b\})$ is not.

As far as we can see, any resolution of the inclusion–exclusion problem must have this feature.

8.3. Relationship to other solution concepts

The literature has proposed a number of weak-dominance analogs to the Pearce (1984) BRS concept. To the best of our knowledge, no paper has provided foundations which fully resolve the inclusion–exclusion problem. Each gives up on one or other criterion.

In terms of inclusion: Samuelson (1992) provided foundations for the consistent pairs concept. (See also Börgers and Samuelson [1992].) A consistent pair may contain an inadmissible strategy, and so inclusion is not satisfied.[15] Asheim (2001), Asheim and Dufwenberg (2003), and Asheim and Perea (2005) take an interesting different approach: They require that Ann consider every strategy for Bob possible, but not every type for Bob. So, epistemically speaking, they have partial rather than full inclusion.

In terms of exclusion: Dekel and Fudenberg (1990) introduced the $S^\infty W$ concept of one round of deletion of inadmissible strategies followed by iterated deletion of strongly dominated strategies. There is an obvious

---

[14]This is a semi-formal discussion only. To be precise, we would need to include the types, too.

[15]The modified consistent pairs concept (Ewerhart [1998]) has a flavor of solving the inclusion–exclusion problem. We do not know of epistemic foundations for this concept.

BRS-like version of the definition (Brandenburger, Friedenberg, and Keisler [2008, Section 11B]). The epistemic foundations (Brandenburger [1992], Börgers [1994], and Brandenburger, Friedenberg, and Keisler [2008, Section 11B]) rest on the idea of belief at the 1st level. As mentioned, this concept fails exclusion.

SAS also differs from each of these solution concepts in terms of the strategies that can be played. (See Brandenburger, Friedenberg, and Keisler [2008, Section S.3] for details.)

8.4. *Related PI results*

Proposition 7.1 resembles an earlier result on PI trees, due to Ben Porath (1997, Theorem 2). Ben Porath defines an extensive-form analog to belief at the 1st level, and gives epistemic conditions which yield a solution concept we will call $S^{\infty} CD$. This concept first eliminates all conditionally dominated strategies (Shimoji and Watson [1998]) — i.e., all strategies which are not extensive-form rational. It then iteratively eliminates strategies which are strongly dominated in the matrix (equivalently, strongly dominated at the root of the tree). It is an extensive-form analog to the $S^{\infty}W$ concept, and, indeed, in generic PI trees the two concepts coincide.

The $S^{\infty} CD$ concept need not yield a Nash outcome in PI trees satisfying SPC. For example, in Centipede (Figure 4.1), the only elimination via $S^{\infty} CD$ is the strategy *In* at every node for Ann.

Ben Porath then adds an assumption which does result in Nash outcomes in (generic) PI games. This is a 'grain of truth' condition on his epistemic analysis, which says that each player assigns positive probability to the actual state — and, therefore, to the actual strategies. See Ben Porath (1997, p. 38). Presumably, the admissibility requirement (recall Remark 3.1) plays a similar role in our analysis.

Battigalli and Friedenberg (2009) study the epistemic condition of "rationality and common strong belief of rationality" (RCSBR), due to Battigalli and Siniscalchi (2002). RCSBR does not impose an admissibility requirement. For instance, in simultaneous games, RCSBR is characterized by BRS.[16] Nevertheless, there is a point of connection with this chapter. Battigalli and Friedenberg show that in PI games satisfying no-relevant ties (refer back to the discussion after Definition 7.1 for a definition), RCSBR is characterized by Nash outcomes in extensive-form rational strategies.

---

[16]Here, the definition of a BRS incorporates a maximality condition; see Battigalli and Friedenberg (2009). See also Footnote 4.

8.5. *n-player games*

We have treated two-player games, but the analysis extends to n-player games.

A set $\times_{i=1}^{n} Q^i \subseteq \times_{i=1}^{n} S^i$ is an (*n-player*) *SAS* if, for each player i: (i) each $s^i \in Q^i$ is admissible with respect to $\times_{j=1}^{n} S^j$ (ii) each $s^i \in Q^i$ is admissible with respect to $S^i \times \times_{j \neq i} Q^j$; (iii) for any $s^i \in Q^i$, if $r^i \in \mathrm{su}(s^i)$ then $r^i \in Q^i$.

Of course, conditions (i) and (ii) are equivalent to: (i$'$) each $s^i \in Q^i$ is optimal under some $\mu^{-i} \in \mathcal{M}(\times_{j \neq i} S^j)$, with Supp $\mu^{-i} = \times_{j \neq i} S^j$, given S^i; (ii$'$) each $s^i \in Q^i$ is optimal under some $\nu^{-i} \in \mathcal{M}(\times_{j \neq i} S^j)$, with Supp $\nu^{-i} = \times_{j \neq i} Q^j$, given S^i. Under this definition, all the results in this chapter hold for the n-player case, including Proposition 7.1. (We now use the n-player Minimax Theorem for PI games: $\min_{\times_{j \neq i} S^j} \max_{S^i} \pi^i(\cdot, \cdot) = \max_{S^i} \min_{\times_{j \neq i} S^j} \pi^i(\cdot, \cdot)$.)

8.6. *Invariance continued*

SAS satisfies invariance (Proposition 5.2). In Section 2, we also mentioned the idea of a deeper notion of invariance — that epistemic reasoning should not change across equivalent games. Does SAS satisfy such "epistemic invariance," too?

To answer, we need a formalization of this notion. Here is one possibility. Fix a tree Γ and an associated type structure, where a type t^a for Ann is associated with a conditional probability system (CPS) on $S^b \times T^b$, and the family of conditioning events (at least) includes all events in $S^b \times T^b$ that correspond to an information set in some tree with the same fully reduced strategic form as Γ. Now bring in SAS's and their epistemic underpinnings — viz., lexicographic probability systems (LPS's) à la Brandenburger, Friedenberg, and Keisler (2008). A full-support LPS on $S^b \times T^b$ naturally induces a CPS on $S^b \times T^b$, where the family of conditioning events consists of all nonempty open sets in $S^b \times T^b$ (Brandenburger, Friedenberg, and Keisler [2006, 2008]). This family includes the events that correspond to the information sets. So, SAS can be said to pass one test of epistemic invariance. Still, this is only a sketch, and a full-blown formulation of epistemic invariance is warranted.

8.7. *Relationship to the stability literature*

SAS is a solution concept derived from the epistemic program. In this chapter, the focus was on SAS as a solution concept in its own right. We have covered several general properties of a solution concept — existence, invariance, extensive-form rationality, and projection. Of course, there are other properties one can consider. Starting with Kohlberg and Mertens (1986), the stability literature has developed a long list of potentially desirable properties of solution concepts. One is the difference property (Kohlberg and Mertens [1986, Section 2.6]). Roughly speaking, this requires that for any tree and subtree, an outcome allowed by the solution concept on the original tree is also allowed by the solution concept on the difference tree — i.e., on the tree obtained by pruning the subtree according to the solution concept. Elsewhere (Brandenburger and Friedenberg [2009]), we show that if a solution concept satisfies existence, extensive-form rationality, and difference on the domain of PI trees satisfying SPC, then this concept must yield the BI outcome in these trees. It follows that SAS fails the difference property. Still, we do not view this as a flaw in the SAS concept, since we do not insist on BI. Mertens himself has expressed a similar view in the context of equilibrium analysis: "I had (and still have) some instinctive liking for the bruto Nash equilibrium, or a modest refinement like admissible equilibria" (Mertens [1989, pp. 582–583]).

It would certainly be interesting to conduct a comprehensive examination of which properties various epistemic solution concepts do or do not satisfy. Speaking as epistemic game theorists, we believe that this would help determine which properties are desirable and which ones are not. Of course, this exercise would also serve as a kind of 'audit' of different epistemic concepts. Our investigation of SAS in this chapter is only a first step in this direction.

Appendix

We begin with the proof of Proposition 5.1.

Lemma A.1. *Fix a strategy s^a and some $\varphi^a \in \mathcal{M}(S^a)$ such that $\pi^a(\varphi^a, s^b) = \pi^a(s^a, s^b)$ for all $s^b \in S^b$ Then s^a is optimal under $\mu^b \in \mathcal{M}(S^b)$ if and only if all $r^a \in \mathrm{Supp}\ \varphi^a$ are optimal under μ^b.*

Proof. Routine. □

Corollary A.1. *Fix a strategy s^a that is admissible given $Q^a \times Q^b$. Then, for each $r^a \in \mathrm{su}(s^a)$ with $r^a \in Q^a$, r^a is admissible given $Q^a \times Q^b$.*

The next lemma is Lemma F1 in Brandenburger, Friedenberg, and Keisler (2008), but we give a statement and proof here for ease of reference.

Lemma A.2. *If $s^a \in S_m^a$ then there is a $\mu^b \in \mathcal{M}(S^b)$, with Supp $\mu = S_{m-1}^b$, such that $\pi^a(s^a, \mu^b) \geq \pi^a(r^a, \mu^b)$, for each $r^a \in S^a$.*

Proof. By Remark 3.1, there is a $\mu^b \in \mathcal{M}(S^b)$, with Supp $\mu = S_{m-1}^b$, such that $\pi^a(s^a, \mu^b) \geq \pi^a(r^a, \mu^b)$ for all $r^a \in S_{m-1}^a$. Suppose there is an $r^a \in S^a \backslash S_{m-1}^a$ with

$$\pi^a(r^a, \mu^b) > \pi^a(s^a, \mu^b). \tag{A.1}$$

Then $r^a \in S_l^a \backslash S_{l+1}^a$, for some $l < m - 1$ Choose r^a (and l) so that there does not exist with $q^a \in S_{l+1}^a$ with $\pi^a(q^a, \mu^b) > \pi^a(s^a, \mu^b)$.

Fix some $\nu^b \in \mathcal{M}(S^b)$, with Supp $\nu^b = S_l^b$, and define a sequence of measures $\mu_n^b \in \mathcal{M}(S^b)$, for each $n \in \mathbb{N}$, by $\mu_n^b = (1 - \frac{1}{n})\mu^b + \frac{1}{n}\nu^b$. Note that Supp $\mu_n^b = S_l^b$ for each n. Using $r^a \notin S_{l+1}^a$, and Remark 3.1 applied to the $(l + 1)$-admissible strategies, it follows that for each n there is a $q^a \in S_l^a$ with

$$\pi^a(q^a, \mu_n^b) > \pi^a(r^a, \mu_n^b). \tag{A.2}$$

We can assume that $q^a \in S_{l+1}^a$. (Choose $q^a \in S_l^a$ to maximize the left-hand side of Equation (A.2) among all strategies in S_l^a.) Also, since S_{l+1}^a is finite, there is a $q^a \in S_l^a$ such that Equation (A.2) holds for infinitely many n. Letting $n \to \infty$ yields

$$\pi^a(q^a, \mu^b) \geq \pi^a(r^a, \mu^b). \tag{A.3}$$

From Equations (A.1) and (A.3) we get $\pi^a(q^a, \mu^b) > \pi^a(s^a, \mu^b)$, contradicting our choice of r^a. $\qquad\square$

Proof of Proposition 5.1. We show that the IA set is an SAS. Fix $s^a \in S_M^a$ Certainly, $s^a \in S_1^a$ and so s^a is admissible with respect to S^b, establishing condition (i). Since $S_M^i = S_{M+1}^i$, we know that s^a is admissible with respect to $S_M^a \times S_M^b$. Using Lemma A.2 and Remark 3.1, it follows that s^a is admissible with respect to $S^a \times S_M^b$, establishing condition (ii). For condition (iii), we show, by induction on m, that if $r^a \in \mathrm{su}(s^a)$, then $r^a \in S_m^a$. The result is immediate for $m = 0$, so assume $r^a \in S_m^a$. Then, using the fact that $S^a \in S_{m+1}^a$ and Corollary A.1, $r^a \in S_{m+1}^a$. $\qquad\square$

To prove Proposition 5.2, we need two lemmas.

Lemma A.3. *If $q^a \in \mathrm{su}(r^a)$ and $r^a \in \mathrm{su}(s^a)$, then $q^a \in \mathrm{su}(s^a)$.*

Proof.   Immediate.   □

Lemma A.4.   *Fix $s^a \in S^a$ and $\varphi^a \in \mathcal{M}(S^a)$ with $\pi^b(s^b, \varphi^a) = \pi^b(s^b, s^a)$ for all $s^b \in S^b$. Fix also $X \subseteq S^a$ with Supp $\varphi^a \subseteq X$, and some $Y \subseteq S^b$. Then s^b is admissible with respect to $X \times Y$ if and only if it is admissible with respect to $(X \cup \{s^a\}) \times Y$.*

Proof.   We can obviously assume that $s^a \notin X$. Now, if s^b is admissible with respect to $X \times Y$, there is a $\mu^a \in \mathcal{M}(S^a)$, with Supp $\mu^a = X$, such that s^b is optimal under μ^a given Y. Define $\nu^a \in \mathcal{M}(S^a)$ by

$$
\nu^a(r^a) = \begin{cases} \varepsilon & \text{if } r^a = s^a, \\ \mu^a(r^a) - \varepsilon\varphi^a(r^a) & \text{otherwise,} \end{cases}
$$

where $\varepsilon > 0$ is chosen small enough that every $\mu^a(r^a) - \varepsilon\varphi^a(r^a) > 0$. (This is possible since $\varphi^a(r^a) > 0$ implies $\mu^a(r^a) > 0$.) Then Supp $\nu^a = X \cup \{s^a\}$, and $\pi^b(r^b, \nu^a) = \pi^b(r^b, \mu^a)$ for all $r^b \in S^b$. Thus, s^b is admissible with respect to $(X \cup \{s^a\}) \times Y$.

Conversely, if s^b is admissible with respect to $(X \cup \{s^a\}) \times Y$, there is a $\mu^a \in \mathcal{M}(S^a)$, with Supp $\mu^a = X \cup \{s^a\}$, such that s^b is optimal under μ^a given Y. Define $\nu^a \in \mathcal{M}(S^a)$ by $\nu^a(r^a) = \mu^a(r^a) + \mu^a(s^a)\varphi(r^a)$ for $r^a \in X$. Then Supp $\nu^a = X$, and $\pi^b(r^b, \nu^a) = \pi^b(r^b, \mu^a)$ for all $r^b \in S^b$. Thus, s^b is admissible with respect to $X \times Y$.   □

Proof of Proposition 5.2. Begin with part (a). It is immediate that each $s^a \in \bar{Q}^a \backslash \{q^a\}$ satisfies conditions (i)–(iii) of an SAS. So we will turn to Bob.

Since each $s^b \in \bar{Q}^b$ is admissible with respect to $(S^a \cup \{q^a\}) \times S^b$ (condition (i) of an SAS applied to $\bar{G}$), Lemma A.4 implies that each $s^b \in \bar{Q}^b$ is admissible with respect to $S^a \times S^b$. Next, note that $s^b \in \bar{Q}^b$ is admissible with respect to $\bar{Q}^a \times S^b$ (condition (ii) of an SAS applied to $\bar{G}$). It suffices to consider the case when $q^a \in \bar{Q}^a$. Then Supp $\varphi^a \subseteq \bar{Q}^a$ (condition (iii) of an SAS applied to $\bar{G}$). It follows from Lemma A.4 that s^b is admissible with respect to $(\bar{Q}^a \backslash \{q^a\}) \times S^b$, establishing condition (ii) of an SAS for Bob.

For condition (iii) of an SAS, suppose r^b supports $s^b \in \bar{Q}^b$, via $\rho^b \in \mathcal{M}(S^b)$, in the game G. We have to show that $r^b \in \bar{Q}^b$. This will follow from condition (iii) applied to $\bar{G}$, provided $\pi^b(\rho^b, q^a) = \pi^b(s^b, q^a)$. Notice that

$$
\pi^b(\rho^b, q^a) = \sum_{u^b \in S^b} \pi^b(u^b, q^a)\rho^b(u^b)
$$

$$
= \sum_{u^b \in S^b} \sum_{s^a \in S^a} \pi^b(u^b, s^a)\varphi^a(s^a)\rho^b(u^b)
$$

$$= \sum_{s^a \in S^a} \sum_{u^b \in S^b} \pi^b(u^b, s^a) \rho^b(u^b) \varphi^a(s^a)$$

$$= \sum_{s^a \in S^a} \pi^b(s^b, s^a) \varphi^a(s^a) = \pi^b(s^b, q^a)$$

as required.

For part (b) of the proposition, first suppose that q^a does not support any strategy in Q^a. Any $s^a \in Q^a$ is admissible with respect to $S^a \times S^b$ (resp. $S^a \times Q^b$) among strategies in S^a. It follows from Lemma A.1 that each $s^a \in Q^a$ is also admissible with respect to $(S^a \cup \{q^a\}) \times S^b$ (resp. $(S^a \cup \{q^a\}) \times Q^b$). This establishes conditions (i) and (ii) of Definition 3.4 for $s^a \in Q^a$. Condition (iii) is immediate for this case. Next, because each $s^b \in Q^b$ is admissible with respect to $S^a \times S^b$, it is also admissible with respect to $(S^a \cup \{q^a\}) \times S^b$ by Lemma A.4. Condition (ii) is immediate. Finally, notice that if r^b supports s^b in $\bar{G}$, then certainly it does in G, so condition (iii) follows.

Next, suppose q^a supports some $s^a \in Q^a$, and write $\bar{Q}^a = Q^a \cup \{q^a\}$. It is immediate that each $r^a \in Q^a$ satisfies conditions (i) and (ii) of Definition 3.4. Lemma A.1 implies that q^a also satisfies (i) and (ii), since s^a does. Condition (iii) is clearly satisfied for any $r^a \in Q^a$, since $q^a \in \bar{Q}^a$. Condition (iii) is also satisfied for q^a. To see this, use Lemma A.3 to get that if u^a supports q^a, then u^a also supports s^a (since q^a supports s^a). Applying condition (iii) to G then implies $u^a \in Q^a$.

Next, consider some $s^b \in Q^b$. Conditions (i) and (ii) are as above (i.e., as in the case where q^a does not support any strategy in Q^a). Turn to condition (ii). Using condition (iii) already established for q^a, $\mathrm{Supp}\phi^a \backslash \{q^a\} \subseteq Q^a$. So, by Lemma A.4, any $s^b \in Q^b$ is admissible with respect to $\bar{Q}^a \times S^b$. $\square$

The IA set is one SAS (Proposition 5.1). We have just proved that SAS's are invariant to the fully reduced strategic form of the game. So, we have: Fix two matrices with the same fully reduced strategic form, and let $S^a_M \times S^b_M$ be the IA set for one of these matrices. Then, $S^a_M \times S^b_M$ induces an SAS of the second matrix. But does it induce the IA set of the other matrix? That is, is the IA set itself invariant? The answer is yes, in the following (restricted) sense.

Let $\bar{S}^i_m$ denote the set of m-admissible strategies for player i (where $i = a, b$) in the game $\bar{G}$.

Proposition A.1. *For all m,*

(a) *if* $\mathrm{Supp}\ \varphi^a \subseteq S^a_m$, *then* $\bar{S}^a_m \times \bar{S}^b_m = (S^a_m \cup \{q^a\}) \times S^b_m$,
(b) $\bar{S}^a_m \times \bar{S}^b_m = S^a_m \times S^b_m$ *otherwise.*

Proof. The proof is by induction on m. For $m = 0$, the result is immediate. Assume the claim holds for m. We show it then holds for $m+1$. This is immediate from the induction hypothesis if $\bar{S}_m^a = S_m^a$. So, we will suppose that $\bar{S}_m^a = S_m^a \cup \{q^a\}$.

We first show that if Supp $\varphi^a \subseteq S_{m+1}^a$, then $\bar{S}_{m+1}^a = S_{m+1}^a \cup \{q^a\}$; and $\bar{S}_{m+1}^a = S_{m+1}^a$ otherwise. Certainly, if $s^a \in \bar{S}_{m+1}^a$ and $s^a \neq q^a$, then s^a is admissible given $S_m^a \times \bar{S}_m^b$ So, by the induction hypothesis, $\bar{S}_{m+1}^a \subseteq S_{m+1}^a \cup \{q^a\}$. Fix $s^a \in S_m^a$ that is admissible given $S_m^a \times S_m^b$. By Lemma A.1, s^a is also admissible given $(S_m^a \cup \{q^a\}) \times S_m^b$. Again using Lemma A.1, Supp $\varphi^a \subseteq S_{m+1}^a$ if and only if $q^a \in S_{m+1}^a$. The claim then follows from the induction hypothesis.

Next, we show that $\bar{S}_{m+1}^b = S_{m+1}^b$. Since $\bar{S}_m^a = s_m^a \cup \{q^a\}$, Supp $\varphi^a \subseteq S_m^a \subseteq \bar{S}_m^a$. (This is the induction hypothesis.) The result now follows from Lemma A.4. □

Proof of Lemma 7.1. Fix an admissible s^a, and also some $\varphi^a \in \mathcal{M}(S^a)$ with $\pi^a(\varphi^a, s^b) = \pi^a(s^a, s^b)$ for all $s^b \in S^b$. Without loss of generality, take $s^a \in$ Supp φ^a. Suppose, contra hypothesis, there exists $r^a, q^a \in$ Supp ϕ^a with $\pi^a(r^a, r^b) > \pi^a(q^a, r^b)$ for some $r^b \in S^b$. Let h_1 be the last common predecessor of the terminal nodes $\zeta(r^a, r^b)$ and $\zeta(q^a, r^b)$ and note that Ann moves at h_1. Of course, there may be many choices of profiles (r^a, r^b) and (q^a, r^b). If so, choose profiles so that there does not exist another with a last common predecessor (strictly) following h_1.

We will first argue that there is some q^b that allows h_1, with $\pi^a(q^a, q^b) > \pi^a(r^a, q^b)$. If not, then $\pi^a(r^a, s^b) \geq \pi^a(q^a, s^b)$ for all s^b that allow h_1, with strict inequality for some s^b that allows h_1 (in particular, for r^b). Note that we can construct a strategy that allows h_1, agrees with r^a at node h_1 and onwards, but otherwise agrees with q^a. As such, q^a must be inadmissible. But this implies that s^a is inadmissible (Corollary A.1), a contradiction.

For the remainder of the proof, we will take $\pi^a(r^a, r^b) \neq \pi^a(r^a, q^b)$. If this is not the case, then $\pi^a(q^a, q^b) > \pi^a(r^a, q^b) = \pi^a(r^a, r^b) > \pi^a(q^a, r^b)$, and a corresponding argument can then be made with respect to the pairs (q^a, q^b) and (q^a, r^b).

Let h_2 be the last common predecessor of $\zeta(r^a, r^b)$ and $\zeta(r^a, q^b)$ and note that Bob moves at h_2. Refer to Figure A.1 and note that, since Ann moves at h_1 and Bob moves at h_2, these nodes are distinct. Moreover, since (r^a, r^b) passes through both these nodes, they must be (strictly) ordered. In particular, h_2 must strictly follow h_1, since q^b allows h_1.

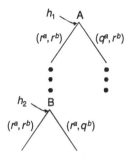

Figure A.1.

Now pick strategies $\hat{r}^b$ and $\hat{q}^b$, each of which allows h_2. The two strategies agree in all but one respect: At and after h_2, the strategy $\hat{r}^b$ agrees with r^b and the strategy $\hat{q}^b$ agrees with q^b.

Fix a strategy $u^a \in \operatorname{Supp} \varphi^a$ that allows h_2. Then, for every strategy s^b that allows h_2, we have $\pi^a(u^a, s^b) = \pi^a(r^a, s^b)$. (If not, we contradict our choice of (r^a, r^b) and (q^a, r^b).) Given this, we can write

$$\pi^a(s^a, \hat{r}^b) = \phi^a(S^a(h_2))\pi^a(r^a, \hat{r}^b) + c,$$
$$\pi^a(s^a, \hat{q}^b) = \phi^a(S^a(h_2))\pi^a(r^a, \hat{q}^b) + c,$$

where

$$c = \sum_{w^a \notin S^a(h_2)} \varphi^a(w^a)\pi^a(w^a, \hat{r}^b) = \sum_{w^a \notin S^a(h_2)} \varphi^a(w^a)\pi^a(w^a, \hat{q}^b).$$

(This last equality uses the fact that, by construction, $\zeta(w^a, \hat{r}^b) = \zeta(w^a, \hat{q}^b)$ whenever w^a does not allow $S^a(h_2)$.)

Since $(\phi^a(S^a(h_2)) > 0$, we have $\pi^a(s^a, \hat{r}^b) \neq \pi^a(s^a, \hat{q}^b)$. By choice of $\hat{r}^b$ and $\hat{q}^b$, it follows that s^a allows h_2. So, given that we chose $s^a \in \operatorname{Supp} \varphi^a$ (again using the facts established in the previous paragraph), we have $\pi^a(s^a, \hat{r}^b) = \pi^a(r^a, \hat{r}^b)$ and $\pi^a(s^a, \hat{q}^b) = \pi^a(r^a, \hat{q}^b)$. This implies

$$(1 - \varphi^a(S^a(h_2)))\pi^a(s^a, \hat{r}^b) = c = (1 - \varphi^a(S^a(h_2)))\pi^a(s^a, \hat{q}^b).$$

Since $\pi^a(s^a, \hat{r}^b) \neq \pi^a(s^a, \hat{q}^b)$, this can occur only if $\varphi^a(S^a(h_2)) = 1$. But, q^a does not allow h_2 (refer to Figure A.1) and $q^a \in \operatorname{Supp} \varphi^a$. So, $\varphi^a(S^a(h_2)) < 1$, a contradiction. $\qquad\square$

Finally, we prove that in a PI tree satisfying SPC, there is an admissible BI strategy profile. This was used in the proof of Proposition 7.2.

Fix a PI tree Γ and let N^a (resp. N^b) be the set of nodes at which a (resp. b) moves. It will also be useful to fix the following terminology, tailored to a PI tree.

Definition A.1. Say r^a *weakly dominates* s^a at node $n \in N^a$ if s^a, r^a allow n and:

(i) for each s^b that allows n, $\pi^a(r^a, s^b) \geq \pi^a(s^a, s^b)$;

(ii) for some s^b that allows n, $\pi^a(r^a, s^b) > \pi^a(s^a, s^b)$.

Say s^a is *weakly dominated* at n if there is an r^a that weakly dominates s^a at n. Say s^a is *admissible* at n if it is not weakly dominated at n.

Note, if s^a precludes n, then s^a is admissible at n. The following lemma is immediate by finiteness of S^a.

Lemma A.5. *If s^a is inadmissible at n, then there is an admissible strategy r^a that weakly dominates s^a at n.*

Lemma A.6. *Fix a PI tree. A strategy s^a is admissible if and only if it is admissible at each $n \in N^a$.*

Proof. Suppose s^a is admissible. Then, applying Proposition 3.1 in Brandenburger and Friedenberg (2003), s^a is admissible at each $n \in N^a$. For the converse, suppose s^a is admissible at each node $n \in N^a$. Then, applying Lemma 4 in Battigalli (1997) and Proposition 3.1 in Brandenburger and Friedenberg (2003), we get that s^a is admissible. $\square$

Proposition A.2. *Fix a PI tree satisfying SPC. There is an admissible BI strategy profile.*

Proof. Fix a BI profile (s^a, s^b). We will construct a new BI profile, viz. (r^a, s^b), so that r^a is admissible. Once we do this, we can apply the same argument to construct a new BI profile, viz. (r^a, r^b), so that r^b is also admissible. This will complete the proof.

If s^a is admissible, we are done. So, suppose s^a is inadmissible. Then, by Lemma A.6, there is a node $n \in N^a$ at which s^a is weakly dominated. Let $n_1, \ldots, n_K$ be a list of all nodes (for a) where: (i) s^a is weakly dominated at each n_k; and (ii) s^a is admissible at each node n that precedes n_k. (Note, the nodes $n_1, \ldots, n_K$ cannot be ordered. Also, each $n_k \in N^a$.) We will use these nodes to construct inductively strategies $f(s^a, n_k), k = 1, \ldots, K$, for a.

Begin with node n_1. By Lemmas A.6 and A.5, there is an admissible strategy q^a that weakly dominates s^a at n_1. Construct a strategy $f^a(s^a, n_1)$

as follows: Let $f^a(s^a, n_1)$ coincide with q^a at n_1 and each node that follows n_1, provided q^a allows the node. Otherwise, let $f^a(s^a, n_1)$ coincide with s^a. Now suppose $f^a(s^a, n_k)$ is defined. Consider the node n_{k+1} and an admissible strategy q^a (possibly different from earlier q^a's, of course) that weakly dominates s^a at n_{k+1}. (Again, we use Lemmas A.6 and A.5.) Construct $f^a(s^a, n_{k+1})$ analogously: Let $f^a(s^a, n_{k+1})$ coincide with q^a at n_{k+1} and each node that follows n_{k+1}, provided q^a allows the node. Otherwise, let $f^a(s^a, n_{k+1})$ coincide with $f^a(s^a, n_k)$. Denote by r^a the resulting strategy $f^a(s^a, n_K)$.

Note that r^a is admissible. Indeed, by the above construction and Lemma A.6, r^a is admissible at each node $n \in N^a$. Now apply Lemma A.6 to get the result. Next, we turn to showing that (r^a, s^b) is a BI profile. Here is the idea. First, we show that, for each node n, both a and b are indifferent between any (r_n^a, s_n^b) and (s_n^a, s_n^b) where we write r_n^a (resp. s_n^a or s_n^b) for a strategy that allows n and thereafter agrees with r^a (resp. s^a or s^b). From this we will conclude that (r^a, s^b) is a BI profile.

Step I: For each node $n, \pi^a(r_n^a, s_n^b) = \pi^a(s_n^a, s_n^b)$ and $\pi^b(s_n^b, r_n^a) = \pi^b(s_n^b, s_n^a)$. To show this, it suffices to consider a node n at which a moves. If r^a coincides with s^a at n and each node that follows n, certainly $\pi^a(r_n^a, s_n^b) = \pi^a(s_n^a, s_n^b)$. If not, then r^a must weakly dominate s^a at some node that (weakly) proceeds n. It follows that $\pi^a(r_n^a, s_n^b) \geq \pi^a(s_n^a, s_n^b)$. Moreover, given that (s^a, s^b) is a BI profile, it follows that $\pi^a(s_n^a, s_n^b) \geq \pi^a(r_n^a, s_n^b)$. So, $\pi^a(r_n^a, s_n^b) = \pi^a(s_n^a, s_n^b)$. Now, note that a moves at the last common predecessor of $\zeta(r_n^a, s_n^b)$ and $\zeta(s_n^a, s_n^b)$. So, by SPC, $\pi^b(s_n^b, r_n^a) = \pi^b(s_n^b, s_n^a)$ as desired.

Step II: For each node $n \in N^a$ (resp. $n \in N^b$), r_n^a (resp. s_n^b) maximizes $\pi^a(\cdot, s^b)$ (resp. $\pi^b(s_n^b, \cdot)$) among all strategies that allow n. Notice, this statement is immediate from Step I and the fact that (s^a, s^b) is a BI profile, provided n is a penultimate node (i.e., a "last move" in the tree). Then, assuming the statement holds for all nodes that follow n, the result again follows (for n) from Step I and the fact that (s^a, s^b) is a BI profile.

It is immediate from Step II that (r^a, s^b) is a BI profile. $\square$

References

Asheim, G (2001). Proper rationalizability in lexicographic beliefs. *International Journal of Game Theory*, 30, 453–478.

Asheim, G and M Dufwenberg (2003). Admissibility and common belief. *Games and Economic Behavior*, 42, 208–234.

Asheim, G and A Perea (2005). Sequential and quasi-perfect rationalizability in extensive games. *Games and Economic Behavior*, 53, 15–42.

Aumann, R (1998). On the Centipede game. *Games and Economic Behavior*, 23, 97–105.

Battigalli, P (1997). On rationalizability in extensive games. *Journal of Economic Theory*, 74, 40–61.

Battigalli, P and A Friedenberg (2009). Context-dependent forward-induction reasoning. Available at http://www.public.asu.edu/~afrieden.

Battigalli, P and M Siniscalchi (2002). Strong belief and forward-induction reasoning. *Journal of Economic Theory*, 106, 356–391.

Ben Porath, E (1997). Rationality, Nash equilibrium and backwards induction in perfect-information games. *Review of Economic Studies*, 64, 23–46.

Bernheim, D (1984). Rationalizable strategic behavior. *Econometrica*, 52, 1007–1028.

Blume, L, A Brandenburger, and E Dekel (1991). Lexicographic probabilities and choice under uncertainty. *Econometrica*, 59, 61–79.

Börgers, T (1994). Weak dominance and approximate common knowledge. *Journal of Economic Theory*, 64, 265–276.

Börgers, T and L Samuelson (1992). Cautious utility maximization and iterated weak dominance. *International Journal of Game Theory*, 21, 13–25.

Brandenburger, A (1992). Lexicographic probabilities and iterated admissibility. In Dasgupta, P, D Gale, O Hart, and E Maskin (Eds.), *Economic Analysis of Markets and Games*, pp 282–290. Cambridge, MA: MIT Press.

Brandenburger, A and A Friedenberg (2003). The relationship between rationality on the matrix and the tree. Available at http://www.stern.nyu.edu/~abranden.

Brandenburger, A and A Friedenberg (2009). Are admissibility and backward induction consistent? Available at http://www.stern.nyu.edu/~abranden.

Brandenburger, A, A Friedenberg, and HJ Keisler (2006). Notes on the relationship between strong belief and assumption. Available at http://www.stern.nyu.edu/~abranden.

Brandenburger, A, A Friedenberg, and HJ Keisler (2008). Admissibility in games. *Econometrica* 76, 307–352.

Brandenburger, A, A Friedenberg, and HJ Keisler (2008). Supplement to "Admissibility in games." *Econometrica*, 76, Available at http://econometricsociety.org/ecta/Supmat/5602_extensions.pdf.

Dalkey, N (1953). Equivalence of information patterns and essentially determinate games. In Kuhn, H and A Tucker (Eds.), *Contributions to the Theory of Games, Vol. 2*, pp 217–244, Princeton, NJ: Princeton University Press.

Dekel, E and D Fudenberg (1990). Rational behavior with payoff uncertainty. *Journal of Economic Theory*, 52, 243–267.

Elmes, S and P Reny (1994). On the strategic equivalence of extensive form games. *Journal of Economic Theory*, 62, 1–23.

Ewerhart, C (1998). Rationality and the definition of consistent pairs. *International Journal of Game Theory*, 27, 49–59.

Kohlberg, E and J-F Mertens (1986). On the strategic stability of equilibria. *Econometrica*, 54, 1003–1038.

Kuhn, H (1950). Extensive games. *Proceedings of the National Academy of Sciences*, 36, 570–576.

Kuhn, H (1953). Extensive games and the problem of information. In Kuhn, H and A Tucker (Eds.), *Contributions to the Theory of Games, Vol. 2*, pp 193–216, Princeton, NJ: Princeton University Press.

Marx, L and J Swinkels (1997). Order independence for iterated weak dominance. *Games and Economic Behavior*, 18, 219–245.

Mertens, J-F (1989). Stable equilibria — A reformulation. *Mathematics of Operations Research*, 14, 575–625.

Pearce, D (1984). Rationalizable strategic behavior and the problem of perfection. *Econometrica* 52, 1029–1050.

Rosenthal, R (1981). Games of perfect-information, predatory pricing, and the chain-store paradox. *Journal of Economic Theory*, 25, 92–100.

Samuelson, L (1992). Dominated strategies and common knowledge. *Games and Economic Behavior*, 4, 284–313.

Selten R (1978). The chain store paradox. *Theory and Decision*, 9, 127–159.

Shimoji, M and J Watson (1998). Conditional dominance, rationalizability, and game forms. *Journal of Economic Theory*, 83, 161–195.

Stuart, H (1997). Common belief of rationality in the finitely repeated prisoners' dilemma. *Games and Economic Behavior*, 19, 133–143.

Thompson, F (1952). Equivalence of games in extensive form. Research Memorandum RM-759. Santa Monica: The RAND Corporation.

Subject Index

Author Index

Printed in the United States
By Bookmasters